EXCITATION ENERGY AND ELECTRON TRANSFER
IN PHOTOSYNTHESIS

Excitation Energy and Electron Transfer in Photosynthesis

Dedicated to Warren L. Butler

edited by

GOVINDJEE
University of Illinois at
Urbana-Champaign
Urbana, Illinois, U.S.A.

J. BARBER
Imperial College of Science
and Technology
London, U.K.

W.A. CRAMER
Purdue University
Lafayette, Indiana, U.S.A.

J.H.C. GOEDHEER
Rijksuniversiteit Utrecht
Utrecht, The Netherlands

J. LAVOREL
Association Recherche en
Bioenergie Solaire
Cadarche, Saint Paul
Lez Durance, France

R. MARCELLE
Research Station of Gorsem West
Sint Truiden, Belgium

B. ZILINSKAS
Cook College
Rutgers University
New Brunswick, New Jersey
U.S.A.

Reprinted from Photosynthesis Research, Volume 10, Number 3

1987 **MARTINUS NIJHOFF PUBLISHERS**
a member of the KLUWER ACADEMIC PUBLISHERS GROUP
DORDRECHT / BOSTON / LANCASTER

Distributors

for the United States and Canada: Kluwer Academic Publishers, P.O. Box 358, Accord Station, Hingham, MA 02018-0358, USA
for the UK and Ireland: Kluwer Academic Publishers, MTP Press Limited, Falcon House, Queen Square, Lancaster LA1 1RN, UK
for all other countries: Kluwer Academic Publishers Group, Distribution Center, P.O. Box 322, 3300 AH Dordrecht, The Netherlands

Library of Congress Cataloging in Publication Data

Excitation energy and electron transfer in photosynthesis.

Bibliography: p.
1. Photosynthesis. I. Govindjee, 1933–
II. Title: Electron transfer in photosynthesis.
QK882.E96 1986   581.1'3342   86–23776
ISBN-13:978-94-010-8076-7   e-ISBN-13:978-94-009-3527-3
DOI:10.1007/978-94-009-3527-3

ISBN-13:978-94-010-8076-7

Copyright

Contents

I. Photosynthetic Unit; the Antenna System; and the Photosynthetic Pigments

* Indicates overviews, others are original papers.

Contributors

Numbers in square brackets indicate the pages on which the authors' contributions begin.

G. Akoyunoglou   [25],   Biology Department Nuclear Research Center, "Demokritos", Athens, GREECE

S.I. Allakhverdiev   [209],   Institute of Soil Science and Photosynthesis, U.S.S.R. Academy of Sciences, Pushchino, Moscow-Region-142292, U.S.S.R.

J. Amesz   [191],   Department of Biophysics, Huygens Laboratory of the State University, P.O. Box 9504, 2300 RA Leiden, THE NETHERLANDS

G.P. Anderson   [291],   Department of Biochemistry, Ohio State University, 484 W. 12th Avenue, Columbus, Ohio 43210, U.S.A.

J.H. Argyroudi-Akoyunoglou   [25],   Biology Department, Nuclear Research Center "Demokritos", Athens, GREECE

Mordhay Avron   [259],   Biochemistry Department, Weizmann Institute of Science, Rehovot-76100, ISRAEL

G.T. Babcock   [277],   Department of Chemistry, Michigan State University East Lansing, Michigan 48824, U.S.A.

J. Barber   [97],   Imperial College of Science and Technology, Department of Pure and Applied Biology, London, SW7 2BB, U.K.

John Biggins   [137],   Division of Biology and Medicine, Brown Univ., Providence, Rhode Island 02912, U.S.A.

W. Bilger   [157],   Institut für Botanik und Pharmazeutische Biologie der Universität Würzburg, Mittlerer Dallenbergweg 64, D-8700 Würzburg, F.R.G.

Norman I. Bishop [1], Department of Botany and Plant Pathology, Oregon State University, Corvallis, Oregon 97331-2901, U.S.A.

Salil Bose [181], Department of Plant Biology, Carnegie Institute of Washington, 290 Panama Street, Stanford, California 94305-1297, U.S.A.

K. Brettel [307], Max-Volmer Institut für Biophysikalische und Physikalische Chemie, Technische Universität Berlin, Strasse des 17 Juni 135, 1000 Berlin 12, PC14, F.R.G.

Jacques Breton [87], Service de Biophysique, Départment de Biologie, Centre d'Etudes Nucleaires de Saclay, 91191-Gif-sur-Yvette Cedex, FRANCE

J-M. Briantais [173], Laboratoire de Photosynthèse, C.N.R.S., 91190-Gif-Sur-Yvette, FRANCE

Doug Bruce [137], Division of Biology and Medicine, Brown University, Providence, Rhode Island 02912, U.S.A.

Sharon Campbell [63], Department of Biochemistry and Microbiology, Cook College, Rutgers University, New Brunswick, New Jersey 08903, U.S.A.

O. Canaani [145], Biochemistry Department, Weizmann Institute of Science, Rehovot-76100, ISRAEL

T.K. Chandrashekar [277], Department of Chemistry, Michigan State University, East Lansing, Michigan 48824, U.S.A.

W.A. Cramer [247], Department of Biological Sciences, Lilly Hall of Life Sciences, Purdue University, West Lafayette, Indiana 47907, U.S.A.

Karoly Csatorday   [63],   Department of Biochemistry and
                                 Microbiology, Cook College, Rutgers
                                 University, New Brunswick, New
                                 Jersey 08903, U.S.A.

*A. Faludi-Daniel   [71],   Department of Plant Physiology, Bio-
                                 logical Research Center, Hungarian
                                 Academy of Sciences, P.O. Box 521,
                                 Szeged-6701, HUNGARY

W. Draber   [235],   Agrochemicals, Chemical Research,
                                 Bayer AG, Monheim, F.R.G.

J.E. Draheim   [291],   Department of Biochemistry, Ohio State
                                 University, 484 W. 12th Avenue, Colum-
                                 bus, Ohio 43210, U.S.A.

L.N.M. Duysens   [191],   Department of Biophysics, Huygens
                                 Laboratory of the State University, P.O.
                                 Box 9504, 2300 RA Leiden, THE
                                 NETHERLANDS

J.J. Eaton-Rye   [219],   Department of Plant Biology, Universi-
                                 ty of Illinois at Urbana-Champaign, 289
                                 Morrill Hall, 505 South Goodwin Ave-
                                 nue, Urbana, Illinois 61801, U.S.A.

David C. Fork   [181],   Department of Plant Biology, Carnegie
                                 Inst. of Washington, 290 Panama Street,
                                 Stanford, California 94305-1297, U.S.A.

Giorgio Forti   [131],   Centro CNR Biologia Cellulare e
                                 Molecolare delle Pianta, Dipartimento
                                 di Biologia, Universita di Milano, Via
                                 Celoria 26, 20133-Milano, ITALY

Elisabeth Gantt   [55],   Environmental Research Center, Smith-
                                 sonian Institution, 12441 Parklawn
                                 Drive, Rockville, Maryland 20852-1773,
                                 U.S.A.

Gy. I. Garab   [71],   Department of Plant Physiology, Bio-
                                 logical Research Center, Hungarian

* deceased

Academy of Sciences, P.O. Box 521, Szegad-6701, HUNGARY

Nicholas E. Geacintov [87], Chemistry Department, New York University, New York, New York 10003, U.S.A.

Demetrios F. Ghanotakis [337], Division of Biological Sciences, The University of Michigan, Ann Arbor, Michigan 48109-1048, U.S.A.

Govindjee [5, 219], Departments of Physiology & Biophysics and Plant Biology, University of Illinois at Urbana-Champaign, 289 Morrill Hall, 505 S. Goodwin Avenue, Urbana, Illinois 61801, U.S.A.

E.L. Gross [291], Department of Biochemistry, Ohio State University 484 W. 12th Avenue, Columbus, Ohio 43210, U.S.A.

A. Grover [77], Bioenergetics Laboratory, School of Life Sciences, Jawaharlal Nehru University, New Delhi-110067, INDIA

Paola M.G. Grubas [131], Centro CNR Biologia Cellulare e Molecolare delle Pianta, Dipartimento di Biologia, Universita de Milano, Via Celoria 26, 20133-Milano, ITALY

Lucia E. Hancock [137], Department of Physics and Astronomy, Rochester University, Rochester, New York 14627, U.S.A.

Cheryl A. Hanzlik [137], Department of Physics and Astronomy, University of Rochester, Rochester, New York 14627, U.S.A.

M. Havaux [145], Biochemistry Department, Weizmann Institute of Science, Rehovot, 76100, ISRAEL

Stephen K. Herbert [181], Department of Plant Biology, Carnegie Institute of Washington, 290 Panama Street, Stanford, California 94305-1297, U.S.A.

G. Hervo   [173],   Service de Biophysique, Centre d'Etudes Nuclèaires de Saclay, BP2, 91191-Gif-Sur-Yvette Cedex, FRANCE

M. Hodges   [173],   Laboratoire de Photosynthèse, C.N.R.S., 91190 Gif-Sur-Yvette, FRANCE

A.R. Holzwarth   [163],   Max-Planck Institut für Strahlenchemie, Stift-strasse 34−36, D-4330 Mülheim a.d. Ruhr, F.R.G.

Peter H. Homann   [351],   Institute of Molecular Biophysics and Department of Biological Science, Florida State University, Tallahassee, Florida 32306-3015, U.S.A.

P. Horton   [151],   Department of Biochemistry, and Research Institute for Photosynthesis, The University, Sheffield S10 2TN, U.K.

Yorinao Inoue   [285],   Solar Energy Research Group, The Institute of Physical and Chemical Research (RIKEN), Hirosawa, Waiko, Saitama-351-01, JAPAN

S.L. Ketchner   [291],   Department of Biochemistry, Ohio State University, 484 W. 12th Avenue, Columbus, Ohio 43210, U.S.A.

J.G. Kiss   [71],   Clinic of ENT, Medical University, P.O.-Box 422, Szeged-6701, HUNGARY

Masao Kitajima   [369],   Research Laboratories, Asaka, Fuji Photo Film Co., Ltd., Asaka-Shi, Saitama-Ken 351, JAPAN

V.V. Klimov   [209],   Institute of Soil Science and Photosynthesis, USSR Academy of Sciences, Pushchino, Moscow Region-142292, U.S.S.R.

David B. Knaff   [361],   Department of Chemistry and Biochemistry, Texas Technical University, Lubbock, Texas 79409-4260, U.S.A.

Robert S. Knox   [87, 137],   Department of Physics and Astronomy and Laboratory for Laser Energetics, The University of Rochester, New York 14627, U.S.A.

V.G. Ladygin [209], Institute of Soil Science and Photosynthesis, U.S.S.R. Academy of Sciences, Pushchino, Moscow Region-142292, U.S.S.R.

P. Lee [151], Department of Biochemistry and Research Institute for Photosynthesis, The University, Sheffield S10 2TN, U.K.

Arthur C. Ley [43], Department of Cellular and Developmental Biology, 16 Divinity Avenue, Harvard University, Cambridge, Massachusetts 02138, U.S.A.

R. Malkin [51], Division of Molecular Plant Biology, Hilgard Hall, University of California, Berkeley, California 94720, U.S.A.

S. Malkin [145], Biochemistry Department, Weizmann Institute of Science, Rehovot-76100, ISRAEL

P. Mathis [201], Service de Biophysique, Centre d'Etudes Nuclèaires de Saclay, 91191-Gif-Sur Yvette Cedex, FRANCE

David Mauzerall [17], The Rockefeller University, 1230 York Avenue, New York, New York 10021, U.S.A.

Mamoru Mimuro [55], Environmental Research Center, Smithsonian Institution, 12441 Parklawn Drive, Rockville, Maryland 20852-1773, U.S.A.

M. Miyao [343], National Institute for Basic Biology, Myodaiji, Okazaki-444, JAPAN

P. Mohanty [77], Bioenergetics Laboratory, School of Life Sciences, Jawaharlal Nehru University, New Delhi-110067, INDIA

I. Moya   [173],   Laboratoire de Photosythèse, CNRS, 91190-Gif-Sur-Yvette, FRANCE

N. Murata   [343],   National Institute for Basic Biology, Myodaiji, Okazaki-444, JAPAN

P.J. O'Malley   [277],   Department of Chemistry, Michigan State University, East Lansing, Michigan 48824, U.S.A.

G.C. Papageorgiou   [299],   Department of Biology, Nuclear Research Centre, Demokritos, Aghia Paraskevi, Attiki, Athens, GREECE

I. Rodriguez   [277],   Department of Chemistry, Michigan State University, East Lansing, Michigan 48824, U.S.A.

S.C. Sabat   [77],   Bioenergetics Laboratory, School of Life Sciences, Jawaharlal Nehru University, New Delhi-110067, INDIA

D.A. Sanderson   [291],   Department of Biochemistry, Ohio State University, 484 W.12th Avenue, Columbus, Ohio 43210, U.S.A.

Kimiyuki Satoh   [35],   Department of Biology, Faculty of Science, Okayama University, Okayama 700, JAPAN

O. Saygin   [307],   Max-Volmer Institut für Biophysikalische und Physikalische Chemie, Technische Universität Berlin, Strasse des 17 Juni 135, 1000 Berlin 12, PC14, F.R.G.

G.H. Schatz   [163],   Max-Planck Institut für Stahlenchemie, Stiftstrasse 34–36, D-4330 Mülheim a.d. Ruhr, F.R.G.

E. Schlodder   [307],   Max-Volmer Institut für Biophysikalische und Physikalische Chemie, Technische Universität Berlin, Strasse des 17 Juni 135, 1000 Berlin 12, PC 14, F.R.G.

U. Schreiber [157], Institut für Botanik und Pharmazeutische Biologie der Universität Würzburg, Mittlerer Dallenbergweg 64, D-8700 Würzburg, F.R.G.

P. Setif [201], Service de Biophysique, Centre d'Etudes Nuclèaires de Saclay, 91191-Gif-Sur-Yvette Cedex, FRANCE

Yosepha Shahak [259], Biochemistry Department, Weizmann Institute of Science, Rehovot-76100, ISRAEL

G. Sotiropoulou [299], Department of Biology, Nuclear Research Centre, Demokritos, Aghia Paraskevi, Attiki, Athens, GREECE

Reto J. Strasser [109], Department of Bioenergetics, Institute of Biology, University of Stuttgart, Ulmerstrasse 227, D-7000 Stuttgart 60, F.R.G.

S.M. Theg [247], Department of Biological Sciences, Lilly Hall of Life Sciences, Purdue University, West Lafayette, Indiana 47907, U.S.A.

Robert K. Togasaki [269], Department of Biology, Indiana University, Bloomington, Indiana 47401, U.S.A.

Zs.M. Toth [71], Department of Plant Physiology, Biological Research Center, Hungarian Academy of Sciences, P.O. Box 521, Szeged 6701, HUNGARY

A. Trebst [235], Department of Biology, Ruhr University, Bochum, F.R.G.

Imre Vass [285], Department of Theoretical Physics, Jozsef Attila University, Szeged, HUNGARY

M.H. Vidal [201], Department de Physico-Chimie, Service de Chimie Moleculaire, Centre d'Etudes Nuclèaires de Saclay; UA331 (CNRS),

91191-Gif-Sur-Yvette Cedex, FRANCE

C. John Whitmarsh   [269],   Department of Plant Biology, University of Illinois at Urbana-Champaign, 289 Morrill Hall, 505 S. Goodwin Avenue, Urbana, Illinois 61801, U.S.A.

W.R. Widger   [247],   Department of Biological Sciences, Lilly Hall of Life Sciences, Purdue University, West Lafayette, Indiana 47907, U.S.A.

H.T. Witt   [307],   Max-Volmer Institut für Biophysikalische und Physikalische Chemie, Technische Universität Berlin, Strasse des 17 Juni 135, 1000 Berlin 12, PC 14, F.R.G.

T. Yamashita   [327],   Institute of Biological Sciences, Tsukuba University, Sakura-Mura, Ibaraki-305, JAPAN

Charles F. Yocum   [337],   Division of Biological Sciences, The University of Michigan, Ann Arbor, Michigan 48109-1048, U.S.A.

Barbara A. Zilinskas   [63],   Department of Biochemistry and Microbiology, Cook College, Rutgers University, New Brunswick, New Jersey 08903, U.S.A.

Preface

The present volume "Excitation Energy and Electron Transfer in Photosynthesis" is dedicated to a colleague and dear friend Warren L. Butler. I first met Warren when he visited the University of Illinois at Urbana during the early sixties; he left an indelible impression on me as a person with warmth and enthusiasm. Initially, he was someone I looked to for guidance, but later we also became friends. Whenever I passed through Los Angeles, I always telephoned Warren and often ended up taking a plane to San Diego to stay with two wonderful people, Warren and his wife Lila. His invitations could never be refused. Below I reproduce the words of Herbert Stern on Warren L. Butler's life; these words express my sentiments as well as those of many of Warren's friends: "A lifetime of acedemic creativity criss-crossed by streaks of highbrow and lowbrow fun. There is no summary to this adventure because we can neither make nor proclaim an end. Warren has bequeathed us his garden of academic treasures. It is ours to keep and tend. There is lots of joy in our many recollections of Warren's life and sorrow's foil can only brighten the brightness that the joy radiates."

This volume, containing 39 chapters, is a collection of overviews and original papers on: (I) Photosynthetic Unit, the Antenna System, and Photosynthetic Pigments (9 chapters); (II) Excitation Energy Migration, Regulation of Energy Transfer, State Transitions and Chlorophyll *a* Fluorescence (11 chapters); (III) Reaction Centers, Primary Photochemistry and Early Acceptors and Donors (3 chapters); (IV) Electron Transfer (9 chapters); (V) Oxygen Evolution (5 chapters) and, (VI) Photosynthetic Bacteria: Metabolism (2 chapters). Originally, the collection was to be only a special volume of the regular issues of *Photosynthesis Research*. However, due to its large size, coherence of topics, and importance, it was decided to publish it also as a seperate volume. This volume was edited by J. Barber, W.A. Cramer, J.H.C. Goedheer, J. Lavorel, R. Marcelle, B. Zilinskas and me.

There have been several dedications[1] to Warren. The current volume differs from others in that it contains both overviews and original papers by 90 international authorities in the field of "Excitation Energy and Electron Transfer in Photosynthesis". Our dedication pages are by a friend and colleague of Warren's, Norman I. Bishop. We have also included a list of Warren's photosynthesis publications. The volume contains discussions on challenging and controversial questions on photosynthesis. It is up to the reader to keep and tend Warren's garden of academic treasures.

Since both background (in overview chapters, marked with asterisks in the table of contents) and current research are included here, this volume will serve not only the researchers but also the beginning students in biochemistry, biophysics, cell biology and plant physiology.

I hope that the readers will find this volume to be exciting and useful and a worthy memorial to Warren.

Urbana, Illinois   GOVINDJEE

---

[1] *Photochemistry and Photobiology 42* (6) December, 1985 – special issue on *Photomorphogenesis in Plants,* an area not covered in the current book; *"Warren L. Butler",* printed in 1985 at Matsusaka Press, Matsue, Japan, – this is a collection of the reprints of most of Warren's scientific publications; and *"Light Emission by Plants and Bacteria,* Academic Press, Orlando, 1986 – this is a collection of chapters written for students as well as researchers on many facets of light emission in the living system.

Figure 1. Warren L. Butler (1925–1984)

Photosynthesis Research 10: 147–149 (1986)
© *Martinus Nijhoff Publishers, Dordrecht*

WARREN L. BUTLER; A TRIBUTE TO A FRIEND AND FELLOW SCIENTIST

"**The talent of success is nothing more than doing what
you can do well; and doing well whatever you do, without
a thought of fame.**"
...Henry Wadsworth Longfellow...

It was a deep and personal shock to learn that my friend and
colleague of many years, Warren Butler, died on June 21, 1984. For those
of us who were so fortunate to have known Warren personnaly as a fellow
scientist and as a warm understanding person, this tragic information was
difficult to comprehend. Few of us realized that the occassion of the
VIth International Congress on Photosynthesis in Brussels in August of
1983 would be our last opportunity to share his gifted insight into
mechanisms of photobiology, his warm, friendly, and often pixish
personality, his tolerance of the less gifted of his colleagues and his
forebearance of his own physical disability.
Warren was unique in many ways including the fact that he was a
native to the Northwest(USA), having been born in Yakima, Washington on
January 28, 1925. He completed the earlier phases of his education in
Portland, Oregon culminating with a B.A. degree in physics from Reed
College in 1949. His Ph.D. was received in 1955 at the University of
Chicago where he was the initial graduate in the then new field of
biophysics and the last student of the Nobel Laureate, James Franck.
I came into Warren Butler's life in August, 1955 when I began a
postdoctoral program with Hans Gaffron in the same laboratory. At that
time Warren was putting the finishing touches to his dissertation and had
agreed to remain for an additional year of research in this laboratory.
His fascination with instrumentation was clearly apparent even then for
his small laboratory-office was filled with self-designed gadgets for
monitoring early transients in carbon dioxide metabolism of intact
leaves. For an absolute neophyte family to the complexities of urban
living in the environs of the University, Warren and his charming wife,
Lila, immediately served as welcome beacons. Soon we were impressed to
learn that Warren and his young family lived near enough to Lake Michigan
to have their own "private" beach; we were delighted to be invited to
share this facility with them on a typical muggy summer day of Chicago.
It was then learned that their house and its beach were remnants of the
housing units provided for the construction workers of the Chicago
World's Columbian exposition of 1893. Warren's ingenuities in protecting
the Butler family, and the plumbing, from the often inclement weather of
Chicago abounded in this ancient abode. Throughout our year together in
the Franck-Gaffron laboratories our families shared many moments
together which served to maintain our friendship over the years. It would
be consiserably later, when I first visited Warren and his family in
their new California home overlooking the Pacific Ocean and recalled
their earlier situation in Chicago, that I recognized how far Warren had
progressed over the years of our acquaintance.
Warren's decision to join the research group at the USDA (United
States Department of Agriculture) laboratory in Beltsville, Maryland

where he began a rewarding period with S. B. Hendricks, K.H. Norris, H. A. Borthwick, H. W. Siegelman and others, was fundamental in the development of his scientific career in photobiology. It was there that his and Karl Norris's abilities for developing the right kind of instrumentation for the special problem at hand led to the initial, in vivo, detection and assay of phytochrome by absorption spectroscopy. A publication describing their finding appeared in 1959 and became central to a rapidly expanding field of research in light control of plant development. Indeed other contemporary scientists, including H. Lundegardh, B. Chance, L. Duysens, H. Kautsky and others had used absorption and emission spectroscopy as a non-destructive analytical device for assaying biochemical processes. However, Warren's early training in physics and his exposure to biophysical aspects of photosynthesis and contemporary experimetal biology during his Ph.D studies awakened in him a well-grounded curiosity about a broad range of photobiological phenomena which motivated his scientific curiosity throughout his life. The optical instrumentation developed for phytochome detection in intact plant tissue was applied effectively by Warren to measure changes in energy transfer during the development of the chloroplast, to detect a new long wavelength absorption band in chloroplasts and to measure the lifetime of chlorophyll fluorescence at low temperature. Following a successful sabbatical program at the Johnson Foundation in 1964, a year which intensified his interest in photosythesis research, Warren, and his expanding family, moved to La Jolla where further important achievements awaited to be made in the broadening field of photobiology. From his association with the Beltsville group nearly fifty, fundamental publications resulted, a pace of productivity which would continue in his own laboratory at La Jolla.

In his twenty year tenure in the Department of Biology of the University of California San Diego, Warren Butler's special talents would further manifest themselves in an every expanding search for clarification and understanding of effects of light on algae and higher plants. Through collaboration with a unique selection of postdoctoral researchers and students, including W. Cramer, T. Yamashita, L.Pratt, B. Epel, H. Ninneman, D. Hopkins, K. Erixon, H. Oelze-Karow, S. Okayama, R. Lozier, K. Poff, M. Kitajima, R. Strasser, and others, an additional one hundred and ten publications would result. His continued work on energy distribution between the two photosystems culminated in the still accepted "tripartite" model; his meaurements on the redox potential of Q, on two-light effects on fluorescence, on the redox status of cytochrome b-559, on the role of C-550, on the effect of TRIS extraction on PS-II activity and on many other reactions of the chloroplast all represent pioneering contributions to the field of photosynthesis research. Between these activities additional important contributions were made to our knowledge of phytochome, of photoinhibition of microsomal and mitochondrial electron transport, of phototaxis, of blue-light mediated photocontrol of plant development and other related areas of photobiology. Warren's responsibilities for these major contributions did not go unrecognized. Early in his career he received numerous plaudits, citations and awards including the C.F. Kettering Research Award and the Sterling Hendricks Memorial Lecturer' Award. Later he was elected a Fellow of the National Academy of Science and of the American Association of Arts and Sciences; subsequently he was named as a Foreign Associate of the French Academy of Sciences. The awareness and apprecation of his role in advancing the frontiers of photobiology by his many friends and colleagues is apparent through the many tributes and dedications presented since his death. (See, for example, "Warren L. Butler, 1925-1984" by H. Ninneman, <u>Photochem</u>. <u>Photobiol</u>. 42 (6), 619-624,1985.)

To have been asked to write the dedicatory remarks for this commemorative issue of **Photosynthesis Research** has been an honor for me. As a career-long friend and colleague, Warren L. Butler occupies a special place in my memory for he was a warm and caring individual who was devoted to his wife and children and was a remarkably persistent and dedicated scientist. For those of us who knew him personally, we will miss his marvelous insight into problems of photobiology, his special charismatic wit and his momentously raised, black eyebrows backgrounded by an abundance of silver hair. Younger members of our limited scientific community who did not have this special privelege nevertheless have the opportunity to read his papers, examine his ideas and be positively influenced by his contributions. The cadre of young scientists who trained in his shadow will continue to transfuse our branch of science with Warren's gifts for years to come.

Norman I. Bishop
DEPARTMENT OF BOTANY AND PLANT PATHOLOGY
OREGON STATE UNIVERSITY
CORVALLIS, OREGON 97331-2902
U.S.A.

Photosynthesis Research 10: 151–159 (1986)
© *Martinus Nijhoff Publishers, Dordrecht*

PUBLICATIONS OF WARREN L. BUTLER ON PHOTOSYNTHESIS

BY GOVINDJEE
Departments of Physiology and Biophysics and Plant Biology,
University of Illinois at Urbana-Champaign, 289 Morrill Hall
505 S. Goodwin Avenue, Urbana, Il 61801 (U.S.A.)

A.   Photosynthetic Pigments; Orientation and Their Function

(1) Butler, W.L. (1960) A Far-red Absorbing Form of Chlorophyll. Biochem. Biophys. Res. Commun. 3:685-688.

(2) Butler, W.L. (1961) A Far-red Absorbing Form of Chlorophyll in vivo. Arch. Biochem. Biophys. 93:413-422.

(3) Olson, R.A., Butler, W.L. and Jennings, W.H. (1961) The Orientation of Chlorophyll Molecules in vivo: Evidence from Polarized Fluorescence. Biochim. Biophys. Acta 54:615-617.

(4) Olson, R.A., Butler, W.L. and Jennings, W.H. (1962) The Orientation of Chlorophyll Molecules in vivo: Further Evidence from Dichroism. Biochim. Biophys. Acta 58:144-146.

(5) Butler, W.L. and Baker, J.E. (1963) Low Temperature Spectra of Chloroplast Fragments. Biochim. Biophys. Acta 66:206-211.

(6) Olson, R.A., Jennings, W.H. and Butler, W.L. (1964) Molecular Orientation: Spectral Dependence of Dichroism of Chloroplasts in vivo. Biochim. Biophys. Acta 88:318-330.

(7) Olson, R.A., Jennings, W.H. and Butler, W.L. (1964) Molecular Orientation: Spectral Dependence of Bifluorescence of Chloroplasts in vivo. Biochim. Biophys. Acta 88:331-337.

(8) Butler, W.L., Olson, R.A. and Jennings, W.H. (1964) Oriented Chlorophyll in vivo. Biochim. Biophys. Acta 88:651-653.

(9) Butler, W.L. (1966) Spectral Characteristics of Chlorophyll in Green Plants. IN: The Chlorophylls (edited by L.P. Vernon and G.R. Seeley), pp. 343-379. Academic Press, New York.

(10) Cramer, W.A. and Butler, W.L. (1968) Further Resolution of Chlorophyll Pigments in Photosystem 1 and 2 of Spinach Chloroplasts by Low-temperature Derivative Spectroscopy. Biochim. Biophys. Acta 153:889-891.

(11) Butler, W.L. and Hopkins, D.W. (1970) Higher Derivative Analysis of Complex Absorption Spectra. Photochem. Photobiol. 12:439-450.

(12) Ley, A.C. and Butler, W.L. (1977) Isolation and Function of Allophycocyanin B of Porphyridium cruentum. Plant Physiol. 59:974-980.

B. 1. Chlorophyll a Fluorescence; Relation to Photosynthesis

(13) Butler, W.L. (1962) Effects of Red and Far-Red Light on the Fluorescence Yield of Chlorophyll in vivo. Biochim. Biophys. Acta 64:309-317.

(14) Butler, W.L. (1963) Effect of Light Intensity on the Far-red Inhibition of Chlorophyll a Fluorescence in vivo. Biochim. Biophys. Acta 66:275-276.

(15) Butler, W.L. and Bishop, N.I. (1963) Action of the Two Pigment System on Fluorescence Yield of Chlorophyll a. IN: Photosynthetic Mechanisms of Green Plants. NAS-NRC Publications #1145, pp. 91-100, Washington, D.C.

(16) Butler, W.L. (1966) Fluorescence Yield in Photosynthetic Systems and Its Relation to Electron Transport. IN: Current Topics in Bioenergetics (edited by D.R. Sanadi), Volume 1, pp 49-73. Academic Press, New York.

(17) Lozier, R.H. and Butler, W.L. (1972) The Effects of Dibromo-thymoquinone on Fluorescence and Electron Transport of Spinach Chloroplasts. FEBS Lett. 26:161-164.

(18) Epel, B.L. and Butler, W.L. (1972) Spectroscopic Analysis of a High Fluorescent Mutant of Chlamydomonas reinhardtii. Biophysic. J.12:922-929.

(19) Epel, B.L., Butler, W.L. and Levine, R.P. (1972) A Spectro-scopic Analysis of Low-Fluorescent Mutants of Chlamydomonas reinhardtii Blocked in their Water-Splitting Oxygen-Evolving Apparatus. Biochim. Biophys. Acta 275:395-400.

(20) Butler, W.L. (1977) Chlorophyll Fluorescence: A Probe for Electron Transfer and Energy Transfer. IN: Encyclopedia of Plant Physiology New Series Vol. 5. Photosynthesis I.(edited by A. Trebst and M. Avron), pp. 149-167. Springer Verlag, Berlin.

(21) Butler, W.L. and Strasser, R.J. (1977) Does the Rate of Cooling Affect Fluorescence Properties of Chloroplasts at -196 C? Biochim. Biophys. Acta 462:283-289.

(22) Strasser, R.J. and Butler, W.L. (1977) Fluorescence Emission Spectra of Photosystem I, Photosystem II and the Light-harvesting Chlorophyll a /b Complex of Higher Plants. Biochim. Biophys. Acta 462:307-313.

(23) Satoh, K. and Butler, W.L. (1978) Low Temperature Spectral Properties of Subchloroplast Fractions Purified from Spinach. Plant Physiol. 61:373-379.

(24) Satoh, K. and Butler, W.L. (1978) Competition Between the 735nm Fluorescence and the Photochemistry of Photosystem I in Chloroplasts at Low Temperature. Biochim. Biophys. Acta 502:103-110.

B. 2. Chlorophyll a Fluorescence: Lifetimes

(25) Butler, W.L. and Norris, K.H. (1963) Lifetime of the Long-wavelength Chlorophyll Fluorescence. Biochim. Biophys. Acta 66:72-77.

(26) Butler, W.L., Tredwell, C.J., Malkin, R. and Barber, J., (1979) The Relationship Between the Lifetime and Yield of the 735nm Fluorescence of Chloroplasts at Low Temperatures. Biochim. Biophys. Acta 545:309-315.

(27) Magde, D., Berens, S.J. and Butler, W.L. (1982) Picosecond Fluorescence in Spinach Chloroplasts. Proc. Soc. Photo-Opt. Instrum. Eng. (SPIE) 322: 80-86.

(28) Butler, W.L., Magde, D. and Berens, S.J. (1983) Fluorescence lifetimes in the Bipartite Model of the Photosynthetic Apparatus with α , β Heterogeneity in Photosystem II. Proc. Natl. Acad. Sci. U.S.A. 80:7510-7514.

(29) Berens, S.J., Scheele, J., Butler, W.L. and Magde, D. (1985) Kinetic Modeling of Time-resolved fluorescence in Spinach Chloroplasts. Photochem. Photobiol. 42:59-68.

C. Energy Transfer; Energy Distribution; Bipartite and Tri - partite Models

(30) Butler, W.L. and Kitajima, M. (1975) A tripartite Model for Chloroplast Fluorescence., IN: Proceedings of the 3rd International Congress on Photosynthesis Rehovot 1974 (edited by M. Avron), pp. 13-24. Elsevier Publishers, Amsterdam.

(31) Butler, W.L. and Kitajima, M. (1975) Energy Transfer Between Photosystem II and Photosystem I in Chloroplasts. Biochim. Biophys. Acta 396:72-85.

(32) Kitajima, M., and Butler, W.L. (1975) Excitation Spectra for Photosystem I and Photosystem II in Chloroplasts and the Spectral Characteristics of the Distribution of Quanta Between the Two Photosystems. Biochim. Biophys. Acta 408:297-305.

(33) Butler, W.L. (1976) Energy Distribution in the Photosynthetic Apparatus of Plants. Brookhaven Symposia in Biology 28: 338-346.

(34)Ley, A. C. and Butler, W.L. (1976) Efficiency of Energy Transfer from Photosystem II to Photosystem I in Porphyridium cruentum. Proc. Natl. Acad. Sci. U.S.A. 73:3957-3960.

(35) Satoh, K., Strasser, R., and Butler, W.L. (1976) A Demonstration of Energy Transfer from Photosystem II to Photosystem I in Chloroplasts. Biochim. Biophys. Acta 440:337-345.

(36) Ley, A.C and Butler, W.L. (1977) The Distribution of Excitation Energy Between Photosystem I and Photosystem II in Porphyridium cruentum. IN: Special issue of Plant and Cell Physiol (edited by S. Miyachi, S. Katoh, Y. Fujita and K. Shibata), pp. 33-46. Japan Soc. of Plant Physiol.,Tokyo.

(37) Butler, W.L. and Strasser, R.J. (1977) Tripartite Model for the Photochemical Apparatus of Green Plant Photosynthesis. Proc. Natl. Acad. Sci. U.S.A. 74:3382-3385.

(38) Ley, A.C. and Butler, W.L. (1977) Energy Transfer from Photosystem II to Photosystem I in Porphyridium cruentum. Biochim. Biophys. Acta 462:290-294.

(39) Strasser, R.J. and Butler, W.L. (1977) Energy Transfer and the Distribution of Excitation Energy in the Photosynthetic Apparatus of Spinach Chloroplasts. Biochim. Biophys. Acta 460:230-238.

(40) Butler, W.L. (1978) Energy Distribution in the Photochemical Apparatus of Photosynthesis. Annl. Rev. Plant Physiol. 29:345-378.

(41) Butler, W.L. and Strasser, R.J. (1978) Effect of Divalent Cations on Energy Coupling Between the Light-harvesting Chlorophyll a/b Complex and Photosystem II. Proceedings of the 4th International Congress on Photosynthesis 1977, pp 11-20. Biochem. Soc. London.

(42) Butler, W.L. (1979) Tripartite and Bipartite Models of the Photochemical Apparatus of Photosynthesis. IN: Chlorophyll Organization and Energy Transfer in Photosynthesis. Ciba Foundation Symposium 61 (new series) pp.237-256. Excerpta Medica Elsevier, Holland.

(43) Butler, W.L. (1980) Energy Transfer Between Photosystem II Units in a Connected Package Model of the Photosynthetic Apparatus of Photosynthesis. Proc. Natl. Acad. Sci. U.S.A. 77:4697-4701.

(44) Ley, A.C. and Butler, W.L. (1980) Energy Distribution in the Photochemical Apparatus of Porphyridium cruentum in State I and State II. Biochim. Biophys. Acta 592:349-363.

(45) Butler, W.L. (1981) A Simplified Approach to a Model for the Photochemical Apparatus of Photosynthesis which Includes a Partial Connection Between Photosystem II Units. IN: Photosynthesis I. Photophysical Processes-Membrane Energization (edited by G.Akoyunoglou) pp 273-279. Balaban International Science Services, Philadelphia.

(46) Berens, S.J. and Scheele, J. Butler, W.L. and Magde, D. (1985) Time-resolved Fluorescence Studies of Spinach Chloroplasts-Evidence for Heterogeneous Bipartite Model.

Photochem. Photobiol. 42:51-57.

D. Primary Photochemistry of Photosystem II

(47) Butler, W.L. (1972) The Relationship Between P-680 and C-550. Biophys. J. 12:851-857.

(48) Butler, W.L. (1972) The Influence of Membrane Potential on Measuremnts of C-550 at Room Temperature. FEBS Lett. 20:333-338.

(49) Okayama, S. and Butler, W.L. (1972) The Influence of Cyto-chrome b 559 on the Fluorescence yield of Chloroplasts at Low Temperature. Biochim. Biophys. Acta 267:523-529.

(50) Butler, W.L. (1972) On the Primary Nature of Fluorescence Yield changes Associated with Photosynthesis. Proc. Natl. Acad. Sci. U.S.A. 69:3420-3422.

(51) Butler, W.L. (1972) Primary Photochemistry of Photosystem II of Photosynthesis. Accnts. Chem. Res. 6:177-184.

(52) Butler, W.L., Erixon K. and Okayama, S. (1972) The primary Photochemical Reaction of Photosystem II. IN: Proceedings of the IInd International Congress on Photosynthesis Stresa 1971, (edited by Forti, G., Avron, M. and Melandri, A.), pp 73-80. Dr. W. Junk N.V. Publishers, The Hague.

(53) Butler, W.L., Visser, J.W.M. and Simons, H.L. (1973) The Kinetics of Light-induced Changes of C-550, Cytochrome b 559 and Fluorescnece Yield in Chloroplasts at Low Temperature. Biochim. Biophys. Acta 292:140-151.

(54) Butler, W.L., Visser, J.W.M. and Simons, H.L. (1973) The Back Reaction in the Primary Electron Transfer Couple of Photosystem II of Photosynthesis. Biochim. Biophys. Acta 325:539-545.

(55) Kitajima, M. and Butler, W.L. (1973) C-550 in Photosystem II in Subchloroplast Paticles. Biochim. Biophys. Acta 325:558-564

(56) Kitajima, M. and Butler, W.L. (1975) Quenching of Chlorophyll Fluorescence and Priamry Photochemistry in Chloroplasts by Dibromothymoquinone. Biochim. Biophys. Acta 376:105-115.

(57) Butler, W.L. and Kitajima, M. (1975) Fluorescence Quenching in Photosystem II of Chloroplasts. Biochim. Biophys. Acta 376:116-125.

(58) Mathis, P., Butler, W.L., and Satoh, K. (1979) Carotenoid Triplet State and Chlorophyll Fluroescence Quenching in Chloroplasts and Subchloroplast Particles. Photochem. Photobiol. 30:603-614.

(59) Butler, W.L. (1984) Exciton Transfer Out of Open Photosystem II Reaction Centers. Photochem. Photobiol. 40:513-518.

E. Redox Potentials of Intermediates

(60)   Cramer, W.A. and Butler, W.L. (1969) Potentiometric Titration of the Fluorescence Yield of Spinach Chloroplasts. Biochim. Biophys. Acta 172:503-510.

(61)   Erixon, K., Lozier, R. and Butler, W.L. (1972) The Redox State of Cytochrome b 559 in Spinach Chloroplasts. Biochim. Biophys. Acta 267:375-382.

(62)   Lozier, R.H. and Butler, W.L. (1974) Redox Titration of the Primary Electron Acceptor of Photosystem I in Spinach Chloroplasts. Biochim. Biophys. Acta 333:460-464.

F. Absorbance Changes; Cytochromes; Cytochrome b 559; C-550

(63)   Levine, R.P., Gorman, D.S., Avron, M., and Butler, W.L. (1966) Light-induced Absorbance Changes in Wild-type and Mutant Strains of Chlamydomonas reinhardtii. Brookhaven Symposia in Biology 19:143-148.

(64)   Cramer, W.A. and Butler, W.L. (1967) Light-induced Absorbance Changes of Two Cytochrome b Components in the Electron Transport System of Spinach Chloroplasts. Biochim. Biophys. Acta 143:332-339.

(65)   Erixon, K. and Butler, W.L. (1971) Light-induced Absorbance Changes in Chloroplasts at -196°C. Photochem. Photobiol. 14:427-433.

(66)   Erixon, K. and Butler, W.l. (1971) The Relationship Between Q, C-550 and Cytochrome b 559 in Photoreactions at -196°C in Chloroplasts. Biochim. Biophys. Acta 234: 381-389.

(67)   Butler, W.L. and Okayama, S. (1971) The Photoreduction of C-550 in Chloroplasts and its Inhibition by Lipase. Biochim. Biophys. Acta 245:237-239.

(68)   Butler, W.L. (1971) The Relationship Between C-550 and Delayed Fluorescence. FEBS Lett. 19:125-127.

(69)   Erixon, K. and Butler, W.L. (1971) Destruction of C-550 by Ultraviolet Radiation. Biochim. Biophys. Acta 253:483-486.

(70)   Okayama, S., Epel, B.L. , Erixon, K., Lozier, R. and Butler, W.L. (1971) The Effects of Lipase on Spinach and Chlamydomonas Chloroplasts Biochim. Biophys. Acta 253:476-482.

(71)   Lozier, R.H. and Butler, (1974) Light-induced Absorbance Changes in Chloroplasts Mediated by Photosystem I and Photosystem II at Low Temperature. Biochim. Biophys. Acta 333:465-480.

(72)   Butler, W.L. (1978) On the Role of Cytochrome b 559 in Oxygen Evolution in Photosynthesis. FEBS Lett. 95:19-25.

(73) Matsuda, H. and Butler, W.L. (1983) Restoration of High Potential Cytochrome $\underline{b}$ 559 in Liposomes. Biochim. Biophys. Acta 724:123-127.

(74) Matsuda, H. and Butler, W.L. (1983) Restoration of High Potential Cytochrome $\underline{b}$ 559 in Photosystem II Particles in Liposomes. Biochim. Biophys. Acta 725:320-324.

(75) Butler, W.L. and Matsuda H. (1983) Possible Role of Cytochrome $\underline{b}$ 559 in Photosystem II. IN: The Oxygen Evolving System of Photosynthesis (edited by Y .Inoue, Crofts, A.R., Govindjee, Murata, N. Renger, G. and Satoh, K.), pp 113-122. Academic Press Japan, Inc. Tokyo.

(76) Butler, W.L. and Matsuda, H. (1984) Reconstitution of Photosynthetic Oxygen Evolution in Lipsomes. IN: Advances in Photosynthesis Research. Volume I (edited by C. Sybesma), pp. 749-754. Martinus Nijhoff/Dr. W. Junk Publishers, The Hague, The Netherlands.

G.   Inactivation and Reactivation of Photosystem II

(77) Yamashita, T. and Butler, W.L. (1968) Donation of Electrons to Photosystem II in Chloroplasts by p-Phenylene Diamine. IN: Comparative Biochemistry and Biophysics of Photosynthesis (edited by Shibata, K., Takamiya, A., Jagendorf, A. T., and Fuller, R.C.), pp 179-185 University of Tokyo Press, Tokyo.

(78) Yamashita, T. and Butler, W.L. (1968) Photoreduction and Photophosphorylation with Tris-washed Chloroplasts. Plant Physiol. 43: 1978-1986.

(79) Yamashita, T. and Butler, W.L. (1968) Inhibition of Chloroplasts by UV-Irradiation and Heat Treatment. Plant Physiol. 43: 2037-2040.

(80) Yamashita, T. and Butler, W.L. (1969) Inhibition of the Hill Reaction by Tris and Restoration of Electron Donation to Photosystem II. Plant Physiol. 44:435-438.

(81) Yamashita, T. and Butler, W.L. (1969) Photooxidation by Photosystem II of Tris-washed chloroplasts. Plant Physiol. 44: 1342-1346.

(82) Yamashita, T. and Butler, W.L. (1969) Electron Donation to Tris-washed Chloroplasts by Artifical Electron Donors. Progress in Photosynthesis Research 3:1236-1240.

(83) Lozier, R., Baginsky, M. and Butler (1971) Inhibition of Electron Transport in Chloroplasts by Chaotropic Agents and the Use of Manganese as an Electron Donor to Photosystem II. Photochem. Photobiol. 14:323-328.

(84)   Okayama, S. and Butler, W.L. (1972) Extraction and Reconstitution of Photosystem II. Plant Physiol. 49:769-774.

(85)   Lozier, R.H. and Butler, W.L. (1973) Effects of Photosystem II Inhibitors on Electron Paramagnetic Resonance Signal II of Spinach Chloroplasts. Photochem. Photobiol. 17:133-137.

(86)   Okada, M., Kitajima, M. and Butler, W.L. (1976) Inhibition of Photosystem I and Photosystem II in Chloroplasts by UV Radiation. Plant and Cell Physiol. 17:35-43.

H.   Development; Greening

(87)   Butler, W.L. (1960) Energy Transfer in Developing Chloroplasts. Biochem. Biophys. Res. Commun. 2:419-422.

(88)   Butler, W.L. (1961) Chloroplast Development: Energy Transfer and Structure. Arch. Biochem. Biophys 92:287-295.

(89)   Butler, W.L. (1965) Development of Photosynthetic Systems I and II in a Greening Leaf. Biochim. Biophys. Acta 102:1-8.

(90)   Butler, W.L. and Briggs, W.R. (1966) The Relation Between Structure and Pigments During the First Stages of Proplastid Greening. Biochim. Biophs. Acta 112:45-53.

(91)   de Greef, J., Butler, W.L. and Roth, T.F. (1971) Greening of Etiolated Bean Leaves in Far-Red Light. Plant Physiol. 47:457-464.

(92)   Oelze-Karow, H. and Butler, W.L. (1971) The Development of Photophosphorylation and Photosynthesis in Greening Bean Leaves. Plant Physiol. 48:621-625.

(93)   Oelze-Karow, H. and Butler, W.L. (1972) Cyclic Photophosphorylation in Developing Bean Leaves. IN: Proceedings of the IInd International Congress on Photosynthesis Stresa 1971 (edited by Forti, G., Avron, M. and Melandri, A.) pp 2401-2406. Dr. W. Junk N.V. Publishers, The Hague.

(94)   Butler, W.L., de Greef, J., Roth, T.F. and Oelze-Karow, H. (1972) The Influence of Carbonylcyanide-m-chlorophenylhydrazone and 3-(3,4-Dichlorophenyl)-1,1-dimethylurea on the Fusion of Primary Thylakoids and the Formation of Cyrstalline Fibrils in Bean Leaves Partially Greened in Far Red Light. Plant Physiol. 49:102-104.

(95)   Baker, N.R. and Butler, W.L. (1976) Development of the Primary Photochemical Apparatus of Photosynthesis During Greening of Etiolated Bean Leaves. Plant Physiol. 58:526-529.

(96)   Strasser, R. and Butler, W.L. (1976) Correlation of Absorbance Changes and Thylakoid Fusion with the Induction of Oxygen Evolution in Bean Leaves Greened by Brief Flashes.

Plant Physiol. 58:371-376.

(97)   Strasser, R.J. and Butler, W.L. (1976) Energy Transfer in the Photochemical Apparatus of Flashed Bean Leaves. Biochim. Biophys. Acta 449:412-419.

(98)   Strasser, R.J. and Butler, W.L. (1977) The Yield of Energy Transfer and the Spectral Distribution of Excitation Energy in the Photochemical Apparatus of Flashed Bean Leaves. Biochim. Biophys. Acta 462:295-306.

(99)   Strasser, R.J. and Butler, W.L. (1978) Energy Coupling in the Photosynthetic Apparatus During Development.   Proceedings of the 4th International Congress on Photosynthesis 1977, pp 527-536. Biochem. Soc. London.

(100) Ley, A.C. and Butler, W.L. (1980) Effects of Chromatic Adaptation on the Photochemical Apparatus of Photosynthesis in Porphyridium cruentum. Plant Physiol. 65:714-722.

I. Gas Exchange

(101)   Butler, W.L. (1955) Measurement of Photosynthetic Rates and Gas Exchange Quotients During Induction Periods. Ph.D. Thesis, University of Chicago.

(102)   Butler, W.L. (1957) Transient Phenomena in Leaves as Recorded by a Gas Thermal Conductivity Meter. IN: Research in Photosynthesis, edited by Gaffron, H., Brown, A.H., French, C.S., Livingston, R., Rabinowitch, E.I., Strehler, B.L. and Tolbert, N.E., pp 399-405. Intersience Publishers, New York.

(103)   Butler, W.L. (1960) A Secondary Photosynthetic Carboxylation. Plant Physiol. 35:233-237.

I.   **Photosynthetic Unit; The Antenna System; and the Photosynthetic Pigments**

Figure 2. Norman I. Bishop (author of dedication) and James Franck (Professor of Warren), 1963

Figure 3. The Butler family, Vancouver, Canada.

.

Photosynthesis Research 10: 163–170 (1986)
© Martinus Nijhoff Publishers, Dordrecht

THE OPTICAL CROSS SECTION AND ABSOLUTE SIZE OF A PHOTOSYNTHETIC UNIT

David MAUZERALL

The Rockefeller University, 1230 York Avenue, N.Y., N.Y. 10021

KEY WORDS: Photosynthetic unit, Optical cross section, Poisson distributions, Traps or reaction centers, Single turnover flashes, Quantum requirement.

ABSTRACT
    The concepts of a photosynthetic unit (PSU) and of an optical cross section are defined. The various estimates of sizes of photosynthetic units are described, and it is shown how an unambiguous measurement of the size of a unit can be obtained by measurement of its optical cross section via the saturation response to a single turnover light flash. The Emerson-Arnold unit must be divided by the quantum requirement for oxygen to obtain the true size of the unit. The size so obtained is the average number of chlorophylls per trap or reaction center. The effects of escape from open and closed traps are considered and it is shown that when these escape probabilities are equal, their effect on the saturation curve vanishes, leaving the simple cumulative one hit Poisson distribution.

CONCEPT OF THE PHOTOSYNTHETIC UNIT
    The concept of the photosynthetic unit (PSU) originated with the extraordinary experiments of Emerson and Arnold [1, 2, 3,] and from the prescient calculations of Gaffron and Wohl [4]. The former found that only one molecule of oxygen was produced per 2500 chlorophyll (Chl) molecules when Chlorella cells were illuminated with brief ($< 10^{-5}$ s) but saturating flashes of light. The latter pointed out that a steady state production of oxygen could be reached well before each chlorophyll molecule in the sample absorbed a photon in weak continuous light. These insights lead to the view that many chlorophyll molecules contribute absorbed photons to a reaction center where the photochemistry takes place. Following some confusion, the development of the theory of molecular energy transfer by Forster [5] put the hypothesis on a firm footing. Thus the "size" of the Emerson-Arnold PSU could be definitely stated as 2500 Chl/O_2 but its meaning became obscure as the evidence for two photosystems and multi-trapped units arose.

SIZE OF THE PHOTOSYNTHETIC UNIT
    Over the years it became customary to divide the Emerson-Arnold PSU by four (the number of electrons necessary to make oxygen assuming one electron per photon), by eight (doubling the above because of the two serial photosystems), or by ten (the measured quantum requirement for formation of oxygen) to obtain the "actual" size of the PSU. As we will see, the last choice is correct for the size of the photosynthetic unit that forms O_2 (PSU-O_2).
    With the discovery by Kok [6] of the reaction center of one of the photosystems, P700, and the finding of other components of the photosystems present in small numbers such as cyt b_{559}, it has become popular to divide the total number of chlorophylls by the number of molecules of component z to obtain a PSU-z size. While useful as an internal measure

of changes as in light adaption or development, these numbers tend to be assigned a value which they simply do not possess. The total chlorophyll lumps together systems I and II plus possible cyclic reactions and there is no way to separate them. Yet they are often referred to as the size of system I or II. The problem is aggravated by observations that the ratio of systems I and II can be far from unity and in either direction [7, 8, 9, 10, 11]. However, Whitmarsh and Ort (12) with careful analysis claim the ratio in spinach chloroplasts is 1 ± 0.1. There is also the problem of determining the precise number of these components which may be present in a center. The average number need not even be integral.

Attempts have been made to deduce PSU sizes from rate measurements in steady state light. Unless carefully analyzed, such measurements can intermingle cross sections, yields and turnover times. Many measures have bypassed the idea of a PSU size and stressed the fraction of absorbed light which excites system I or II, *e.g.* the α and β of Butler's analysis [13]. The terms are simply related; *e.g.*, β the fraction of absorbed light captured by system II is: $\beta = \sigma_{II} / (\sigma_{II} + \sigma_{I} + \sigma_{x})$, where σ_{x} (see below) is the cross section of systems I and II (O_2) and of cyclic systems or losses (x). Many analyses simply assume σ_{x} to be zero.

It was consideration of the need for equal excitation of the serially linked systems I and II that led Myers and Graham [14] to first postulate "spillover" from system II to system I. However, the finding of unequal numbers of the two photosystems in many organisms has considerably complicated this simple picture. Butler attempted to bypass these problems by using low temperature fluorescence at specific wavelengths as a measure of (relative) excitation of systems I and II. His value for β in chloroplasts was 0.3, whereas measurements using fluorescence induction at room temperature gave a value of 0.5 [15]. More recently, Malkin *et al.* [16] proposed a method using this fluorescence induction to obtain PSU-II sizes in leaves and other optically dense systems. However in using this method it is simply assumed that β = 0.5. Thus all these approaches leave much to be desired in determining the absolute size of a PSU.

OPTICAL CROSS SECTION AND THE ABSOLUTE SIZE OF A PSU

Absorption of light in solutions is usually described by Beer's law which relates concentration of molecules, path length and absorbancy index to the absorbance of the solution. The uncritical transfer of this description to heterogeneous systems such as algae leads to considerable confusion over "concentration" and "path length." These difficulties are readily avoided by the use of an optical cross section, σ with units of area [17]. When multiplied by fluence E, photons per area, one simply has the average number of photons absorbed by the sample. In solution σ per molecule in A^2 is equal to the decadic molar absorbancy index times 3.82×10^{-5}. Because of heterogeneous absorption and scatter, σ in a cell is unlikely to be equal to that in solution, particularly in regions of strong absorption (see below). The *in vivo* absorption must be directly measured and compared to the same amount of pigment (s) in solution. Assuming the sample isotropic, the above relation can be used to convert absorbance index to cross sections. If one can measure the cross section of a PSU, then dividing by the *in vivo* cross section of the pigments at this wavelength results in the absolute size of a PSU.

The direct way to determine σ of a PSU is by measuring the yield of a a product or intermediate (with due care for kinetics) as a function of the energy of a homogeneous, saturating, single turnover flash of monochromatic light on an optically thin sample. The energy of the flash can

be expressed as a fluence: photons per unit area. For a simple, Poissonian system, the yield, Y_i of a photoproduct, i, is given by:

$$Y_i = \phi_i \ (1 - e^{-\sigma_i E}) \tag{1}$$

where ϕ_i is the yield of product per hit, σ_i is the optical cross section for the unit producing i and E is the energy fluence in photons per
unit area [18]. Then $\sigma_i E$ is the average number of hits per target. It
is assumed that E is uniform and constant across the sample $i.e.$, the
sample is optically thin. A "thick" sample requires integration of eq (1)
over the optical path. It is also assumed that the sample absorbs iso-
tropically. Energy transfer in the PSU with molecules distributed over a
range of inter-transition moment angles will guarantee this. The extensive depolarization of emission observed at most wavelengths confirms iso-
tropy in photosynthetic systems. The measured cross section σ_i is the
effective cross section, $i.e.$, it contains any losses between absorption
and hitting or closing of a photochemical trap. Since the quantum yield
of elementary reactions in photosynthetic systems is $\geqslant$ 95%, the distinc-
tion is often ignored. By a simple Poissonian system we mean one which
responds linearly and where one or more hits to the system produce the
same effect as one hit: thus the cumulative one hit Poisson distribution.
Since $\exp(-\sigma_i E)$ is that fraction of targets not hit, eq (1) follows. The
1/e point of the normalized yield gives the $1/\sigma$ point on the fluence
axis. This was clear to Arnold in the thirties but conclusive experiments
had to await the availability of pulsed lasers. Kohn [19] attempted the
measurement using a flash lamp and filters but the results were uncertain.
Weaver and Weaver [20] measured σ by using the ESR signal of P_{700} as
the indicator. They obtained a value of 30 A^2 at 694 nm corresponding to
120 Chl molecules. Sorokin [21] determined the size of system II unit by
fluorescence measurement on pea chloroplasts as 150 Chl molecules, with a
σ_{II} at 436 nm of 300 A^2.

It may be thought that the Poisson distribution applies only to the
single trap or "puddle" arrangement of traps in a unit and not to the
multi-trapped or "lake" model. In the latter model the excitations are
assigned a probability of escaping from a hit with a closed trap and
moving on to find an open trap [22]. This thought is incorrect for the
case where the probabilities of escape from an open or a closed trap are
equal [23]. For this, possibly not uncommon, case one obtains the simple
Poisson distribution. However, the measured cross section is the average
per trap $i.e.$ σ/T, where T is the number of traps per connected unit.

It is important to have a clear understanding of the term "probability
of escape". It is the probability that an excitation moving out of a trap
reaches a point where it has equal probability of being captured by
another (open or closed) trap as by the original trap. An excitation
which only "hangs around" a trap affects (and complicates) only the micro-
kinetics, it has no effect on the statistics of hits. If one makes the
restrictive assumption that the fraction of closed traps is constant ir-
respective of closing (or opening) of traps during the process, one can
sum the geometric series describing the repeated trials and obtain [24]:

$$P = \frac{(1 - A) \ C/T}{1 - B + (B - A) \ C/T} \tag{2}$$

where P is the probability that an excitation is not trapped in a unit
containing C closed traps out of T total Traps, A is the probability of
escape from closed traps, and B is that from open traps. If B = 0,

$$P_A = \frac{(1 - A)\ C/T}{1 - AC/T} \qquad (3)$$

we recover the Joliots' equation [22], with A = p, C/T = q in their nota-
tion.
If A = 0,

$$P_B = \frac{C/T}{1 - B(1 - C/T)} \qquad (4)$$

and if A = B,

$$P_{AB} = C/T \qquad (5)$$

there is no observable effect of the "escapes"!

Unfortunately the restrictive assumption of constant fraction of closed
traps does not take into account the stochastic, consecutive nature of
trapping and escape. The fraction of closed traps, C/T changes during the
repeated trials by an excitation until it is "fully" trapped or lost.
This effect enters as more than one photon per trap hits the units, an absolute necessity for a flash to be saturating. We can assume multi-photon
effects are not present $i.e.$ the excitation rate if < 1 ns $^{-1}$. Thus the
above widely used equations are inadequate. A three line algorithm (see
appendix) suffices to do the calculation more correctly. The numerical
result is obtained that again when A = B, there will be small effect of
escapes on the observed shape of the saturation curve: it will be close
to Poissonian. When A > B, the saturation curve will be sharper, and when
B > A it will be more shallow than Poissonian [23]. The shape of the
curve is also a weak function of T, the number of traps per unit.

It must also be clear that these equations in no way supplant the use
of Poisson distributions. One must do the calculations for units hit
1,2 ...n times, each weighted by the n hit Poisson factor for the average
number of hits at the given flash energy. The Poisson statistics are man-
dated by the experiment: a very large number of targets each of area much
smaller than the illumination area, together with a large number of
photons.

One can obtain similar conclusions following the far more elaborate
Master Equation, or micro-kinetic approach of Paillotin $et\ al.$ [25] or the
macro-kinetic approach of Sonneveld $et\ al.$ [26]. The objection to these
treatments when applied to yields is the unnecessary kinetics and surfeit
of parameters. The yields are given by ratios of sums and products of
individual rate constants and many combinations of these will produce similar yields. Moreover the kinetics of the trapping and detrapping are almost certainly in the subnanosecond time range. Thus the kinetics observed on the longer time scales are simply the convolution of the yield
function with the time shape of the light pulse, and contain no novel information. Even the multi-photon effects on yield are readily treated by
the statistical approach [23].

Arthur Ley and I [27] used this method and analysis on the yield of
oxygen evolution from Chlorella. We found that for standard Chlorella,
σ_{O_2} was 90 A^2 at 596 nm. This corresponds to 300 Chl molecules (a + b),
since the optical cross section of a Chl molecule (σ_{chl}) in Chlorella at
596 nm is 0.3A^2. There are several unique properties to this measurement.
Nowhere does the total Chl content of the sample enter. In fact the presence of 'dead' cells has no effect as long as the sample is optically
thin! One measures only those chlorophylls which contribute an excitation
to the oxygen forming apparatus. If a Chl molecule delivers a fraction of

its excitation to the O_2 system, we measure only that fraction of a molecule. The measured σ is <u>independent</u> of the <u>efficiency</u> of this apparatus as long as it is constant <u>during a measurement of the yield</u> as a function of flash energy. The efficiency of O_2 evolution is in the ϕ term in eq (1) and it cancels in the normalization of the measured O_2 yields. Since the measurement is carried out in the steady state, the σ measured is the <u>average</u> of all that of the S-states of the O_2 cycle. It can be extended to measure σ of individual S_i state, by varying the energy of the ith flash.

THE EMERSON-ARNOLD UNIT AND THE CROSS SECTION OF THE O_2 UNIT

These concepts can be used to interrelate the classical Emerson-Arnold size of the Photosynthetic Unit, PSU and the quantum requirement (QR, reciprocal of the quantum yield) for O_2 formation. Under optically thin conditions, the total quanta absorbed by a sample containing n Chl molecules and illuminated by a single turnover flash of energy E is n $\sigma_{cell} E$. In steady state, the O_2 produced by each flash is $(1 - \exp(-\sigma_a E))n/PSU_{O_2}$, where n/PSU_{O_2} is the yield, ϕ_{O_2}, for that particular sample: total Chl divided by Chl per O_2 per flash. The minimum quantum requirement occurs at low energy, when $(1 - \exp(-\sigma_a E)) \rightarrow \sigma_a E$. Thus

$$QR = \frac{\text{Photons absorbed}}{O_2 \text{ out}} = \frac{\sigma_{cell}}{\sigma_{O_2}} \cdot PSU_{O_2} \qquad (6)$$

Thus to obtain the true size of the unit for O_2 production, $\sigma_{O_2}/\sigma_{cell}$, one divides the Emerson Arnold Unit size by the quantum requirement for O_2. Equation (6) summarizes the relationship between parameters which have been the subject of 50 years of research and controversy. While it may be overly optimistic to predict the end of the controversy, it is nevertheless encouraging that placing our values of σChl, PSU$_{O_2}$ and σ_{O_2} into equation (6) gives a quantum requirement of 10 ± 1, just the number Emerson and Lewis [28] obtained for <u>Chlorella</u> some time ago. Thus equation (6) has been verified.

The fit of the data on O_2 formation in <u>Chlorella</u> to eq (1) was quite good (±2%) and σ_{O_2} can be determined to about ±5%, part of the error being that of estimating area and light energy. However, this does not mean that each oxygen unit in a Chlorella has precisely this σ_{O_2}. A square distribution of σ over a factor of three may fit the data, but the same distribution over a factor of 10 clearly does not [27]. Similarly, a bimodal distribution of equal amplitudes and $\sigma_x \geqslant 3\sigma_1$, could be detected, but a 15% contribution with $\sigma_2 = 2\sigma_1$, could not. A nonuniform illumination of the sample will broaden the saturation curve in the same way as a bimodal distribution of cross sections. Claims for bimodal distributions in the literature should be evaluated in terms of uniformity of illumination.

The claims of heterogeneity in size of system II centers (see *e.g.* [9]) have been based on absorption changes at 325nm and on fluorescence induction curves. The former measurements have been corrected by Whitmarsh and Ort [12] and by Jursinic and Dennenberg [29], and the latter method has been rightly criticized by Bell and Hipkins [30]. In any case, the small heterogeneity postulated in <u>Chlorella</u> would be within the error of our measurements, and thus of no great consequence to the organism.

We have shown that σ_{O_2} does <u>not</u> increase on closing traps with a background of continuous light [24] contrary to the common belief of a high probability of escape on hitting closed traps. The data is consistent only with a modest escape probability ($A \leqslant 0.25$ if $B = 0$), otherwise $(1 - B)/(1 - A) < 1.5$). The usually quoted value of escape is about 0.5 [15, 22]. Previous measurements however, involve the use of fluorescence to determine the fraction of closed traps, and the use of eq 3, both of which are objectionable as described earlier. Our recent measurements show that fluorescence yields in five species of eucaryotic algae measured by pump and probe flashes are a much more complex function of irradiance than is the steady state O_2 yield [31]. Again the optical cross section was constant within ±10%. Thus we believe there is no strong evidence that excitations have a much higher probability of escaping from closed traps than from open traps.

THE CROSS SECTION OF SYSTEM I

There remains the problem of determining the cross section (σ_I) of system I. Unlike the oxygen forming system, there is usually no stable product from system I. Thus the formation of an intermediate such as P_{700}^+ has been followed by optical [6] or ESR [20] spectroscopy. A stable photoproduct can be formed by utilizing the observation of Gaffron and Rubin [32] that many green algae adaptively form a hydrogenase under an-aerobiosis. Greenbaum [33] has carried out such a measurement and has shown that the PSU size à la Emerson Arnold are about equal for system I and II. An indirect method is to use an acceptor which immediately re-acts with O_2 on reduction, e.g. methyl viologen. If the formation and up-take cannot be separated in time, the O_2 system would have to be poisoned. A method that bypasses many of the difficulties associated with the above procedures has been developed by Ley [in preparation]. It utilizes time dependent changes in O_2 uptake by intact cells and so system II and I can be measured under identical conditions [34].

THE USE OF CROSS SECTIONS

There are several reasons for wishing to obtain the values of σ_i. The first is the solution of eq (6). Now that this equation is estab-lished, one can obtain the true PSU-O_2 size ($\sigma_{O_2}/\sigma_{chl}$) in other algae by simply determining the Emerson-Arnold O_2 unit size and the quantum re-quirement for O_2. The former is easy, the latter is more difficult, but less daunting than our absolute determination of σ_{O_2}. Ley has already shown that the absolute O_2 unit size varies with the light environment of the algae [35]. Second, one can now obtain the true number of active O_2 systems per cell (N_{O_2}/cell) since in steady state O_2 production ($\dot{O}_2$/cell) in the linear light intensity (L) regime:

$$\dot{O}_2/cell = (N_{O_2}/cell)\, \sigma_{O_2}\, L \qquad (7)$$

This is important in comparing the number of supposed system I and II centers in algae and in comparisons with steady state measurements such as in [13, 14]. Third, one can now determine the fraction, f, of these O_2 units that are active in a single, saturating turnover flash since

$$PSU_{O_2} = \frac{N_{chl}/cell}{O_2/cell} = \frac{N_{chl}/cell}{f\, N/cell} \qquad (8)$$

where N_{chl}/cell is the number of chlorophylls per cell. Finally there is the honesty criterion:

$$\sigma_{chl} \cdot N_{chl}/cell = \sigma_{O_2} \cdot N_{O_2}/cell + \sigma_I \cdot N_I/cell + \sigma_x \qquad (9)$$

Where σ_I is the cross section of system I and N_I the number of system I units per cell, determined by these pulse methods. By combining a good determination of σ_I with σ_{O_2} and total yields we can see how close one can account for the total Chl in a cell. For example if 20% of the Chl remains for $\sigma_{\bar{x}}$, then the "extra" 2 photons in the quantum requirement (10 versus 8 for 4 electrons and 2 systems) could be assigned to a cyclic pathway. If on the other hand, $\sigma_{\bar{x}}$ = 0 within error, then the 20% extra photons must be assigned to inefficiency in the trapping.

ACKNOWLEDGEMENT

I thank Arthur Ley for many clarifying discussions of these ideas. This research was supported by the National Science Foundation (PCM 83-16373).

APPENDIX

Algorithm for calculation of yields given the probability of escape from closed (A) and open (B) traps [23, 36].

(1) $E(1 - B)$ $(T - C)/$ $T \longrightarrow H.$
(2) $EAC/$ $T + BE(T - C)/$ $T \longrightarrow E.$
(3) $H + C \longrightarrow C.$

The first line calculates the new closed traps H as the excitation E times the probability of closing $(1 - B)$ times the fraction of open traps $(T - C)/$ T, where B is the probability of escape from open traps, T is the total number of traps per unit and C is the number of closed traps. The second line calculates the residual excitations: the excitation E times the probability of escape A times the fraction of closed traps plus those reflected in the open traps, $BE(T - C)$ T. The third line adds the newly closed traps to those already closed. The initial conditions are $C = 1$ (from the first hit) and $E_{\bar{n}}$ 1 (the nth hit). This calculation is cycled until E is small (e.g., 10^-). C is then multiplied by the Poisson weight for the average hits $x = P_n$ and summed to a yield. Then n is incremented, and a new cycle begins, using the calculated C and E = 1. When the contributions from P_n and C are negligible, the yield at a given x is available.

REFERENCES

1. Emerson R and Arnold W (1932) J Gen Physiol 15: 391-420
2. Emerson R and Arnold W (1933) J Gem Physiol 16: 191-205
3. Arnold W and Kohn HI (1934) J Gen Physiol 18: 109-112
4. Gaffron H and Wohl W (1934) Naturwiss 24: 81-90
5. Forster T (1948) Ann Physik 2: 55-75
6. Kok B (1956) Biochim Biophys Acta 22: 399-401
7. Kawamura M, Mimuro M and Fujita Y (1979) Plant Cell Physiol 20: 697-705
8. Melis A and Theilen A (1980) Biochim Biophys Acta 589: 275-286
9. Melis A and Brown JS (1980) Proc Natl Acad Sci USA 77: 4712-4716
10. Falkowski PG, Ownes TG, Ley AC ad Mauzerall DC (1981) Plant Physiol 68: 969-973
11. Myers J and Graham JR (1983) Plant Physiol 71: 440-442
12. Whitmarsh J and Ort DR (1984) Arch Biochem Biophy 231: 378-389
13. Butler W (1978) Ann Rev Plant Physiol 29: 345-378
14. Myers J and Graham JR (1963) Plant Physiol 38: 105-116
15. Malkin S and Kok B (1966) Biochim Biophys Acta 126: 413-432
16. Malin S, Armond PA, Mooney HA and Fork DC (1981) Plant Physiol 67: 570-579
17. Herron HA and Mauzerall D (1972) Plant Physiol 50: 141-148
18. Mauzerall D (1980) Adv Biol Med Phys 17: 173-198
19. Kohn H.I. (1936) Nature 137: 706
20. Weaver EC and Weaver HE (1969) Science 165: 906-907

21. Sorokin EM (1969) Fiziol Rast 18: 473-479
22. Joliot A and Joliot P (1964) C R Acad Sci: Paris 258: 4622-4625
23. Mauzerall D (1982) in : Biological Events Proved by Ultrafast Laser Spectroscopy (Alfano R, ed) pp. 215-235, Academic Press, New York
24. Ley A and Mauzerall D (1985) Biochim Biophys Acta, in press
25. Paillotin G, Gaecintov NE and Breton (1983) Biophys J 44: 65-78
26. Sonneveld A, Rademaker H and Duysens LNM (1979) Biochim Biophys Acta 548: 536-551
27. Ley A and Mauzerall D (1982) Biochim Biophys Acta 680: 95-106
28. Emerson R and Lewis CM (1943) Am J Bot 30: 165-178
29. Jursinic P and Dennenberg R (1985) Arch Biochem Biophys 241: 540-549
30. Bell DH and Hipkins MF (1985) Biochim Biophys Acta 807: 255-262
31. Falkowski PG, Wyman K, Ley AC and Mauzerall D (1986) Biochim Biophys Acta, in press
32. Gaffron H and Rubin J (1942) J Gen Physiol 26: 219-240
33. Greenbaum E (1977) Science 196: 878-879
34. Greenbaum N and Mauzerall D (1986) Abstract, Meeting of American Society for Photobiology, June 23-27, Los Angeles
35. Ley, A (1984) Plant Physiol 74: 451-454
36. Mauzerall D (1981) in Proceedings of the 5th International Congress on Photosynthesis (Akoyunoglou G, ed) pp. 47-58, Balaban International Science Services, Philadelphia

Photosynthesis Research 10: 171–180 (1986)
© Martinus Nijhoff Publishers, Dordrecht

ORGANIZATION OF THE PHOTOSYNTHETIC UNITS, AND ONSET OF ELECTRON TRANSPORT
AND EXCITATION ENERGY DISTRIBUTION IN GREENING LEAVES

G. AKOYUNOGLOU AND J.H. ARGYROUDI-AKOYUNOGLOU
Biology Department, Nuclear Research Center "Demokritos", Athens, Greece-

ABSTRACT
    The development and organization of the Photosynthetic units follow a
step-wise assembly process. First the core complexes of the PSI and PSII
units are formed, followed by their light-harvesting components; then an
assembly process of these components into supramolecular structures takes
place. Parallel to this, the control of excitation energy distribution between the two photosystems is established. This control is attributed to
the modulation of the PSI unit effective cross section, which is possible
only when LHC-I is formed and assembled into CPIa. Parallel to the formation of PSI and PSII, the electron carriers are synthesized and the electron transport chain is assembled. The number of PSII units operating per
electron transport chain remains constant throughout development and equal
to that of the mature chloroplast, but the number of PSI units per chain
varies with PSII unit size. During development, when the rate of Chl$\underline{a}$ synthesis is low, relative to the other thylakoid components, or is completely
stopped, then the newly formed or preexisting LHC-I and LHC-II proteins are
digested and their Chl$\underline{a}$ is used for the formation of PS core complexes.

1. INTRODUCTION
    In photosynthetic membranes electrons flow from H$_2$O to NADP$^+$ via two se-
parate photoreactions, PSI and PSII, connected in series, according to the
generally accepted Z scheme. Light is absorbed by two pigment assemblies
(photosynthetic units) consisting of proteins, pigments, lipids and electron
carriers. All pigments are organized in the PS units, associated with proteins and lipids in pigment-protein complexes. Each PS unit consists of (a)
the "core" of the unit (the minimum components to carry out the primary photochemical event), a small number of Chl$\underline{a}$ and b-car, which act as antennae,
and the enzymes and cofactors that stabilize the primary charge separation;
(b) the light-harvesting (LHC) pigments (Chl$\underline{a}$, Chl$\underline{b}$ and car) bound to protein. Light excitation reaching a unit can be either used for photochemistry
or dissipated as fluorescence or heat. The efficiency of the photosynthetic
process can be studied by means of the fluorescence emitted. At 20°C the
fluorescence is emitted mostly from the Chl$\underline{a}$ of PSII (main band max at 685
nm) (44); at 77°K the fluorescence is emitted from both photosystems (band
max at 683 and 695 nm from PSII, and 720 to 735 nm from PSI) (25, 40). Upon
illumination of dark adapted chloroplasts at 20°C, the fluorescence intensity rises from an initial level, Fo, to a steady state level, Fm. The increase in the fluorescence yield is thought to be related to the photoreduction
of the primary electron acceptor Q (37). In the presence of DCMU, the rate
of fluorescence rise represents the rate of Q reduction, which depends on
the rate by which the absorbed photons reach the reaction center of PSII.·
Thus, it depends on the intensity and quality of the actinic light, the number of functioning LHC Chl per PSII unit, and the efficiency of the unit to

utilize the absorbed photons. Assuming that the efficiency remains constant, and under constant actinic light, the half-rise time ($t\frac{1}{2}$) of the fluorescence induction is a measure of the relative size of the PSII unit (3,35). The fluorescence yield at 20°C, and the F735/F685 ratio at 77°K, are affected by cations (42,47), and by preillumination with PSI or PSII light (26,48). Cations added to "low-salt" chloroplasts increase the variable fluorescence yield and decrease the F735/F685 ratio. Similarly, preillumination with PSI light induces an increase in the variable fluorescence yield at 20°C, and a decrease in the F735/F685 ratio (State I), while preillumination with PSII light reverses the effect (State II). It has been shown in algae, that State I corresponds to the state with Mg^{2+}, and State II to the state without Mg^{2+} (23). Recent work has shown, however, that cations and PSI light may not act in the same way (43,45), since synchronous cultures of *Scenedesmus obliquus* show the light-induced State transition in the 16th h, but not in the 8th h of development, while the cation-induced increase in the fluorescence yield is observed in both developmental stages (45).

The changes in the fluorescence yield of chloroplasts caused by cations and PSI-PSII light have been attributed to a control mechanism which regulates the distribution of excitation energy between the two photosystems, in order to optimize its utilization. This control mechanism operates through conformational changes of the thylakoid membrane. Two types of mechanisms have been proposed: that of spillover of excitation energy from the antennae of PSII to the reaction center of PSI (block of spillover in the presence of cations) (28,47,48); and that of modulation of the absorption cross section of the PS units (increase of the PSI unit cross section upon illumination with PSII light) (26,28). Recently it has been proposed that the changes in the absorption cross section of the units is due to changes in the supramolecular organization of the pigment-protein complexes, and especially those of the PSI unit (7, 16,31).

2. PIGMENT-PROTEIN COMPLEXES

The Chl in the thylakoid is complexed with intrinsic membrane proteins to form specific associations of pigment-protein complexes. This was first demonstrated when thylakoids solubilized in SDS were electrophoresed by PAGE (49,53,54); in those early studies 30% of the thylakoid Chl was detached from its protein counterpart and ran as free pigment; the rest remained complexed on two pigmented bands: CPI, rich in Chla with apparent Mr of 110 kDa and CPII, containing equal amount of Chla and Chlb, with apparent Mr of 35 kDa. CPI was thought to be derived from PSI, since it contained P700 and was missing from the PSI-less mutant of *Scenedesmus* (41). CPII, although at first considered to represent the PSII unit, was later shown to contain only the light-harvesting complex of PSII (LHC-II); this was suggested from the finding that the Chlb-less agranal intermittent light leaves, lacking CPII (20), are fully active in CO_2 fixation (5) and in partial PSII photoreactions (10). These plants, after transfer to continuous light, form large PSII units, parallel to the formation of Chlb and CPII (3). Recent milder SDS-PAGE procedures have been successful in separating more than 90% of the thylakoid Chl in association with proteins (11,15), allowing the separation of six major pigment-protein complexes from SDS-solubilized mature thylakoids: CPIa, CPI, LHCP1, LHCP2, CPa and LHCP3 (7,15), with apparent Mr of 190, 110, 90, 65,50, and 35 kDa, respectively (7). These complexes are real entities existing *in vivo*, as shown by C^{14}-labelling experiments (18). Separation of the pigment-protein complexes from developing thylakoids, which contain only the core of the PSI and PSII units (15), from barley mutants (11) or from subchloroplast particles enriched in PSI or PSII (14,15), have shown that CPIa and CPI originate in the PSI unit, while the LHCPs and CPa in the PSII unit. CPI comes from the core of the PSI unit, and contains Chla, P700 (30 Chl/P700), b-car and two polypeptides of about 69 kDa (7,38); its fluorescence emission peak at 77°K is at 725 nm (7,15,19). CPIa is the PSI unit in a solubilized form, as suggested from its identical polypeptide composition to that of the Triton X-100-extracted PSI particles (it contains the 69 kDa components of CPI, and polypeptides of 24, 23, 21, 20, 16-15, 10.5-9 kDa) (12,46); it contains Chla, Chlb, P700 (100 Chl/P700), lutein and b-car and its emission band at 77°K is at 732 nm (12). CPa comes from the core of the PSII unit, and the LHCP1, LHCP2 and LHCP3 are the trimeric, dimeric and monomeric form of LHC-II (7,15). CPa contains Chla, the RC of PSII, b-car, the 47 and 43 kDa polypeptides and a number of low Mr ones. Its emission band is at 695nm at 77°K (51). LHC-II contains Chla, Chlb (Chla/Chlb=1.2), xanth and a 25 kDa polypeptide; its emission band at 77°K is at 685nm (19). CPIa can be dissociated into its components, CPI and LHC-I after reelectrophoresis on gels containing Na+ in the stacking gel (11); LHC-I has an apparent Mr of 75 kDa, and contains Chla, Chlb (Chla/Chlb=3.4), xanth and a 21 kDa polypeptide; its emission at 77°K ranges from 717 to 730 nm, depending on the isolation procedure (7,12,46). Its electrophoretic mobility is identical to that of LHCP2; thus, the LHCP2 band isolated from SDS-solubilized thylakoids is a mixture of LHC-I and -II; and it is enriched in LHC-I whenever the thylakoid starting material is a fraction derived from PSI-rich stroma lamellae (12). The pigment/pigment ratio in each complex is constant from the beginning of their biosynthesis; however, the pigment/protein ratio in each complex depends on the developmental stage of the chloroplast, increasing as greening proceeds (7).

As mentioned above, cations induce the dissociation of CPIa into CPI and LHC-I; moreover they induce the dissociation of the oligomeric forms of the LHC-II into its monomer (16). The degree of dissociation depends on cation concentration present in the solubilization buffer or in the gel. The dissociation is accompanied by drastic changes in the 77°K emission spectra of

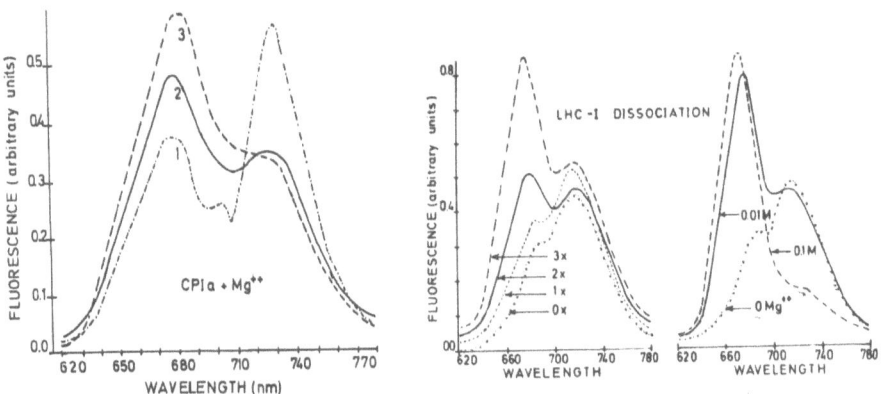

FIGURE 1. a) The effect of Mg^{2+} on the 77°K fluorescence spectra of CPIa, isolated from SDS-solubilized thylakoids by sucrose density gradient centrifugation. (1):0.09 mM Mg^{2+}, (2): 1 mM Mg^{2+}, (3): 1.6 mM Mg^{2+}. b) 77°K fluorescence spectra of LHC-I separated from CPIa by SDS-PAGE; spectra recorded *in situ* as affected by freeze thawing (left) or Mg^{2+}(right). Left: the gel slice was frozen in liquid N_2 and the spectrum recorded (0x), then was thawed and refrozen for 1 to 3 more times (1x-3x). Right: the gel slice was immersed for 1 min at 25°C in 0.01 M or 0.1 M Mg^{2+}, and the spectra recorded.

TABLE 1. PSI, PSII, P700 and cyt$\underline{f}$ content of plastids obtained from 6-d etiolated bean leaves exposed first to intermittent light (LDC), and then transferred to continuous light (CL).

Sample	Chl (a+b) (ug/g fresh wt)	P700	Chla Chlb	Activity/h g fresh wt PSI	PSII	PSII PSI	cytf (nmol/ g fr wt)
28 LDC	210	1.45	high	102	200	2.4	1.68
57 LDC	393	2.83	26.0	188	373	2.4	3.18
86 LDC	610	4.26	14.0	289	543	2.3	4.72
28 LDC+49h CL	2765	6.96	3.0	553	470	1.0	4.70
57 LDC+49h CL	1529	3.84	3.9	344	463	1.7	4.67
86 LDC+49h CL	1214	3.10	4.7	279	466	2.0	4.73
Green Control	2800	7.00	3.0	560	476	1.0	4.70

PSII activity:umoles DCIP reduced; PSI activity:umoles oxygen consumed. The PSII/PSI ratio was estimated on the assumption that in mature plastids it is equal to 1.0.

the detergent solubilized thylakoids, and of the isolated complexes, especially of those of the PSI unit (12, 16 , 48, 46). For example, the CPIa complex has a 77^0K emission band at 732nm; upon cation addition the band shifts to 678nm. Similarly, the LHC-I emits in the absence of cations at 719nm, and in their presence at 685nm (see fig. 1a and 1b). Since these changes are similar to those observed when cations are added to "low-salt" chloroplasts, and follow similar cation concentration dependence, it has been proposed that they may reflect the modulation of the PSI unit cross section, which occurs *via* the process of organization-dissociation of the PSI unit components (7, 16, 31).

3. FORMATION OF THE PSI AND PSII UNITS AND ONSET OF THE ELECTRON FLOW.

PSI activity appears soon after the etiolated leaves are exposed to light, while the time of PSII appearance varies, depending on the etiolated tissue age (10,24,27,36). The PSI activity on a Chl basis is high at the beginning, then it decreases and finally levels off. On the contrary, the PSII activity is initially low, increasing with exposure to light and finally reaches a plateau (4,10). In continuous illumination of normal light intensity, complete PSI and PSII units are formed from the beggining of greening; prolonged illumination results only in increasing the number of the units per chloroplast. However, under conditions where the rate of Chla formation is low, relatively to that of the other thylakoid components, small-sized PSI and PSII units are formed, containing only the core of the unit. When the rate of Chl formation is enhanced, the units increase in size by incorporation of LHC. This step-wise formation and growth of the units has been observed (a) in etiolated leaves exposed first to intermittent light (4,7,13), ms flashes (9, 52), far-red light (29,34,39), or vary low-light intensity (8,50) and then transferred to continuous light of normal intensity; and (b) in etiolated leaves transferred to darkness after a short preexposure to continuous light (1,2,7,17).

Leaves exposed to intermittent light: Greening in intermittent light (2 min light+98 min dark) results in limited Chla formation, since its synthesis is allowed to occur only during the 2 min period of each cycle (13). The "protochloroplasts" formed under these conditions have single "primary" thylakoids and no grana stacks (14). SDS-PAGE of "primary" thylakoids shows that only CPI and CPa (the core complexes of the PSI and PSII units) are present;

the LHC-I and LHC-II are missing (15). The existence of small-sized PSI and PSII units in intermittent light leaves is also supported by photosynthetic activity measurements: "protochloroplasts" show high PSI/Chl, PSII/Chl, P700/ /Chl ratios and require high light intensity for saturation of their PSI and PSII activities (3,4). In these plastids, the rate of the PSII unit formation is higher than that of the PSI, so that 2.4 times higher PSII/PSI ratios are attained than in control green chloroplasts. After transfer of the leaves to continuous light, all complexes are formed and the PSI/Chl, PSII/Chl, P700/ /Chl, the light requirement for saturation and the PSII/PSI ratio reach either values of mature chloroplasts, or intermediate ones, depending on preexposure to intermittent light (see Table 1). All these reflect the gradual increase of the PSI and PSII unit size, and suggest that the final stage of thylakoid development is a function of the mode of the PS unit formation and growth(3,4)

Parallel to the PSI and PSII unit formation all electron carriers are also synthesized, so that the entire electron transfer chain from H_2O to $NADP^+$is assembled and functions properly from the beginning of greening. The change in cytf concentration during greening is shown in Table 1. The rate of cytf formation follows closely the rate of the PSII unit formation. This indicates that throughout development, the number of PSII units operating on the same electron transport chain (PSII/cytf) remains constant and equal to that of mature chloroplasts, but the number of PSI units per electron transport chain (PSI/cytf) varies, and depends on the PSII unit size.

During greening in ms flashes (one flash every 15 min dark), far-red light or very low-light intensity, the rate of PChlide a to Chlide a phototransform-ation is slowed down, either because the time of illumination is very short (ms flash) or the light intensity too low; accordingly, the rate of Chla syn-thesis is much lower than the rate of the other thylakoid components (8,9,29, 34,39,50,52). Plastids formed under these conditions resemble in structure, composition and function those formed in intermittent light.

Leaves transferred to darkness after short preexposure to continuous light: Contrary to the situation observed with intermittent light plants, where Chla synthesis is low, exposure of etiolated leaves to continuous light indu-ces formation of Chla at a high rate from the onset of exposure; all pigment-protein complexes are formed under these conditions, the PSI and PSII units are of normal size, and the thylakoids have the composition of those of matu-re chloroplasts. The number of thylakoids per chloroplast increases gradually and stops after reaching the stage of the mature chloroplast (4). When leaves with chloroplasts still in the process of development are transferred to dark-ness, the following are observed (1,2,7,17): In the dark the Chla/leaf remains constant, the Chlb/leaf and the relative concentration of LHC-II and CPIa de-crease, while that of CPI and CPa increase; the 21 and 25 kDa polypeptides of LHC-I and LHC-II, respectively, decrease, the 42 and 47 kDa ones of CPa, and the 67 kDa of CPI increase. This suggests that new, small-sized PSI and PSII units are formed in darkness, as also demonstrated by the increased PSI/Chl, PSI/leaf, P700/Chl, PSII/Chl and PSII/leaf, and the increased light intensity requirement for PSI and PSII saturation. The rate of PSII formation is higher than that of PSI, and thus, the PSII/PSI ratio, which is constant throughout the continuous illumination, increases by about 2.3 times. These findings suggest that upon transfer of the leaves to darkness, Chla stops to be formed but the thylakoid polypeptides continue to be synthesized and inserted into the developing thylakoid. Since no more Chla is formed, the reaction center polypeptides remove the Chla from LHC-I and LHC-II to form new small-sized PS units (i.e., CPI and CPa). The LHC-I and LHC-II devoid of Chla are disorgani-zed; their Chlb is degraded and their polypeptides digested. After transfer-ring the leaves again to continuous light, new Chl is formed, the LHC comple-xes are reassembled and all the units increase in size.

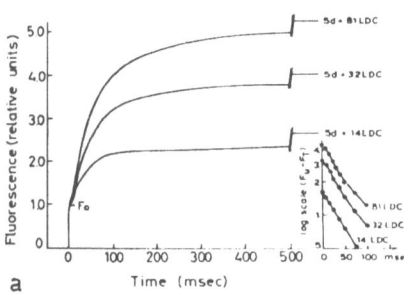
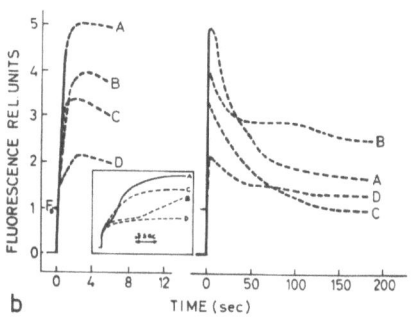

a   Time (msec)   b   TIME (sec)

FIGURE 2. a): Fluorescence induction in the presence of 12 uM DCMU of developing protochloroplasts. Exciting light intensity 575 uW/cm². Inset:log plot of
Fm-Ft for the same curves. Plastids isolated from 5-d etiolated bean leaves,
exposed to light-dark cycles (LDC, 2 min light-98 min dark). Fluorescence measurements at 5 ug Chl/ml. The curves are actual oscilloscope traces normalized
to the same Fo. b):Time course of the Chla fluorescence emission at 685 nm *in
vivo* of 5-d etiolated bean leaves exposed to LDC as above. (A): 87 LDC, (B):
27 LDC, (D): 13 LDC, (C): mature green bean leaf. Leaves kept in darkness for
8 min prior to fluorescence measurements.

4. ASSEMBLY AND INTERACTIONS OF THE PHOTOSYNTHETIC UNITS AS MONITORED BY FLUO-RESCENCE MEASUREMENTS

The appearance, assembly and interactions of the PS units has been also
studied by fluorescence techniques (3,6,19,24,27,35).

Young etiolated leaves show variable fluorescence very soon after exposure
to continuous light (3,24,27); the Fm/Fo ratio of the plastids is low and the
fluorescence rise in the presence of DCMU is slow, but soon the values reach
those of mature chloroplasts. The low Fm/Fo early in greening reflects the presence of newly formed and still unorganized Chl, which contributes to the non
variable fluorescence emission (Fo). Since the relative concentration of unorganized Chl decreases as greening proceeds, its contribution to the Fo level
is gradually decreased.

Plastids of intermittent light leaves, which contain only the core of the
PS units, show also the typical biphasic fluorescence induction kinetics. In
this case, however, the Fm/Fo and Fv/Fm ratios are higher, the fluorescence
yield (Fm/Chl) is lower and the t½ of the fluorescence induction in the presence of DCMU is larger (at least sevenfold) than those of mature plastids
(fig. 2a). All these reflect in fact the absence of LHC-II from these plastids.
The high Fv/Fm, which represents the yield of primary photochemistry, indicates that the Chl of the PSII unit core is more efficient in utilizing the absorbed energy than the Chl of LHC-II, which is missing. The low Fm/Chl and the
high t½ also reflect the absence of LHC-II and the small size of the PSII unit.
In the beginning of greening in intermittent light the slow rise of the fluorescence kinetics in the prescence of DCMU is exponential, and later becomes
sigmoidal (fig. 2a). The sigmoidicity becomes more pronounced as exposure to
intermittent light proceeds, indicating the gradual development of the excitation energy transfer between PSII units (3,4,6). The *in vivo* fluorescence
transient kinetics of intact intermittent light leaves are similar to those
of intact green leaves and algae, i.e., they show a fast increase to the Fo
level and then a slower rise to the steady state level through transient variations resulting in three peaks (P-to-S transient) (Fig. 2b) (6). This shows
that LHC-II and grana are not required for the P-to-S transition.

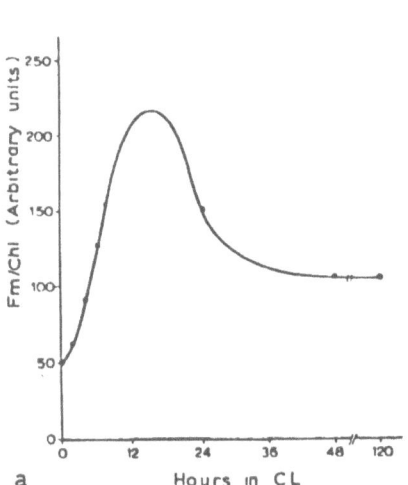
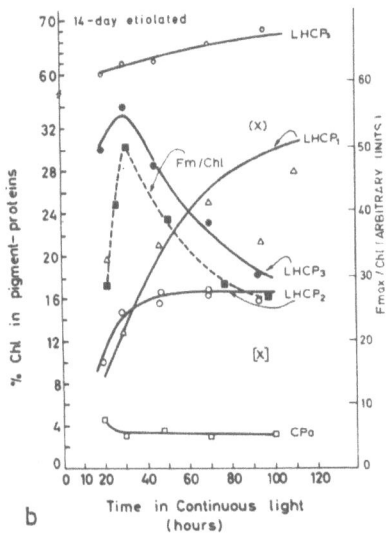

FIGURE 3. a): Changes in the fluorescence yield (Fm/Chl) of chloroplasts isolated from 6-d etiolated bean leaves exposed to 28 light-dark cycles and then transferred to continuous light. b): Distribution of Chl among the pigment-protein complexes of PSII in thylakoids of 14-d etiolated bean leaves exposed to continuous light (solid lines) and Fmax/Chl in chloroplasts (---). For fluorescence measurements, Chl was at 5 ug/ml.

Upon transfer of these leaves to continuous light, the Fm and Fo increase, while the $t\frac{1}{2}$ decreases. The final values depend on preexposure time to intermittent light; thus, the Fm/Fo and $t\frac{1}{2}$ reach either values of the green control (short preexposure) or higher values (long preexposure). Tne decrease in $t\frac{1}{2}$ correlates with the increase in the PSII unit size. The changes in Fm//Chl observed (fig. 3a) (initial increase to a max and subsequent decrease to a plateau level twice as high as the starting value) follow closely the changes in the biosynthesis and organization of the LHC-II. The initial increase parallels the formation of LHC-II, and indicates that the Chla of the LHC-II has a higher fluorescence yield than the antennae Chla of the PSII unit core. The rest of the changes follow closely the changes in the relative concentration of the monomeric and oligomeric forms of LHC-II occuring during greening; thus, the Fm/Chl is high when the monomers predominate, and low when the monomers are organized into oligomeric forms (fig. 3b) (4,31)

Since room temperature fluorescence changes show only the changes occuring in the PSII units, the 77^OK fluorescence emission spectra have been used to monitor changes occuring in the PSI unit as well. Early in greening in continuous light the 77^OK fluorescence emission spectra show a major peak at 680 nm and a satellite band at 735 nm, so that the F680/F735 ratio is high. Gradually the F680/F735 ratio decreases (the F735 increases and the F680 decreases) and after some time it levels off. At the same time the 680nm peak shifts to 683 nm and a shoulder at 695nm appears (55). The increase at 735nm indicates the formation of the PSI unit complexes (CPI and LHC-I) and their organization into the CPIa.

The absence of LHC-I and accordingly of CPIa, from protochloroplasts" and their formation after transfer of the intermittent light plants to continuous light, is also evident from the 77^OK fluorescence emission spectra of the

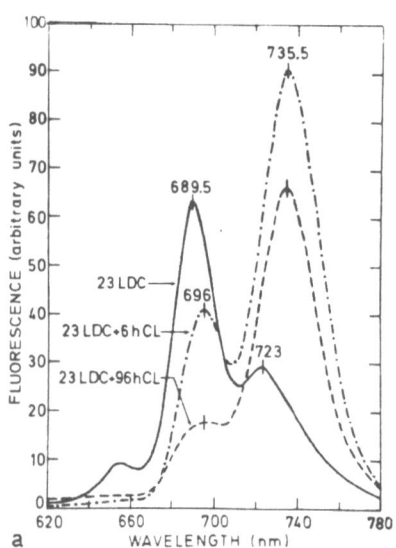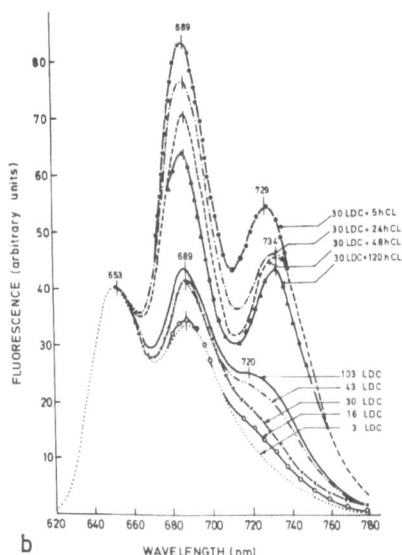

FIGURE 4. 77°K fluorescence emission spectra of 6-d etiolated bean leaves (a), or of their plastids (b), exposed to intermittent light (LDC) and then transferred to continuous light (CL). The spectra of the plastids were recorded in the presence of the internal fluorescence standard phycocyanin (PC). PC/Chl=0.8 (w/w). The normalization of the fluorescence intensity at 653 nm was done by changing the excitation light intensity.

leaves and the isolated plastids (7,19,31). As shown in fig. 4a, in intermittent light leaves the main emission band is at 690 nm. The 735 nm band, due to the antennae of PSI (30) is absent, but a band at 725nm, probably due to the core of the PSI unit, is noticed. After transfer to continuous light, the 725nm peak is red-shifted to 735nm and increases reaching values higher than those at 690nm. The changes in the absolute fluorescence yield at 77°K of the developing isolated plastids are shown in fig. 4b. In this case, the spectra were recorded in the presence of phycocyanin (internal standard, 7,19)

5. ONSET OF THE CONTROL OF EXCITATION ENERGY DISTRIBUTION BETWEEN PSI AND PSII
    The most characteristic structural change occuring during the phototransformation of etioplasts or "protochloroplasts" to the mature chloroplast stage is the differentiation of prothylakoids and "primary" thylakoids to stroma and grana lamellae. Stroma and grana thylakoids differ not only in their composition, but also in their photochemical activities. Stroma lamellae are enriched in Chla, CPIa, CPI, LHC-I and PSI activity; grana are enriched in Chlb LHC-II, CPa and PSII activity. Soon after etiolated leaves are exposed to continuous light small grana stacks with 2-3 thylakoids per granum appear. The thylakoid number per granum increases as time of illumination is prolonged. At the same time the cation control of excitation energy distribution between PSII and PSI and the capacity for State I/State II transition appear. Plastids of intermittent light leaves do not show the cation control of excitation energy distribution, nor the light-induced State transition. Both effects appear gradually after transfer of the leaves to continuous light. As mentioned above, "protochloroplasts" lack grana stacks, and are deficient in Chlb, LHC-II LHC-I and CPIa. All these components are formed in parallel upon transfer of

the plants to continuous light. Since there is a very good correlation between the formation of Chlb, LHC-II and grana, and the onset of the cation
regulation of the excitation energy distribution between PSII and PSI, it has
been suggested that LHC-II and grana are involved in this process (7,21,33).
The rational for this involvement is that when grana are formed there is a
spatial separation of PSII and PSI units, and accordingly, inhibition of excitation energy spillover from PSII to PSI. During unstacking, it has been
proposed (22) that a lateral movement of PSI and PSII units occurs, leading
to randomization of the units, facilitating spillover. However, during proto-
chloroplast to chloroplast differentiation, LHC-I and CPIa are also formed.
Considering the cation effect on the $77^\circ K$ fluorescence properties of these
two complexes, it has been proposed that they may be involved in the regulation of excitation energy distribution between the two photosystems, by affecting the effective cross section of the PSI unit (7,16,19). This is also
supported by late findings (7), showing that upon low-salt unstacking of thylakoids there is no large scale lateral movement of complexes and thus no intermixing of PSI and PSII units, a prerequisite for spillover of excitation
energy to occur from PSII to PSI. In addition, this proposal offers an explanation as to why the PSII-less mutant of *Chlamydomonas* (F34) shows the Mg^{2+}
effect and the State I/II transition, but not the PSI-less mutant (F14) (23).

REFERENCES

1. Akoyunoglou A and Akoyunoglou G (1984). Israel J Bot 33:149-162.
2. Akoyunoglou A and Akoyunoglou G (1985). Plant Physiol 79:425-431.
3. Akoyunoglou G (1977). Arch Biochem Biophys 183:571-580.
4. Akoyunoglou G (1981). In Akoyunoglou G, ed. Photosynthesis, Vol. V, pp.
   353-366. Philadelphia:Balaban Intern Sci Services.
5. Akoyunoglou G and Argyroudi-Akoyunoglou JH (1971). In Forti G, Ayron M
   and Melandri A, eds. Proc 2nd Intern Congress on Photosynthesis Research,
   Vol. 3, pp. 2427-2436.
6. Akoyunoglou G and Argyroudi-Akoyunoglou JH (1978). In Metzner H, ed. Photosynthetic Oxygen Evolution, pp. 453-488. London:Academic Press.
7. Akoyunoglou G and Argyroudi-Akoyunoglou JH (1985). In Packer L, ed. Recent
   Advances in Biological Membrane Studies, pp. 205-236. New York:Plenum.
8. Akoyunoglou G and Argyroudi-Akoyunoglou JH (1986). In Akoyunoglou G and
   Senger H, eds. Regulation of Chloroplast Differentiation (in press) New
   York: Alan R. Liss.
9. Akoyunoglou G, Argyroudi-Akoyunoglou JH, Michel-Wolwertz MR and Sironval,
   C (1966). Physiol Plant 19: 1101-1104.
10. Akoyunoglou G and Michelinaki-Maneta M (1974) In Avron M, ed, Proc. 3rd
    Intern Congress on Photosynthesis, pp. 1885-1896. Amsterdam: Elsevier.
11. Anderson JM, Waldron JC and Thorne SW (1978). FEBS Lett 99:227-233.
12. Argyroudi-Akoyunogloy JH (1984). FEBS lett 171:47-53.
13. Argyroudi-Akoyunoglou JH and Akoyunoglou G (1970) Plant Physiol 46:247-249.
14. Argyroudi-Akoyunoglou JH and Akoyunoglou G (1973) Photochem Photobiol 18:
    219-228.
15. Argyroudi-Akoyunoglou JH and Akoyunoglou G (1979). FEBS Lett 104:78-84.
16. Argyroudi-Akoyunoglou JH and Akoyunoglou G (1983). Arch Biochem Biophys
    227:469-477.
17. Argyroudi-Akoyunoglou JH, Akoyunoglou A, Kalosakas K and Akoyunoglou G
    (1982). Plant Physiol 70:1242-1248.
18. Argyroudi-Akoyunoglou JH and Castorinis A (1980). Arch Biochem Biophys
    200:326-335.
19. Argyroudi-Akoyunoglou JH, Castorinis A and Akoyunoglou G (1984). Israel J
    Bot 33:65-82.

20. Argyroudi-Akoyunoglou JH, Feleki Z and Akoyunoglou G (1971). Biochem Biophys Res Commun 45:606-614.
21. Argyroudi-Akoyunoglou JH and Tsakiris S (1977). Arch Biochem Biophys 184: 307-315.
22. Barber J (1980). FEBS Lett 118:1-10.
23. Bennoun P (1974). Biochim Biophys Acta 368:141-147.
24. Boardman NK and Anderson JM (1978). In Akoyunoglou G and Argyroudi-Akoyunoglou JH, eds. Chloroplast Development, pp. 1-14. Amsterdam: Elsevier.
25. Boardman NK, Thorne SW and Anderson JM (1966). Proc Natl Acad Sci 56:586.
26. Bonaventura C and Myers J (1969). Biochim Biophys Acta 189:366-383.
27. Butler Wl (1965). Biochim Biophys Acta 102:1-8.
28. Butler WL (1976). Brookhaven Symposia in Biology 28:338-344.
29. Butler WL, DeGreef J, Roth TF and Oelze-Karow H (1972). Plant Physiol 49: 102-104.
30. Butler WL and Kitajima M (1975). Biochim Biophys Acta 396:72-85.
31. Castorinis A, Akoyunoglou G and Argyroudi-Akoyunoglou JH (1982). Photobiochem Photobiophys 4:283-291.
32. Cohen D, Malkin S, Shochat S and Ohad I (1976). Plant Physiol 58:257-267.
33. Davis DJ, Armond PA, Gross EL and Arntzen CJ (1976) Arch Biochem Biophys 175:64-70.
34. DeGreef J, Butler WL and Roth TF (1971). Plant Physiol 47:457-464.
35. Dubertret G and Joliot P (1974). Biochim Biophys Acta 357:399-411.
36. Dujardin E, de Kouchkovsky Y and Sironval C (1970). Photosynthetica 4:223.
37. Duysens LNM and Sweers HE (1963). In Japan Soc Plant Physiol, ed. Microalgae and Photosynthetic Bacteria, pp. 353-372. Tokyo:University Press.
38. Fish LE, Kuck U and Bogorad L (1985). J Biol Chem 260:1413-1421.
39. Frado TM and Stern AI (1982). Z Pflanzenphysiol 105:258-266.
40. Goedheer JC (1964). Biochim Biophys Acta 88:304-317.
41. Gregory RPF, Raps S and Bertsch W (1971). Biochim Biophys Acta 234:330.
42. Homann P (1969). Plant Physiol 44:932-936.
43. Krause GH, Briantais JM and Vernotte C (1983). Biochim Biophys Acta 723: 169-175.
44. Kitajima M and Butler WL (1975). Biochim Biophys Acta 376:105-115.
45. Krupinska K, Akoyunglou G and Senger H (1985). Photochem Photobiol 41: 159-164.
46. Kuang TY, Argyroudi-Akoyunoglou JH, Nakatani HY, Watson J and Arntzen CJ (1984). Arch Biochem Biophys 235:618-627.
47. Murata N (1969). Biochim Biophys Acta 172:171-181.
48. Murata N (1969). Biochim Biophys Acta 172:242-251.
49. Ogawa T, Obata F and Shibata K (1966). Biochim Biophys Acta 112:223-234.
50. Ogawa T and Shibata K (1973). Plant Physiol 29:112-117.
51. Satoh K (1980). FEBS Lett 110:53-56.
52. Sironval C (1974). In Avron M, ed. Proc 3rd Intern Photosynthesis Congress, pp. 2153-2162. Amsterdam: Elsevier.
53. Sironval C, Clijsters H, Michel JM, Bronchart R and Michel-Wolwertz MR (1967). In Sironval C, ed. Le Chloroplaste, pp. 99-123. Paris:Masson et Cie.
54. Thornber JP, Stewart JC, Hatton NWC and Bailey JL (1967). Biochemistry 6:2006-2014.
55. Thorne SW and Boardman NK (1971). Plant Physiol 47:252-261.

Photosynthesis Research 10: 181–187 (1986)
© *Martinus Nijhoff Publishers, Dordrecht*

CHLOROPHYLL-PROTEIN COMPLEXES

KIMIYUKI SATOH

Department of Biology, Okayama University, Tsushima Okayama 700, Japan

Key words: Chlorophyll-protein, Reaction center, Light harvesting, Fluorescence

Abstract. Recent advances in the studies on chlorophyll-protein complexes of higher plants are summarized in this article. Special emphasis is laid on the isolation, pigment composition and the absorption and fluorescence properties of the complexes.

1. INTRODUCTION

> The chemistry of chlorophyll extracted from leaves with alcohol or other organic solvents is well known. However, this extracted chlorophyll is only part of the complex existing in living cells. In its functional state, chlorophyll is combined with proteins as insoluble particles that contain also lipids, carotenoids, and even carbohydrates. The main reason for our ignorance about the chemistry of this green coloring matter which absorbs the sunlight used to drive photosynthesis is that the material itself is not soluble, and hence cannot easily be prepared in pure form for chemical analysis.

> From the lecture of C. S. French at the symposium celebrating the 200th anniversary of the momentous experiment of Joseph Priestley (14)

The period of the past 15 years has seen rapid progress in the manipulation of the above mentioned natural green coloring matter "chlorophyll-protein complex". The techniques have been refined to solubilize and purify chlorophylls in combination with their apoproteins in their native states. Thus presently the two lines of research - one searching for the structural minimum capable of photosynthetic electron transport (5,35), and the other trying to directly extract chlorophyll-proteins which must have important functions in photosynthesis (19,32) - have come to a similar point. It has been established that there are several kinds of chlorophyll-protein complexes with a definite polypeptide compositions, functioning either in light-harvesting and/or charge separation in photosynthesis (3,29,31,36). In addition, the recent molecular genetic approach has opened the way to obtain the amino acid sequence of most of the constituent polypeptides of chlorophyll-protein complexes and thus further provided a new perspective for considering the molecular organization and the chemistry or function of chlorophyll-protein complexes in photosynthetic systems (2,9). The basic situation presently emerging from these studies can be summarized as follows:
1. Almost all of the chlorophyll molecules, and virtually all of the

carotenoids, of thylakoid membranes are present in association with proteins forming supramolecular complexes with definite polypeptide and chemical composition.

2. From a functional point of view, they are divided into two groups: reaction center complexes and light-harvesting complexes. The reaction center complexes are composed of both chlorophyll-binding subunits and chlorophyll-free subunits. Their basic structure and their function are well preserved throughout all of the classes of oxygen-evolving organisms ranging from cyanobacteria to higher plants. On the other hand, the types of light-harvesting complex vary with classes of plants: chlorophyll a/b-proteins of PSI and PSII in higher plants, green algae and Euglena; chlorophyll a/c-proteins in brown algae and diatoms; peridinin-chlorophyll a-protein in Dinoflagellates etc.. Furthermore, the amounts of light-harvesting complexes vary with environmental conditions such as light intensity and light quality.

3. Thus most of the functions of the early processes of photosynthesis can efficiently be analyzed in the isolated chlorophyll-protein complexes on the basis of their polypeptide structure. The other functions can also be understood in terms of the interaction between the chlorophyll-protein complexes in the thylakoid membranes; e.g., regulation of excitation energy distribution between two photosystems via protein phosphorylation of light-harvesting chlorophyll a/b-protein complex.

2. ISOLATION OF CHLOROPHYLL-PROTEIN COMPLEXES

A substantial amount of knowledge on the chlorophyll-protein complexes has been provided by the studies on materials separated by SDS (LiDS)-polyacrylamide gel electrophoresis (3,12,25,37). This was especially true for the light-harvesting complexes and the PSI complex. However, the detergent was often harmful to the pigment systems and the photochemical activities, and the method has not contributed much to the identification of PSII chlorophyll-protein complexes with photochemical activity. The use of mild detergents such as digitonin (39) or Triton X-100 (17,38), in combination with sucrose density gradient centrifugation (10,31,39) and isoelectric focusing (31) for separation, has proven to be promising for the isolation of all of the three chlorophyll-protein complexes; namely the PSI complex consisting of both the PSI reaction center complex and the light-harvesting complex for the photosystem (LHC-I), the PSII reaction center complex and the major antenna complex for the PSII reaction center (LHC-II), in a more intact state; the pI values for these complexes were estimated by the method (27,31) to be about 4.8, 4.6 and 4.3, respectively. These values were nearly consistent with the isoelectric points of the membrane surfaces expected to be enriched in each of the three chlorophyll-protein complexes (1).

Ion exchange chromatography using DEAE-cellulose or hydroxylapatite has also been applied to the separation of chlorophyll-protein complexes (36,39). A simple two-step chromatographic procedure using DEAE-Toyopearl was recently developed to purify all of the three chlorophyll-protein complexes simultaneously from the same extract, in a sufficiently pure state for biochemical and biophysical analysis (Figure 1) (40). At this time, the chlorophyll-protein complexes can easily and reproducibly be prepared in a pure form for chemical analysis with ordinary methods used for water-soluble proteins, based on their particle size, surface charge, etc., when it is provided with low concentrations of appropriate detergents, although there are several different classes of preparations for each complex with different degree of disintegrations. On the other hand, the selection of detergent is still largely a matter of trial and error. The pigment systems are often very sensitive to detergents. And thus, for example, in some detergent-solubilized preparations, flash excitation brings about the formation of chlorophyll triplet state with an unusually high yield even when the pigments are attached to the proteins.

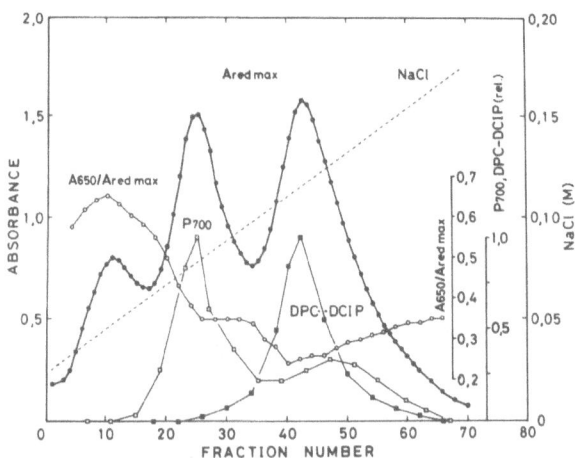

FIGURE 1. Chromatographic separation of the three major chlorophyll-protein complexes of spinach using DEAE-Toyopearl 650 S (From Yamada et al., 1985). A650/A red max, relative contribution of chlorophyll b absorption; P-700, relative amount of PSI reaction center; DPC-DCIP, PSII activity.

3. PROPERTIES OF CHLOROPHYLL-PROTEIN COMPLEXES

   Almost all of the chlorophyll molecules in thylakoid membranes are present in combination with proteins. Probably this is also true for carotenoids; nearly all of them are associated with chlorophyll-protein complexes (41). Chlorophyll a is a component of both reaction center and light-harvesting complexes, whereas chlorophyll b is present exclusively in the light-harvesting complexes. Beta-carotene is present in both PSI and PSII reaction center complexes in a molar ratio of about one to five to chlorophyll a, whereas xanthophyll is the major component of carotenoids in the light-harvesting a/b-protein (LHC-II) of higher plants (41).
   Analysis using the fourth derivative (31) and curve-fitting (7) of the absorption spectra at 77 K of isolated chlorophyll-protein complexes prepared with mild detergents such as digitonin indicates that the chlorophyll forms in vivo are well preserved in the isolated chlorophyll-protein complexes. This is also shown by the fact that when the absorption spectra of the three chlorophyll-protein complexes are added in proportion corresponding to their abundance in the solubilized materials, the sum nearly matches the spectrum of the thylakoids (7).
   The fluorescence emission spectra of chlorophyll in vivo are extremely sensitive to detergent-treatment possibly due to both the sensitivity to

microenvironments of the intramolecular processes competing with fluorescent decay and to the interruption of intermolecular excitation transfer between pigment-protein complexes. Successful fractionation of the thylakoid membranes into the three chlorophyll-protein complexes is shown in Figure 2 which compares the fluorescence emission spectra of thylakoid membranes and digitonin-solubilized chlorophyll-protein complexes (27,31).

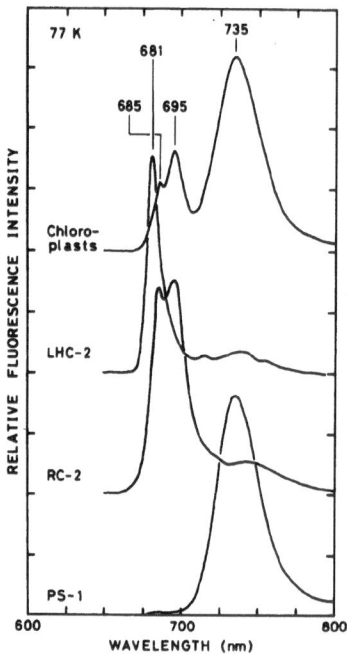

FIGURE 2. Fuluorescence emission spectra at 77 K of spinach chloroplasts and the three major chlorophyll-protein complexes (From Satoh and Butler, 1978;Satoh, 1982). LHC-2, light-harvesting chlorophyll a/b-protein for PSII (LHC-II); RC-2, PSII reaction center complexes; PS-1, PSI complex consisted of both PSI reaction center complex and LHC-I.

Since association of LHC-I to the PSI reaction center complex is much stronger than that between LHC-II and PSII reaction center complex, mild detergent-treatment yields a complex consisting of both LHC-I and PSI reaction center (4). This complex is composed of possibly two large polypeptide subunits of 60-70 kDa encoded by chloroplast DNA and postulated to be the site of primary charge separation in PSI (21,33) and several small polypeptides of 10-25 kDa (4,33). This type of PSI complex retains the secondary electron acceptor (A2) or P-430, in addition to the primary acceptor (A1) and the primary donor chlorophyll a, P-700. This complex also contains about 110 chlorophyll and 20 beta-carotene molecules per P-700 and displays a prominent emission band of F-735, but none of F-685 and F-695 (22). Removal of 2-3 polypeptides in the 20 kDa range from the complex by detergent-treatment, which also accompanies the release of chlorophyll a and b molecules, causes the peak shift from 735 to about 720 nm (15). On the other hand, the isolated chlorophyll a/b-protein complex consisted of polypeptides of about 20 kDa (LHC-I) exhibits F-735 emission (18). The PSI preparation depleted in the LHC-I and consisted of the large subunits and the two small subunits exhibits the electron transport from P-700 to the secondary acceptor (33). The PSI complex prepared by SDS-polyacrylamide gel electrophoresis, which retains both P-700 and A1 and consisted of only the large subunit(s), also displays F-720 band in some preparations, but in most cases the emission peak is shifted to a lower wavelength (670-700 nm). When the chlorophyll content of PSI complex is reduced to only about 10 chlorophyll molecules per P-700, it emits fluorescence at about 700 nm (F-705) at 77 K (16). It is proposed that F-705 originates from the electronically excited state of A1, which is probably a special form of chlorophyll a, and which excited state is produced by charge recombination between P-700+ and A1- (16).

The isolated PSII reaction center complex contains about 50 chlorophyll a and 10 beta-carotene molecules per reaction center chlorophyll a, P-680 (30,40) and emits both F-685 and F-695 at 77 K. Although F-695 becomes insignificant during purification in some PSII preparations, use of 1,10-phenanthroline, which specifically enhances F-695, clearly shows that all the active PSII reaction center preparations display this emission (28). PSII reaction center complex is composed of five kinds of polypeptide of the molecular masses ranging from 7 to 50 kDa; namely a small polypeptide of about 7-10 kDa for the apoprotein of cytochrome b559, two polypeptides at 30-34 kDa (one of which corresponds to the herbicide-binding polypeptide) and two polypeptides at 40-50 kDa. The two large polypeptides of about 43(40 in a cyanobacterium) and 47 kDa have been shown to be chlorophyll-binding (8,12). Both components have been isolated by SDS (or LiDS)-polyacrylamide gel electrophoresis in association with at least a fraction of the chlorophyll molecules (8,12,23,24). The isolated 47 kDa component (CP-47) emits fluorescence peaked at about 695 nm at 77 K which exhibits characteristic temperature dependence (43) sensitivity to 1,10-phenanthroline (34) of F-695 emission in thylakoids, whereas the isolated 43 kDa component (CP-43) emits only F-685 fluorescence. This finding is in accordance with the proposal that CP-47 is the site of charge separation between P-680 and pheophytin (8,10,13,24,42), because F-695 emission is postulated to be derived from electronic excitation of pheophytin a which is produced by charge recombination at the reaction center of PSII (6). However, another line of evidence has recently led to an alternative proposal for the location of the PSII reaction center in the chlorophyll-protein complex that the two polypeptides of 30-34 kDa in the complex form a P-680, pheophytin and quinone-containing core in the PSII reaction center (20). This proposal is based on the sequence homology between the polypeptides of the PSII reaction center complex and those of purple bacterial reaction center complex (20). However, these two polypeptides in the PSII reaction center complex have not yet been isolated in conjunction with chlorophylls.

Isolated LHC-II, which consists of 1-3 polypeptides of about 25 kDa, displays a strong emission at about 680 nm (F-680) at 77K (27,31). Although the fluorescence yield of the emission in the isolated complex is about 10 times higher than those of the PSI and PSII reaction center complexes (27), the F-680 emission in intact thylakoids is insignificant (Fig. 2). This is interpreted by an efficient energy transfer from LHC-II to the PSII reaction center in the thylakoid membranes. At liquid helium temperature, however, we can observe this component as a shoulder in the emission spectrum (26).

4. PERSPECTIVE

In chlorophyll-protein complexes, the polypeptide associations and the network of non-covalent bonding interactions, which anchor pigments and functional groups to polypeptide backbone, are expected to be responsible for the precise positioning which influences the spectroscopic properties of the pigments and ensures the efficient and regulated transformation of energy in photosynthesis. The recent biochemical investigations have succeeded in identifying the polypeptides constituting the major portion of chlorophyll-protein complexes. However, at present, it has not been successful in clearly identifying the site of primary charge separation in the reaction center complex to a specific polypeptide, or to locate the site of chlorophyll- or cofactor-binding to a specific polypeptide structure. The recent remarkable progress on this point is that the chloroplast gene structure for most of the polypeptides of chlorophyll-protein complexes has been identified. These sequence data and calculations of the distribution of hydrophobic residues predict the membrane spanning regions and the position of functional groups on the polypeptides. The data also enable us to consider the structure by examining the homology between different polypeptides from a phylogenic and evolutional point of view. Together with more advanced biochemical investigation, these analyses will provide a new perspective for the understanding of the structure and function of chlorophyll protein complexes at the molecular level. The more recent achievement in the x-ray crystallographic analysis of the purple bacterial reaction center complex has opened an entirely new direction to the study of chlorophyll-protein complexes (11).

References

1. Akerlund H-E, Andersson B, Persson A and Albertsson P-A (1979) Biochim Biophys Acta 552:238-246
2. Alt J, Morris J, Westhoff P and Herrmann G (1984) Current Genetics 8:597-606
3. Anderson JM (1975) Biochim Biophys Acta 416:191-235
4. Bengis C and Nelson N (1975) J Biol Chem 250:2783-2788
5. Boardman NK and Anderson JM (1964) Nature 203:166-167
6. Breton J (1982) FEBS Lett 147:16-20
7. Brown JS (1983) Photosyn Res 4:375-383
8. Camm EL and Green BR (1983) Biochim Biophys Acta 724:291-293
9. Cramar WA, Widger WR, Herrmann RG and Trebst A (1985) Trend Biochem Sci 10:125-129
10. de Vitry C, Wollman F-A and Delepelaire P (1984) Biochim Biophys Acta 767:415-422
11. Deisenhofer J, Epp O, Miki K, Huber R and Michel H (1984) J Mol Biol 180:385-398
12. Delepelaire P and Chua N-H (1979) Proc Natl Acad Sci USA 76:111-115
13. Diner BA and Wollman F-A (1980) Eur J Biochem 110:521-526
14. French CS (1971) Proc Natl Acad Sci USA 68:2893-2897
15. Haworth P, Watson JL and Arntzen CJ (1983) Biochim Biophys Acta 724:151-158
16. Ikegami I and Ke B (1984) Biochim Biophys Acta 764:80-85
17. Ke B, Vernon LP and Chaney T (1972) Biochim Biophys Acta 256:345-357
18. Kuang TY, Argyroudi-Akoyunoglou JH, Nakatani HY, Watson J and Arntzen CJ (1984) Arch Biochem Biophys 235:618-627
19. Markwell JP, Thornber JP and Boggs T (1979) Proc Natl Acad Sci USA 76:1233-1235

20. Michel H and Deisenhofer J (1985) In: Encyclopedia of Plant Physiology, New series Photosynthesis 3 (Staehelin LA and Arntzen CJ, eds.),Springer-Verlag (in press)
21. Moller BL (1981) Carlsberg Res Commun 46:373-382
22. Mullet JE,Burke JJ and Arntzen CJ (1980)PlantPhysiol 65:814-822
23. Nakatani HY (1983) In: The Oxygen Evolving System of Photosynthesis (Inoue Y, Crofts AR, Govindjee, Murata N, Renger G and Satoh K, eds.), pp.49-54, Academic Press, Tokyo
24. Nakatani HY, Ke B, Dolan E and Arntzen CJ (1984) Biochim Biophys Acta 765:347-352
25. Ogawa T, Obata F and Shibata K (1966) Biochim Biophys Acta 112:223-234
26. Rijgersberg CP, Amesz J, Thielen APGM and Swager JA (1979) Biochim Biophys Acta 545:473-482
27. Satoh K (1979) Plant Cell Physiol 20:499-512
28. Satoh K (1980) FEBS Lett 110:53-56
29. Satoh K (1982) In: Methods in Chloroplast Molecular Biology (Edelman M, Hallick RB and Chua N-H, eds.), pp.845-856, Elsevier Biomedical Press, Amsterdam
30. Satoh K (1983) In: The Oxygen evolving System of Photosynthesis (Inoue Y, Crofts AR, Govindjee, Murata N, Renger G and Satoh K, eds.), pp.27-38, Academic Press, Tokyo
31. Satoh K and Butler WL (1978) Plant Physiol 61:373-379
32. Smith EL (1941) J Gen Phys 24:565-582
33. Takahashi Y, Koike H and Katoh S (1982) Arch Biochem Biophys 219:209-218
34. Tang X-S and Satoh K (1984) Plant Cell Physiol 25:935-945
35. Thomas JB, Blaauw OH and Duysens LNM (1953) Biochim Biophys Acta 10:230-240
36. Thornber JP (1975) Ann Rev Plant Physiol 26:127-158
37. Thornber JP, Smith CA and Bailey JL (1966) Biochem J 100:14-15
38. Vernon LP and Shaw ER (1971) In: Methods in Enzymology, Vol. XXIII (San Pietro A ed.), pp.277-289, Academic Press,New York
39. Wessels JSC, van Alphen-van Waveren O and Voorn O (1973) Biochim Biophys Acta 292:741-752
40. Yamada Y, Itoh N and Satoh K (1985) Plant Cell Physiol 26:1263-1271
41. Yamada Y, Takahashi Y and Satoh K (in preparation)
42. Yamagishi A and Katoh S (1983) Arch Biochem Biophys 225:836-846
43. Yamagishi A and Katoh S (1983) In: The Oxygen evolving System of Photosynthesis (Inoue Y, Crofts AR, Govindjee, Murata N, Renger G and Satoh K, eds.), pp.39-48,Academic Press,Tokyo
44. Yamagishi A and Katoh S (1985) Biochim Biophys Acta 807:74-80

Photosynthesis Research 10: 189–196 (1986)
© *Martinus Nijhoff Publishers, Dordrecht*

RELATIONSHIPS AMONG CELL CHLOROPHYLL CONTENT, PHOTOSYSTEM II
LIGHT-HARVESTING AND THE QUANTUM YIELD FOR OXYGEN PRODUCTION
IN CHLORELLA[*]

ARTHUR C. LEY

1. ABSTRACT

Cells of the green alga Chlorella vulgaris were grown under conditions
where total Chl/cell varied by a factor of almost 80; from 0.02 fmol/cell
to nearly 1.6 fmol/cell. The change in Chl/cell was accomplished by an
approximately 11-fold increase in RCII/cell along with a 7-fold increase
in Chl/RCII. The effective absorption cross section per RCII at 596 nm
varied by a factor of 6, increasing with Chl/cell from a minimum of 20 A^2
to a maximum of 116 A^2. In contrast, over the same range of Chl/cell, the
quantum requirement for O_2 production remained relatively constant at
10.4 ± 1.8 quanta absorbed/O_2 evolved. The results are well described by
a simple model in which changes in Chl/cell are produced by coordinated
changes in reaction center and light-harvesting complexes. The model
predicts that between 20 and 40% of the light-harvesting
chlorophyll-protein complexes commonly assigned to PSII, do not function
as antenna for PSII.

2. INTRODUCTION

Photosynthetic organisms often respond to changes in environmental
conditions with changes in the composition of the photosynthetic
apparatus. For example, both total environmental irradiance and its
spectral distribution, have been shown to influence the composition of the
photosynthetic apparatus (1-9). Organisms ranging from cyanobacteria
(4,9) to unicellular eucaryotic algae (2,3,7,8) to ferns (6) and higher
plants (1,5) have all been shown to respond to changes in environmental
light conditions. The responses which have been reported include changes
in photosynthetic rates and turnover times (2,3,5,6), pigment content (1-
9), reaction center (RC) ratios (3-6) and photosystem antenna sizes
(3,4,5-7).
     Advances in biochemical expertise have permitted the quantitative
description of the pigment composition of plants in terms of specific
pigment-protein complexes (10-20) of known compostion and function.
Recent experiments have demonstrated that the relative amounts of the
various pigment-protein complexes can also respond to changes in
environmental conditions (5,6).

---

[*]This paper is dedicated to the memory of Warren L. Butler, a pioneer in
the field of photobiology. His creativity and innovativeness as a
scientist were matched only by his patience as a mentor.

In a previous study (7) it was reported that culture conditions influenced the values of several PSII-related parameters in <u>Chlorella</u>, including cell Chl content, absorption cross sections, and photosynthetic unit sizes. However, under the same conditions, the quantum requirement for O_2 production was essentially unchanged. This report describes the results of a more extensive study of the relationships between cell Chl levels, PSII light-harvesting properties, and the quantum requirement for O_2 production.

3. MATERIALS AND METHODS

Cells of the green alga <u>Chlorella</u> <u>vulgaris</u> were grown in batch cultures exposed to a wide range of irradiances between about 10^6 erg/cm^2-s and 10^2 erg/cm^2-s as previously described (7). Cell cultures were grown with no additional aeration or CO_2 enrichment and it is likely that the cultures growing at the highest irradiances were CO_2-limited (21). In addition, due to their proximity to the metal arc lamp, the high irradiance cultures grew at somewhat higher temperature (25°C) than did the other cultures (20°C). Thus, several environmental factors probably influenced the properties of the cells reported here. Since the aim of this work is to investigate the more fundamental relationships between cell Chl levels and PSII properties, no effort has been to causally relate any cell property to any extrinsic environmental variable.

Chl concentrations were determined from ethanol extracts of cells as described previously (7). When the ratio of Chl-a:Chl-b was greater than 5, a room temperature flourescence assay similar to that of Boardman and Thorne (22) was used. Calibrations using known mixtures of purified Chl-a and Chl-b in 95% ethanol indicate that this procedure is accurate for ratios of Chl-a:Chl-b up to 25. Cell counts were made in quadrupicate using a hemocytometer.

Measurements of the photosynthetic unit size for O_2 production (PSU_{O2}) were determined as described previously (3) using flash rates of 5, 10, 15, and 20 Hz. As has been reported by Myers and Graham (23), the addition of a continuous far red (704 nm) irradiance had no effect on the magnitude of the O_2 flash yields. The number of PSII reaction centers (RCII) per cell were calculated from PSU_{O2} and Chl/cell using the assumption of $4e^-$ per O_2.

Effective absorption cross sections per RCII (σ_{O2}) were determined at 596 nm from laser flash energy saturation curves as described previously (7). The quantum requirement for O_2 production was calculated from PSU_{O2}, σ_{O2}, and the <u>in vivo</u> 596 nm absorption cross section of a Chl molecule, σ_{chl} as (15):

$$\text{Quantum Requirement} = \frac{PSU_{O2} \cdot \sigma_{chl}}{\sigma_{O2}} \qquad (1)$$

4. RESULTS

The culture growth conditions used in this study produced large changes in the pigment and PSII properties of <u>Chlorella</u> cells (Table I, Fig. 1A-E). The data shown in Table I are arranged in order of decreasing Chl-(a+b)/cell, which correlates roughly with increasing growth irradiance. Cells containing more than 1.0 fmole Chl or less than 0.1 fmole Chl grew at the lowest and highest irradiances, respectively.

The values for Chl/cell shown in Table I vary by a factor of almost 80, from about 0.02 fmol/cell to about 1.6 fmol/cell. Over this range, Chl-a/cell increases by a factor of 60, while Chl-b/cell increases more than 800-fold. As a result, the ratio of Chl-a:Chl-b changes by at least an order of magnitude from ≤ 0.04 to about 0.4.

The variation in Chl/cell is produced by changes in both RC/cell and

TABLE 1. Pigment and photochemical characteristics of Chlorella cells having different chlorophyll content.

Chl 1 Cell	Chl a[1] Cell	Chl b[1] Cell	Chl b Chl a	PSU_{O_2}[2]	RCII[3] Cell	Cross[4] Section	Chl RCII	Quantum[5] Requirement
1.57	1.15	0.42	0.37	5480	6.9	113	377	14.5
1.32	0.98	0.34	0.35	4500	7.1	116	387	11.6
0.82	0.60	0.22	0.32	nd	---	95	317	----
0.71	0.54	0.17	0.32	nd	---	96	320	----
0.71	0.52	0.19	0.37	nd	---	98	327	----
0.68	0.51	0.17	0.33	nd	---	90	300	----
0.58	0.43	0.13	0.28	nd	---	85	283	----
0.54	0.43	0.11	0.26	nd	---	74	247	----
0.52	0.41	0.11	0.27	2800	4.5	nd	---	----
0.51	0.39	0.12	0.31	nd	---	95	317	----
0.50	0.39	0.11	0.30	2250	---	80	267	8.4
0.50	0.38	0.12	0.33	3170	3.8	nd	---	----
0.49	0.39	0.10	0.27	2300	5.1	nd	---	----
0.49	0.38	0.11	0.28	2150	5.5	nd	---	----
0.48	0.37	0.11	0.30	nd	---	95	317	----
0.48	0.38	0.10	0.27	2300	5.0	83	277	8.3
0.46	0.35	0.11	0.32	2910	3.8	82	273	10.7
0.44	0.35	0.094	0.27	3230	3.3	nd	---	----
0.44	0.34	0.10	0.30	nd	---	69	230	----
0.43	0.34	0.089	0.26	nd	---	74	247	----
0.43	0.33	0.096	0.29	1970	5.2	nd	---	----
0.41	0.31	0.095	0.31	nd	---	82	273	----
0.40	0.32	0.082	0.26	nd	---	71	237	----
0.40	0.32	0.089	0.24	nd	---	78	260	----
0.38	0.29	0.088	0.30	3000	3.0	nd	---	----
0.37	0.30	0.075	0.25	2240	4.0	70	233	9.6
0.37	0.30	0.074	0.25	nd	---	65	217	----
0.36	0.29	0.073	0.26	2590	3.4	83	277	9.4
0.36	0.29	0.073	0.26	2270	3.8	74	247	9.2
0.34	0.26	0.083	0.32	2930	2.8	nd	---	----
0.34	0.26	0.78	0.30	2370	3.5	67	223	10.6
0.33	0.27	0.064	0.24	2180	3.6	77	257	8.5
0.31	0.25	0.065	0.26	2390	3.1	82	273	8.8
0.30	0.25	0.065	0.26	2010	3.6	nd	---	----
0.27	0.22	0.047	0.22	2600	2.5	nd	---	----
0.26	0.22	0.049	0.23	2600	2.4	nd	---	----
0.19	0.16	0.034	0.20	2090	2.2	nd	---	----
0.19	0.16	0.037	0.22	2260	2.0	nd	---	----
0.17	0.15	0.021	0.14	1480	2.8	42	140	10.6
0.17	0.14	0.030	0.15	1540	2.6	43	143	10.8
0.16	0.14	0.022	0.15	1260	3.1	38	127	9.9
0.13	0.11	0.014	0.13	1150	2.7	nd	---	----
0.12	0.11	0.010	0.09	1340	2.2	nd	---	----
0.10	0.093	0.009	0.10	1300	1.9	nd	---	----
0.092	0.080	0.011	0.14	910	2.4	nd	---	----
0.085	0.080	0.005	0.06	1030	2.0	22	73	14.1
0.053	0.051	0.002	0.04	850	1.4	21	70	12.1
0.039	0.035	0.004	0.10	760	1.2	nd	---	----
0.028	0.027	0.001	0.04	670	1.0	22	73	9.2
0.022	0.021	0.001	0.04	960	0.6	nd	---	----

Units: 1. fmol/cell   2. mole Chl · flash/mole O_2 3. 10^5/cell
         4. A^2 5. photons absorbed/O_2 produced

Chl/RC. Values for PSU_{O2} shown in Table I vary from about 800 Chl/O_2 to about 5500 Chl/O_2. Assuming that the PSU_{O2} measurements count all RCII (7,8,23,24), this variation in PSU_{O2} implies an 11-fold change in RCII/cell, from 0.6×10^5 to 7×10^5, and a 7-fold change in total cell Chl per RCII ($=PSU_{O2} \div 4$), from about 200 to about 1400.

The effective absorption cross section per RCII (σ_{O2}) provides a direct measure of the light-harvesting power of the PSII antenna _in vivo_. (For a more detailed discussion of the measurement and significance of σ_{O2} see ref. 7 and the article by Mauzerall elsewhere in this volume.) Table I presents values for σ_{O2} determined at 596 nm (the wavelength of the laser used for the measurements) in _Chlorella_ cells having greatly different Chl content. Cross sections vary nearly 6-fold with increasing Chl/cell, from a minimum of 21 A^2 to a maximum of 116 A^2. For "bookkeeping" purposes it is useful to convert the measured _in vivo_ cross section values into the specific number of Chl molecules acting as antenna for a RCII. The _in vivo_ absorption cross section for Chl at 596 nm, 0.3 A^2 (7), allows this conversion. The results of this calculation are shown in the eighth column of Table I. As total cell Chl levels change by a factor of 80, the specific size of the antenna available to an individual RCII changes by a factor of 6, from about 70 to nearly 400 Chl functioning as PSII antenna per RCII.

The final column in Table I presents values for the quantum requirement for photosynthetic O_2 production by cells having greatly different Chl content. Remarkably, despite the large variations in all the other parameters listed in Table I, the quantum requirement is essentially constant. The variation between the largest and smallest values shown in Table I is less than a factor of 2 and the average of all the measurements is 10.4 ± 1.8. The quantum requirement of cells having very high or very low Chl/cell may be significantly greater than that of cells having intermediate levels of Chl/cell (12 ± 2 _vs_ 9.6 ± 0.95, respectively).

5. DISCUSSION

The values for the quantum requirement shown in Table I are larger than the theoretical minimum obtained by assuming that photosystems operate with no photochemical losses, that PSI is the sole oxidant of PSII, and that cell Chl is equally divided between PSI and PSII. As shown by Eq. (1), the quantum requirement can be derived from three measured quantities: PSU_{O2}, σ_{Chl}, and σ_{O2}. It has previously been argued (7), that the measured values for σ_{O2} are not greatly decreased by photochemical inefficiencies in PSII. The value for σ_{Chl} is near the lower end of the range of molecular cross sections measured _in vitro_ for Chl-a and Chl-b in various solvents (0.26 A^2 to 0.42 A^2, depending on the Chl species and solvent, unpublished observations). Thus it appears that the quantum requirement is "too high" because, to a large extent, PSU_{O2} values are "too large".

The observation, reported by Myers and Graham (23) and confirmed here, that a far red background irradiance does not increase O_2 flash yields suggests that PSI is not limiting in the PSU_{O2} measurement. If this is the case, the quantum requirement for O_2 production is a direct reflection of the fraction of cell absorbance at 596 nm that contributes to functional PSII light-harvesting. The "elevated" values observed for the quantum requirement imply that this fraction is less than 0.5. Such a situation would arise if a significant fraction of the PSII antenna was associated with non-functional RCII (or reaction centers in which non-photochemical losses occur during the dark processes of O_2 production), or if at 596 nm the total cell PSI absorption exceeds that of PSII by 20 to 30%.

Fig. 1A shows that Chl-b/cell is a remarkably linear function of Chl-a/cell. A linear regression on the data ($r^2 = 0.985$) has a slope of 0.38 (corresponding to an integral ratio Chl-a:Chl-b of 8:3) and an x-axis intercept of 0.08 fmol Chl-a/cell. Cells having less than about 0.08 fmole Chl-a contain essentially no Chl-b. Since Chl-b is thought to be contained only in the light-harvesting complexes of PSI and PSII (10,11,14,16,18,20), it may be that these cells contain only RC complexes. Two aspects of the data shown in Table I support this possiblity. The first is that at low Chl/cell, PSU_{O2} values are similar to those measured in organisms (cyanobacteria and red algae) which lack Chl-b and LHCP (4,9). The second is that $\overline{\sigma_{O2}}$ is constant at 21 A^2 for low values of Chl/cell. Thus for cells containing less than 0.08 fmol Chl-a, changes in Chl/cell would be accomplished solely by changes in RC/cell.

21 A^2 corresponds to an _in vivo_ RCII complex antenna size of about 75 Chl-a. This is larger than the value of 50 Chl-a/RCII observed with isolated complexes (12,19). It may be that the _in vivo_ complex is larger than the isolated particle, or that the value for σ_{Chl} in the complex is larger than the overall average for Chl in the cells.

At low Chl/cell, the ratio of total cell Chl to RCII is about 220:1. Since 75 Chl-a act as antenna per RCII, about 145 Chl-a are associated with PSI per RCII. Furthermore, since both Chl/cell:RCII and $\overline{\sigma_{O2}}$ are constant at low Chl/cell, it seems likely that the ratio RCI:RCII is also constant in this range. The ratio RCII:RCI in _Chlorella_ has been reported to be about 1.4:1 (23). If this is the case at low Chl/cell, the RCI complex contains about 200 Chl-a as antenna. If, on the other hand, the RCI complex is similar to that deduced for cyanobacteria (9) or the PSI "core" particle of higher plants (17,18) and contains about 120 Chl-a, then the ratio RCII:RCI at low Chl/cell is about 0.8.

For values of Chl-a/cell greater than about 0.08 fmole, Chl-b/cell is almost a linear function of Chl-a/cell. This observation implies that increases in cell Chl are achieved by the addition of Chl-a and Chl-b in a fixed molar ratio. Correction of the data shown in Fig. 1A for the measured increase in RCII/cell (with the associated increase in Chl-a only) gives the ratio with which Chl-a and Chl-b are added in light-harvesting complexes: Chl-a:Chl-b = 2.1:1. The data in Table I can be used to calculate that for each increment of 2×10^5 RCII/cell, total cell Chl/RCII also increases by about 210 Chl. Of these 210 Chl, 140 are Chl-a and 70 are Chl-b.

At present, two general classes of light-harvesting pigment-protein complexes can be distinguished: LHCP and LHCI (5,6,10,16,20). LHCPs are characterized by relatively low ratios of Chl-a:Chl-b (between 1:1 and 2:1) and are usually considered to function only as antenna for RCII (5,6,10,11). LHCI (or CP 0) functions in the PSI antenna and has relatively high ratios of Chl-a:Chl-b; 4:1 for LHCI (13,14,16) and about 6:1 for CP 0 (15,20).

The results described above suggest a simple model relating relating cell Chl content, numbers of RCII and their antenna sizes, and quantum requirements. Cells can contain RC complexes (RCI and RCII) and light-harvesting pigment-protein complexes (LHCP and LHCI). The RCII complex contains 75 Chl-a acting as antenna per photochemically active site. An additional 145 chl-a per RCII are associated with PSI (with an unspecified number of RCI). When cells contain less than about 2×10^5 RCII, they contain no LHCP or LHCI and changes in Chl/cell result solely from changes in RC/cell. Cells having more than 2×10^5 RCII/cell contain both reaction center and light-harvesting complexes. Changes in Chl/cell

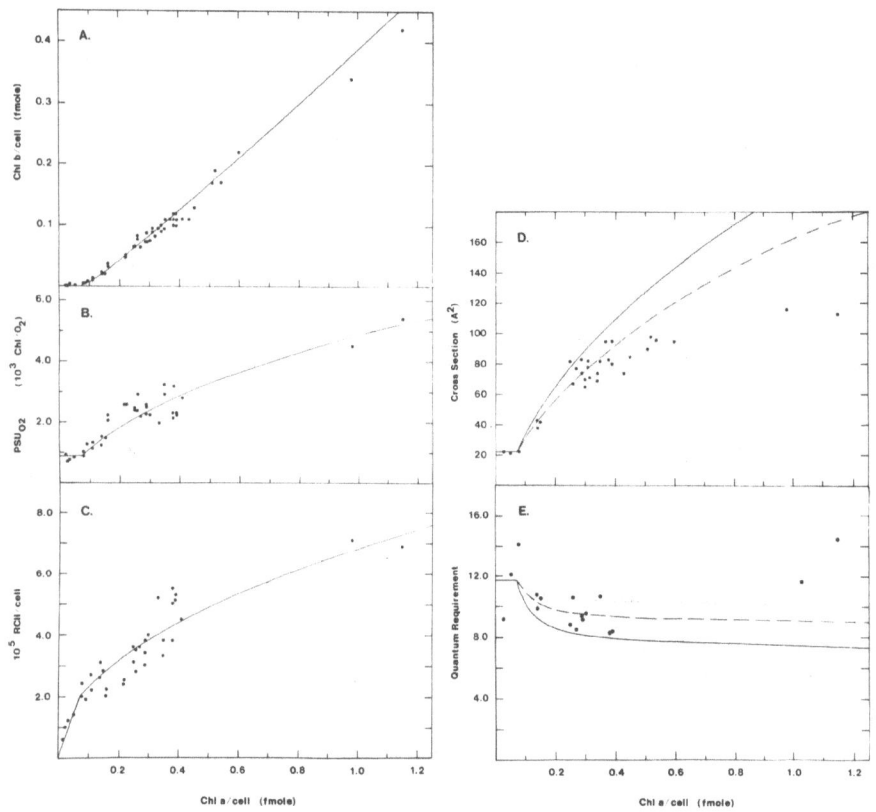

FIGURE LEGEND:
Figure 1. Relationships between cell Chlorophyll-a levels and other
physiologically important parameters in Chlorella. A.) Chl-b/cell
B.) PSU_{O2}, C.) RCII/cell,   D.) σ_{O2} , and E.) Quantum Requirement for
O_2 production. The curves drawn through the data were calculated as
described in the text.

are accomplished by coordinated changes in the numbers of both RC and light-harvesting complexes. Each addition of 10^5 RCII is accompanied by an increase in the ratio of total cell Chl to RCII by the addition of 208 Chl/RCII. 121 of the 208 Chl (67 Chl-a and 54 Chl-b, Chl-a:Chl-b=1.2:1) are associated with LHCP. The remaining 87 Chl (70 Chl-a and 17 Chl-b; Chl-a:Chl-b=4.1:1) function in LHCI.

This model was used to calculate the curves shown in Fig. 1A-E. Considering the oversimplifications inherent in the model, the calculated curves fit the data quite well. For simple bulk properties of the cells (Chl-b vs Chl-a, PSU_{O2} vs Chl-a, and RCII vs Chl-a), the model is adequate as described above. However, calculations of PSII-specific properties such as cross sections or quantum yields require that the distribution of light-harvesting complexes be specified. In Fig. 1D,E, three curves are plotted. The solid curves were obtained by assuming that LHCP associates exclusively with RCII and that LHCI functions solely in PSI. The two other curves in panels D and E were obtained by assuming that some fraction, designated "f", of LHCP did not function as antenna for RCII. In both panels, the dashed and dotted curves were obtained using f=0.2 and f=0.3, respectively. It is clear from the plots shown in Fig. 1 that f=0 does not fit the data.

It is possible to generate fits to the data similar to those shown in Fig. 1 by assuming that LHCI has ratios Chl-a:Chl-b which are greater than 4 (and thus is more similar to CP 0). In these cases, even larger values for f are required (up to 0.4 for Chl-a:Chl-b = 6) to provide fits to the data comparable to those shown in Fig. 1D,E. Finally, it appears that the data shown in Fig. 1 might be best fit if f is allowed to increase with increasing Chl/cell; i.e., as Chl/cell increases, the fraction of LHCP which functions in the PSII antenna decreases.

It is clear from plots such as those shown in Fig. 1D,E that a significant fraction (between 20 and 40%, depending on assumptions concerning the composition of LHCI) of LHCP does not function as light-harvesting antenna for PSII. Presumably, this fraction is available for use by PSI. In this case, the magnitude of f might be controled via the operation of a protein kinase/phosphotase system similar to that described in higher plants (25,26).

6. ACKNOWLEDGEMENTS

This research was supported in part by the United States Department of Agriculture (Contract # 5901941 9-03289) and the National Science Foundation (Grant # PCM 8316373). I thank Dr. David Mauzerall for many helpful discussions and suggestions throughout the course of this work.

7. REFERENCES

1. Boardman NK (1977) Ann. Rev. Plant Physiol. 28: 355-363
2. Falkowski PG and Owens TG (1980) Plant Physiol. 66: 632-635
3. Falkowski PG, Owens TG, Ley AC and Mauzerall DC (1981) Plant Physiol. 68: 969-973
4. Kawamura M, Mimuro M and Fujita Y (1979) Plant Cell Physiol. 20: 697-705
5. Leong T-Y and Anderson JM (1984) Photosyn. Res. 5: 105-115
6. Leong T-Y, Goodchild DJ and Anderson JM (1985) Plant Physiol. 78: 561-567
7. Ley AC and Mauzerall DC (1982) Biochim. Biophys. Acta 680: 95-106
8. Myers J and Graham J-R (1971) Plant Physiol. 48: 282-286
9. Myers J, Graham J-R and Wang RT (1980) Plant Physiol. 66: 1144-1149

w10. Anderson JM (1980) FEBS Lett. 117: 327-331

11. Anderson JM, Waldron JC and Thorne SW (1978) FEBS Lett. 92:27-233

12. Diner BA and Wollman F-A (1980) Eur. J. Biochem. 110: 521-527

13. Dunahay TG and Staehelin LA (1985) Plant Physiol. 78: 606-613

14. Haworth P, Watson JL and Arntzen CJ (1983) Biochim. Biophys. Acta 724:   151-158

15. Ish-Shalon D and Ohad I (1983) Biochim. Biophys. Acta 722: 498-507

16. Lam E, Oritz W, Mayfield S and R. Malkin (1984) Plant Physiol. 74: 650-655

17. Mullet JE, Burke JJ and Arntzen CJ (1980) Plant Physiol. 65: 814-822

18. Mullet JE, Burke JJ and Arntzen CJ (1980) Plant Physiol. 65: 823-827

19. Satoh K and Butler WL (1978) Plant Physiol. 61: 373-378

20. Wollman F-A and Bennoun P (1982) Biochim. Biophys. Acta 680: 352-360

21. Myers J (1951) Ann. Rev. Microbiol. 5: 157

22. Boardman NK and Thorne SW (1971) Biochim. Biophys. Acta 253:221-231

23. Myers J and Graham J-R (1983) Plant Physiol. 73: 440-442

24. Myers J, Graham J-R and Wang RT (1983) Biochim. Biophys. Acta 722: 282-290

25. Bennet J (1983) Phil. Trans. R. Soc. Lon. B302: 113-125

26. Horton P (1983) FEBS Lett. 152: 47-52

Author's address:
      Department of Cellular and Developmental
      Biology, Harvard University, 16 Divinity Avenue,
      Cambridge, Massachusetts 02138 (U.S.A.)

Photosynthesis Research 10: 197—200 (1986)
© *Martinus Nijhoff Publishers, Dordrecht*

A CONSIDERATION OF THE ORGANIZATION OF CHLOROPLAST PHOTOSYSTEM I*

R. MALKIN, DIVISION OF MOLECULAR PLANT BIOLOGY
            UNIVERSITY OF CALIFORNIA, BERKELEY, CA   94720

ABSTRACT
    Procedures that allow the fractionation of a native Photosystem I
complex (PSI-200) into several chlorophyll-containing complexes are now
available. Two complexes, each containing ~50% of the total chlorophyll of the photosystem, can be isolated. One complex contains both
chlorophyll a and b and serves as antenna complex for the reaction center while the reaction center complex contains 100 Chl a molecules per
P700 and has 7 different polypeptides. Only two of the latter (62 and
58 kDa) contain chlorophyll a· and these can be isolated as the photochemically active CPI complex. Based on these fractionation methods, a
model that describes the overall organization of the chlorophyll in
Photosystem is presented.

1. INTRODUCTION
    The recent isolation of three highly resolved thylakoid membrane com-
plexes (PSI, PSII, and the cytochrome b_6-f complex) (2,7,13) in functionally active form has allowed for detailed studies of struc-
ture-function relations in vitro (8). In the case of the case of PSI,
Mullet et al (13), described the purification of a "native" PSI complex
that retains the spectral properties of PSI in vivo. In studies of
this complex, it has been found that there is chlorophyll b associated
with PSI as well as with PSII and that this pigment is present in a
specific PSI antenna complex, known as LHCPI (5,10). The native PSI
complex has also been used to examine the structural organization of
the polypeptide subunits in the complex as well as in thylakoid membranes (15). In the present review, the chlorophyll proteins of PSI
are considered in greater detail. Based on the fractionation of the
native complex into several resolved chlorophyll protein complexes, the
overall organization of Photosystem I is considered and a model describing the photosystem is presented.

2. REVIEW OF RESULTS AND DESCRIPTION OF A MODEL OF PSI
    A flow-diagram for the fractionation of the chlorophyll-protein com-
plexes of PSI from spinach thylakoids is shown in Fig. 1.

*Dedicated to the memory of Warren Butler, who was both a friend and
a colleague.

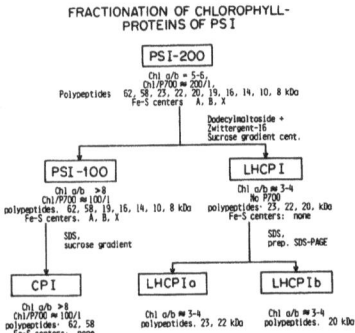

FIGURE 1. Fractionation
of chlorophyll-proteins
of Photosystem I. See
text for further details.

The starting complex, PSI-200, has ~200 Chl molecules per P700, contains Chl a and Chl b (Chl a/b ~5-6) and has approximately 10 polypeptide subunits (10,13). Subsequent fractionation separates this complex into two chlorophyll-containing complexes, one which contains the photochemical reaction center (PSI-100) and the second which contains a Chl a/b light-harvesting antenna complex (16). The reaction center complex contains ~100 Chl/P700 and has seven major polypeptides. This fraction still contains the entire PSI primary electron acceptor complex, based on EPR measurements at cryogenic temperature (3), and is able to utilize either reduced plastocyanin or DCPIP as electron donors in the photoreduction of NADP. The antenna complex contains three polypeptides of molecular weights 23, 22 and 20 kDa and has been fractionated into two chlorophyll-protein complexes: LHCPIa and LHCPIb (9). These have similar Chl a/b ratios, but differ in polypeptide compositions. It should be stressed that the separation of PSI-200 into these two chlorophyll-containing complexes is essentially quantitative in that little free chlorophyll is dissociated by the procedure, and the recovery of chlorophyll in each fraction is approximately 50%.

While PSI-100 is a relatively simple preparation in terms of polypeptide composition, it is possible to fractionate it further to separate the reaction center polypeptides from the low-molecular weight polypeptides in the preparation and obtain a photochemically active fraction known as CPI. This is accomplished by SDS treatment (17). In this case, P700 photooxidation activity has been observed under steady-state illumination even though the bound iron-sulfur centers associated with the stable primary electron acceptor complex are absent. Presumably the early electron acceptors, A_0 and/or A_1, are present in CPI although documentation of this is not complete. CPI has the simplest polypeptide composition of any photochemically active PSI fraction: only the 62 and 58 kDa subunits are present (17). However, the Chl/P700 ratio of 100/1 is the same as in the PSI-100 fraction. This demonstrates that all the chlorophyll a in CPI as well as in PSI-100 is localized in the high-molecular weight subunits, with no chlorophyll in the lower molecular weight subunits, since these are totally absent in CPI. This characterization of CPI also localizes P700 in the high-molecular weight subunits.

The separation of the LHCPI antenna complex into two chlorophyll-containing complexes has been previously described (9). Each contains approximately 50⁰/o of the original chlorophyll b found in PSI-200 and the two fractions are spectrally similar, with a Chl a/b ratio of ~3. However, the polypeptide subunit composition of the two differs in that LHCPIa contains 23 and 22 kDa subunits which LHCPIb contains the 20 kDa subunit.

The fractionation of PSI into several chlorophyll-containing complexes, some of which have an antenna function and some with reaction center activity, which has been described has allowed for a model that delineates that organization and function of chlorophyll in this photosystem. This model is shown schematically in Fig. 2. According to this model, two different chlorophyll b containing complexes are present and serve as light-harvesting antenna complexes for the reaction center complex. Based on the fractionation of PSI-200, the chlorophyll b-containing complexes contain approximately half of the total chlorophyll of PSI, with the remaining 50⁰/o found in a reaction center complex that contains only chlorophyll a. In many regards this situation is analogous to that in the PSII complex where LHCPII, an antenna complex, contains all the chlorophyll b and at least half of the total chlorophyll of this photosystem.

ORGANIZATION OF CHLOROPHYLL IN PHOTOSYSTEM I

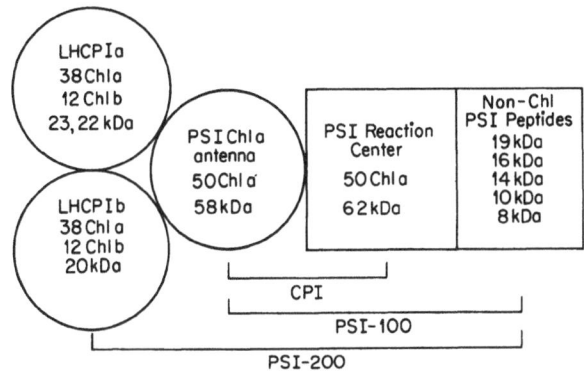

FIGURE 2. Model for the organization of chlorophyll in Photosystem I. Each indicated unit represents a chlorophyll-protein-complex that has been separated by standard biochemical fractionation.

The model shown in Fig. 2 indicates two additional chlorophyll a-containing proteins in PSI and assigns one a function as an antenna complex and the second as the reaction center that contains P700. A separation of these two complexes has not yet been achieved, but there is some indirect evidence that supports this view. It is known that there is a heterogeneity in the high-molecular weight subunits of PSI (17) and in our SDS-PAGE gel system, two subunits (62 and 58 kDa) can be resolved. Both subunits are found in CPI and together they are involved in binding ~100 chlorophyll a molecules. The stoichiometry of these subunits in the complex is not agreed upon as Nelson and co-workers have reported 2 high-molecular weight subunits per PSI complex in Chlamydomonas (14) while Glazer and co-workers have found 4 subunits per complex in a cyan-obacterial PSI (11). In the latter case, this would correspond to 25 chlorophyll a molecules found per ~60 kDa subunit. The recent report of Fish et al (4) has described the cloning of two genes for two high mole-cular weight polypeptides (PSIA1 and PSIA2) and it is likely these two genes are related to the two high-molecular weight subunits. Extensive sequence homology exists between the two peptides and in this regard, the situation is similar to that in PSII where two chlorophyll-protein complexes, one of 47 kDa and one of 43 kDa, are present. In the case of PSII, it is probable that the 47 kDa complex contains the PSII reaction center components while the 43 kDa subunit is a chlorophyll a-antenna complex. Considerable sequence homology also exists between these two proteins (1,6,12). Based on this analogy with PSII, the model of Fig. 2 suggests one PSI subunit (62 kDa) may contain the reaction center com-ponents and approximately 50 chlorophyll a molecules, while the second of 58 kDa may serve as a chlorophyll a-antenna complex with no photo-chemical activity. It is also possible these functions could be re-versed in relation to the two subunits. Until a separation of these two subunits under non-denaturing conditions has been achieved so that chlorophyll binding and photochemical activity can be measured, this model will remain as tentative although its structural analogy with the PSII complex is an attractive feature that argues for a unifying concept of photosystem organization.

ACKNOWLEDGEMENT
    This work was supported in part by a grant from the National Science Foundation. I would like to thank Drs. E. Lam and W. Ortiz for partici-pation in much of the work described.

REFERENCES
1 Alt J, Morris J, Westhoff P and Herrmann RG (1984) Curr. Genetics 8: 597-606
2 Berthold DA, Babcock GT and Yocum CF (1980) FEBS Lett. 134: 231-234
3 Bonnerjea J, Ortiz W and Malkin R (1985) Arch. Biochem. Biophys. 240: 15-20
4 Fish LE, Kuck U and Bogorad L (1985) J. Biol. Chem. 260: 1413-1421
5 Haworth P, Watson JL and Arntzen CJ (1983) Biochim. Biophys. Acta 724: 151-158
6 Holschuh K, Bottomley W and Whitfeld PF (1984) Nucleic Acids Research 12: 8819-8834
7 Hurt E and Hauska G (1981) Eur. J. Biochem. 117: 591-599
8 Lam E and Malkin R (1982) Proc. Natl. Acad. Sci. USA 79: 5494-5498
9 Lam E, Ortiz W and Malkin R (1984) FEBS Lett. 168: 10-14
10 Lam E, Ortiz W, Mayfield S and Malkin R (1984) Plant Physiol. 74: 650-655
11 Lundell DJ, Glazer AN, Melis A and Malkin R (1985) J. Biol. Chem. 260: 646-654
12 Morris J and Herrmann RG (1984) Nucleic Acids Research 12: 2837-2850
13 Mullet JE, Burke JJ and Arntzen CJ (1980) Plant Physiol. 65: 814-822
14 Nechushtai N and Nelson N (1981) J. Biol. Chem. 256:11624-11628.
15 Ortiz W, Lam E, Chollar S, Munt D and Malkin R (1985) 77: 389-397
16 Ortiz W, Lam E, Ghirardi M and Malkin R (1984) Biochim. Biophys. Acta 766: 505-509
17 Vierling E and Alberte RS (1983) Plant Physiol. 72: 625-633

Photosynthesis Research 10: 201–208 (1986)
© *Martinus Nijhoff Publishers, Dordrecht*

A HIGH MOLECULAR WEIGHT TERMINAL PIGMENT ("ANCHOR POLYPEPTIDE") AND A MINOR BLUE POLYPEPTIDE FROM PHYCOBILISOMES OF THE CYANOBACTERIUM NOSTOC SP. (MAC): ISOLATION AND CHARACTERIZATION

MAMORU MIMURO and ELISABETH GANTT

Key words: Terminal pigment, Cyanobacteria, Energy transfer, Photosynthesis, Phycobilisome, Nostoc sp.

ABSTRACT

A 94 kD pigment-polypeptide, which is presumed to be involved in anchoring the phycobilisomes to the thylakoids, was isolated from Nostoc phycobilisomes by gel filtration in 63 mM formic acid. The isolation condition did not require detergents or denaturating reagents, as in previous procedures, and enzymatic degradation was not observed at the low pH of 2.5. The "anchor polypeptide" thus obtained had absorption (Abs) and fluorescence maxima (Em) at 658 and 673 nm, respectively, in 63 mM formic acid at room temperature. The maxima shifted to longer wavelengths in 100 mM potassium phosphate (pH 6.8), Abs 665 and Em 683 nm at room temperature, and Abs 665 and Em 684 nm at liquid nitrogen temperature. The fluorescence maxima at both temperatures correspond to the longest wavelength component resolved in phycobilisomes from second derivative spectra. A minor blue polypeptide was also found by this isolation method. The molecular weight of this polypeptide was ca. 18,000 and is probably similar to a polypeptide which has been found in the phycobilisome core of other cyanobacteria.

INTRODUCTION

Phycobilisomes (PBS), as supra-molecular pigment protein complexes, function as light harvestors in photosynthesis of cyanobacteria and red algae [1, 2]. The light energy captured by PBS is preferentially transferred to pigment system II (PS II) chl $\underline{a}$ in the thylakoid membranes [2, 3]. In many cyanobacteria and some red algae, hemidiscoidally shaped PBS are found [1], which consist of multiple rod structures attached to an allophycocyanin (APC)-core which is adjacent to the photosynthetic membrane. The rods radiating from the APC-core consist of phycocyanin (PC) complexes, or phycocyanin-phycoerythrin (PC-PE) complexes, depending on the pigment composition of the organism. The light energy is transferred from the peripheral complexes to the APC-core, i.e. from the outside to the inside of the PBS. At the PBS-thylakoid attachment site, two polypeptides can function as donors to PS II chl a [2, 3]. One of the donors is the α-subunit of

Abbreviations used: α-APB, α-subunit of allophycocyanin B; APC, allophycocyanin; kD, kilodalton; SDS-PAGE, sodium dodecyl sulfate polyacrylamide gel electrophoresis; PBS, phycobilisome; PE, phycoerythrin; PC, phycocyanin; PS, pigment system.

allophycocyanin B (α-APB), and the other is a high molecular weight terminal pigment protein, also designated as "anchor polypeptide." The former was first reported in Synechococcus sp. and Anabaena variabilis [4] and further characterized from several other species [5-7]. The latter was first found in the red alga Porphyridium cruentum [8, 9] and is present in other PBS preparations including Nostoc [7, 10-12]. The molecular weight of the "anchor polypeptide" ranges from 75,000 to 115,000 depending on the species [12].

The "anchor polypeptide" is a minor PBS component (ca. 1-2%) [13] and is readily degraded to fragments of lower molecular weight [10, 11]. The isolation methods generally used, i.e. column chromatography with Sephacryl S-200 [10] and SDS-PAGE [8, 11], result in a low yield, partly because of apparent proteolytic degradation. We, therefore, used a simple method for the isolation of the "anchor polypeptide" based on a method developed by Fueglistaller et al. [14] for phycobiliprotein subunits. This procedure involves dissociation of PBS in distilled water and separation in 63 mM formic acid on a Bio-Gel column. The results presented here show that the "anchor polypeptide" appears to be undegraded, and that it retained its spectral characteristics with a fluorescence emission at 683 nm (20C). Furthermore, this method was also useful for the identification of a minor blue polypeptide which probably corresponds to the 18.3 kD polypeptide which has been identified as a phycobilisome core component in Synechococcus 6301 [15].

MATERIALS AND METHODS

Algal culture and isolation of PBS

Nostoc sp. (MAC) was grown in a liquid medium [16] at 38C under continuous illumination with daylight fluorescent lamps (ca. 1.5 W/m^2), and supplied with air and 5% CO_2. After 10 to 14 days cells were harvested. Phycobilisomes were isolated by the usual method [17], and after removal from the 1 M sucrose layer were subsequently incubated for at least 15 hr in Triton X-100 (1% v/v) at ca. 20C. This procedure decreased possible contamination of PS II thylakoid components [18]. Following incubation, the sample was centrifuged at 43,000 x g for 30 min, and the top Triton-containing layer was discarded, while the clear violet supernate was diluted three-fold with 0.75 M potassium phosphate (pH 6.8). Phycobilisomes were then pelleted by centrifugation at 254,000 x g for 3 hr.

Column chromatography

A Bio-Gel P-100 (<400 mesh) resin (Bio Rad, Richmond, CA) was hydrated and washed twice in 63 mM formic acid. After packing, the column (2.6 x 90 cm) was thoroughly washed and equilibrated with the same solution. The phycobilisome sample (1 ml) in 0.75 M potassium phosphate (pH 6.8) was diluted with 3 ml of distilled water and dialyzed against 63 mM formic acid (2 l) for 15 hr at 4C. Before loading, insoluble material was removed by centrifugation. The sample (about 20 mg protein per 5 ml) was applied to the column and was then eluted with 63 mM formic acid. The linear flowrate was held at ca. 1 cm/hr, and the elution profile (2.8 ml per fraction) was monitored for absorbance at 278, and 620 nm.

SDS-polyacrylamide gel electrophoresis was carried out according to the method of Laemmli [19] on a 1.5 mm thick slab gel with a gradient of 8 to 15% acrylamide containing 0.1% SDS. The apparent molecular mass of the polypeptides was estimated with marker proteins in the molecular weight range of 14,000 to 200,000 (Bio Rad, Richmond, CA).

Absorption and fluorescence spectra were measured with a Cary 17 spectrophotometer and an SLM 4800 spectrofluorometer, respectively. The spectral sensitivity of the fluorometer was corrected and the spectrum was expressed as quanta per wavelength. The resolution of the fluorometer was better than 0.5 nm.

RESULTS

Fractionation on Bio-Gel P-100

The first protein component in the formic acid elution profile of dissociated phycobilisomes was blue and appeared as a shoulder (fraction 43) on a larger 278 nm absorbance peak (Fig. 1). This blue shoulder contained the 94 kD peptide (Fig. 2, lane 2) which is a long wavelength terminal pigment called the "anchor polypeptide". The 278 nm absorbance peak at fraction 41 results from Triton X-100 micelles which formed during the isolation procedure.

Fractions were spectrally analyzed, from which we then selected those of special interest for SDS-PAGE analysis and comparison with the polypeptide pattern of unfractionated phycobilisomes [13]. The "anchor polypeptide" was only present in fraction 43, and is seen to correspond in molecular weight to presumably the same component in phycobilisomes

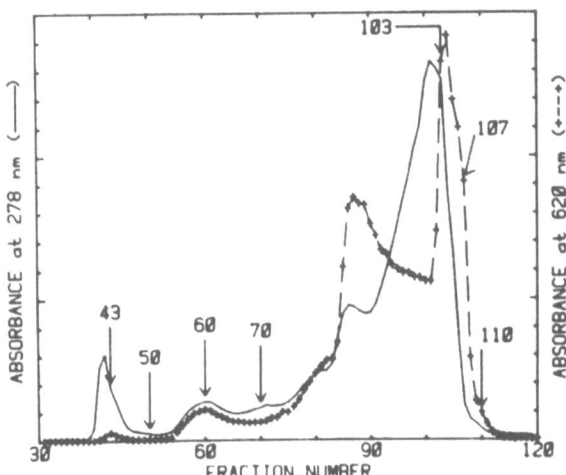

Figure 1. Elution profile of <u>Nostoc</u> phycobilisomes on a Bio Gel P-100 in 63 mM formic acid. Absorbance at 278 nm (___), and at 620 nm (+ ___ +). Arrows indicate comparable fractions analyzed on SDS-PAGE in Fig. 2.

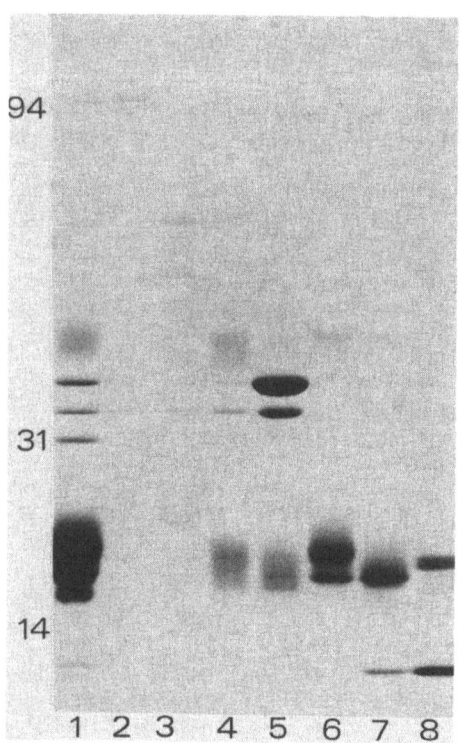

Figure 2. SDS-PAGE of several fractions from a Bio Gel P-100 column.
Lane 1, unfractionated phycobilisomes; lane 2, fraction 43; lane 3,
fraction 53; lane 4, fraction 60; lane 5, fraction 70; lane 6, fraction
103; lane 7, fraction 107; lane 8, fraction 110. On left are the
relative molecular weights.

(lane 1). In the unfractioned PBS the 31-38 kD linker polypeptides are
prominent, as are the subunits of the biliproteins 15 to 20 kD, with PE
near the top of the range and APC near the bottom. A minor peptide
whose molecular mass was about 10 kD was also clearly observed. A
striking feature of the gel pattern was the absence of any apparent
degradation product of the "anchor polypeptide", which had been
frequently observed in SDS-PAGE as an additional band of slightly lower
molecular weight [11, 12].
    Dialysis in formic acid prevented degradation not only of the anchor
polypeptide (lane 2), but it also preserved other polypeptides. The
fractions from 50-53 were pink. They contained several peptides at ca.
21, 36, 50 and 58 kD (lane 3). The 50 kD polypeptide may be a core
component of the PSII reaction center [18], while the subunit at ca. 21
kD is from PE (accounting for the pink color), the 36 kD peptide is a
linker, and 58 kD is unknown. Fraction 60 which was green had a complex

peptide pattern (lane 4). The green color arose from partly denatured
β-subunit of PC, because the color changed to blue when the sample was
placed in 100 mM K-phosphate buffer (pH 6.8). The origin of the 40–43
kD polypeptides is not known. In fraction 70 (blue) the uncolored 38 kD
linker polypeptides was heavily concentrated (lane 5). The color of the
fractions then changed from blue (fraction 85) to violet (fraction 103)
with the elution of a mixture of the subunits of PE, PC and APC as in
lanes 6 and 7. The polypeptide pattern in lane 8 is unusual, because in
addition to the expected enrichment of the low molecular weight
colorless polypeptide at ca. 10 kD, there appeared a distinct blue
polypeptide at ca. 18 kD. This polypeptide has an unusual behaviour on
a Bio-Gel P-100 column, in that it eluted after the smallest
polypeptides, i.e. along with the low molecular weight α–subunit of
APC and the 10 kD colorless polypeptide.

Spectral characterization

The "anchor polypeptide" in 63 mM formic acid had an absorption maximum
(20 C) at 658 nm and two shoulders around 605 and 565 nm (Fig. 3A). The
definitive emission maximum at 673 nm (Fig. 3B) indicates that the
chromophore configuration and its environment are kept relatively intact
in 63 mM formic acid, unlike the loss of fluorescence emission which
occurs on unfolding in 8 M urea [20].

On removal of the 63 mM formic acid, by dialysis against 100 mM
potassium phosphate buffer (pH 6.8), the "anchor polypeptide" formed an
insoluble flocculent blue precipitate. A thin layer of the precipitate
had an emission maximum at 683 nm at room temperature (Fig. 3D) when
measured with front surface optics. When a thicker layer of precipitate
was used for the measurements, the maximum was shifted to longer
wavelengths, typical of reabsorption. The 683 nm maximum is the
shortest obtained and thus is take as the actual fluorescence maximum.
The excitation spectrum (Fig. 3C) (fluorescence monitored at 750 nm)
with the maximum at 665 nm reflects a higher vibrational band than in 63
mM formic acid. At liquid nitrogen temperature, the fluorescence
maximum was at 684 nm with a sharpening of the band-shape (Fig. 3F).
The 684 nm peak was verified by high resolution scanning at 0.25 nm
intervals. The excitation spectrum had the same maximum as that at room
temperature (665 nm) except with a narrower bandwidth (Fig. 3E).

The blue 18 kD polypeptide, most concentrated in fraction 110, is
normally not resolved by SDS-PAGE in unfractionated phycobilisomes
(Fig. 2, lane 1) because of the masking by α and β subunits of the
more prevalent phycobiliproteins. This polypeptide did not seem to be
as hydrophobic as was the "anchor polypeptide" and did not precipitate
in buffer. In 63 mM formic acid fraction 110 had an absorption maximum
at 628 nm (not shown) and a fluorescence emission maximum at 648 nm
(Fig. 4). These maxima closely agree with those found for a 16.2 kD
polypeptide which has been found in <u>Mastigocladus</u> under similar isolation
conditions (14).

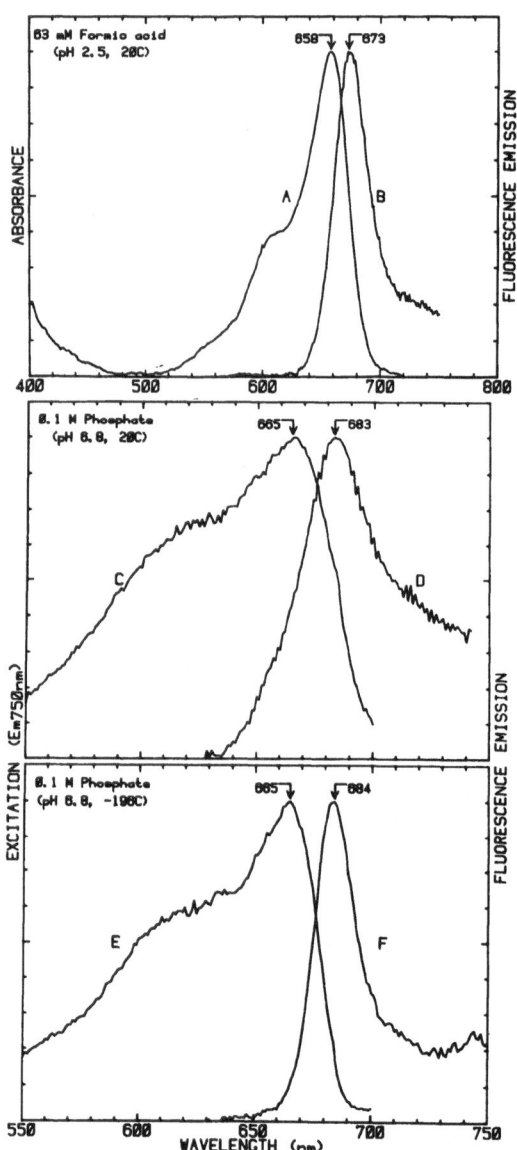

Figure 3. Optical characteristics of the "anchor polypeptide". Absorption (A) and fluorescence emission (B) spectra in 63 mM formic acid at room temperature. Fluorescence excitation (C) and emission (D) spectra in 100 mM potassium phosphate (pH 6.8) at room temperature, and at liquid nitrogen temperature: (E) excitation; (F) emission. For the fluorescence spectra, the sample was excited by 550 nm light (half band width, 8 nm) and detected with a half band width of 4 nm. For the excitation spectra (half band width, 4 nm), the fluorescence was monitored at 750 nm (half band width, 16 nm). Liquid nitrogen spectra were recorded in K-glycerophosphate frozen as a clear "glass".

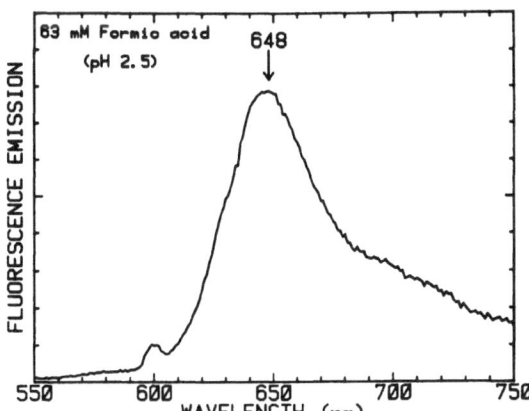

Figure 4. The fluorescence emission spectrum of fraction 110 (in 63 mM formic acid) at room temperature. Excitation was at 600 nm (half band width, excitation 4 nm and emission 2 nm).

DISCUSSION

The high molecular weight "anchor polypeptide" comprises only ca. 1-2% of the total stainable protein in Nostoc [13]. Its relative scarcity and susceptibility to proteolysis [21] have made it difficult to isolate. Furthermore, its hydrophobic nature has required a detergent or a chaotropic agent such as 6 M guanidine-HCl, and SDS-PAGE [10, 11] for its isolation. Its isolation in 63 mM formic acid was simple, and as shown here, was without any indication of proteolylic degradation either in fraction 43, or in the dissociated phycobilisomes (Fig. 2). The yield of ca. 100 μg per column was substantial, especially since contaminating peptides were not detected. Also, as indicated above, the chromophore is stable even when the apoprotein is partly unfolded.

Isolation in 63 mM formic acid has also proven useful for separation of other phycobilisome polypeptides [22]. One of these is the 18 kD blue polypeptide which is spectrally similar to the 16.2 kD polypeptide in Mastigocladus, and probably corresponds functionally to the 18.3 kD polypeptide noted in the APC I complex in Nostoc [11], and is present in the Synechococcus phycobilisome core [15]. The purified peptide from Mastigocladus had an absorption peak at 624 nm, and 643 nm fluorescence emission in 5 mM K-phosphate pH 6.8 (M. M. unpublished).

Isolation of the "anchor polypeptide" without strong protein denaturating agents has provided material for detailed spectral analysis. By its spectral properties it corresponds to the longest wavelength components (of two), which have been resolved in native PBS by second derivative spectroscopy. From these results together with spectral analysis of intact phycobilisomes we have obtained evidence (to be published) which shows that energy transfer from allophycocyanin to the "anchor polypeptide" is the energetically preferred pathway.

ACKNOWLEDGEMENT

We thank Ms. C. A. Lipschultz for her skillful assistance in performing SDS-PAGE. M. Mimuro was recipient of a Smithsonian Institution Fellowship while at the Smithsonian Environmental Research Center. His permanent address is Institute for Basic Biology, Okazaki 444, Japan. This work was supported in part by contract AS05-76ER-04310 from the Department of Energy.

REFERENCES

1. Gantt E (1980) Int Rev Cytol 66: 45-80
2. Gantt E (1982) Ann Rev Plant Physiol 32: 327-347
3. Glazer AN (1984) Biochim Biophys Acta 768: 29-51
4. Glazer AN and Bryant DA (1975) Arch Microbiol 104: 15-22
5. Ley AC, Butler WL, Bryant DA and Glazer AN (1978) Plant Physiol 59: 974-980
6. Lundell DJ and Glazer AN (1981) J Biol Chem 256: 12600-12606
7. Canaani OD and Gantt E (1980) Biochemistry 19: 2950-2956
8. Redlinger T and Gantt E (1981) Plant Physiol 68: 1375-1379
9. Redlinger T and Gantt E (1982) Proc Natl Acad Sci USA 79: 5542-5546
10. Lundell DJ, Yamanaka G and Glazer AN (1981) J Cell Biol 91: 315-319
11. Zilinskas BA (1982) Plant Physiol 70: 1060-1065
12. Gantt E (1985) in Molecular Biology of the Photosynthetic Apparatus. (K. Steinback et al., eds.), Cold Spring Harbor, New York 223-229
13. Canaani OD and Gantt E (1983) Biochim Biophys Acta 723: 340-349
14. Fueglistaller P, Ruembeli R, Suter F and Zuber H (1984) Hoppe-Seyler's Z Physiol Chem 365: 1085-1096
15. Lundell DJ and Glazer AN (1983) J Biol Chem 258: 894-901
16. Van Ballen C (1967) J Phycol 3: 154-157
17. Gantt E, Lipschultz CA, Grabowski J and Zimmerman B (1979) Plant Physiol 63: 615-620
18. Chereskin B and Gantt E (1985) submitted to Arch Biochem Biophys
19. Laemmli VK (1970) Nature 117: 690-685
20. Scheer H (1981) Angew Chem Int Ed Engl 20: 241-261
21. Ruscowski M and Zilinskas BA (1982) Plant Physiol 70: 1055-1059
22. Fueglistaller P, Suter F and Zuber H (1983) Hoppe-Seyler's Z Physiol Chem 364: 691-712

Author's address:
    Smithsonian Institution, Environmental Research Center,
    12441 Parklawn Drive, Rockville, Maryland 20852-1773 (U.S.A.)

Photosynthesis Research 10: 209–215 (1986)
© *Martinus Nijhoff Publishers, Dordrecht*

EXCITON INTERACTIONS IN PHYCOERYTHRIN

KAROLY CSATORDAY, SHARON CAMPBELL, AND BARBARA A. ZILINSKAS

Department of Biochemistry and Microbiology, Cook College, Rutgers University, New Brunswick, NJ 08903, U.S.A.

ABSTRACT
Upon assembly of the phycoerythrin trimer into hexamer and the hexamer into dodecamer, marked spectral changes are observed. The absorption and circular dichroism spectra of the various phycoerythrin aggregates were resolved into Gaussian components representing individual electronic transitions of phycoerythrobilin chromophores within these proteins. While the contribution of a broad, sensitizing band (at 525 nm) is constant, with increasing aggregate size, a short-wavelength pair of bands centered at 555 nm decreases concomitantly with a dramatic increase in the intensity of a long-wavelength pair of chromophore transitions centered at 563 nm. The implications of these spectral changes for efficient energy transfer in the phycobilisome are discussed.

INTRODUCTION
Light energy is harvested in the cyanobacteria and red algae by special multiprotein aggregates called phycobilisomes (PBsomes). These contain phycobiliproteins that absorb radiation and transfer its energy in the form of electronic excitation to the photochemical reaction center. A complete transfer sequence involves phycoerythrin (PE), phycocyanin (PC) and allophycocyanin (APC) [11]. Since this highly efficient transfer is dependent on the distance between donor and acceptor molecules, as well as on the spectral overlap between their respective emission and absorption spectra, the spectral properties as well as the possibility of their delegation to particular chromophore transitions are of interest. Chromophores with short-wavelength absorption bands sensitize the fluorescence of those absorbing at longer wavelengths, and thus, the terms "sensitizing" (˘s˘) and "fluorescent" (˘f˘) chromophores were introduced by Teale and Dale (8). In the present study, we attempt to correlate spectral characteristics of PE with individual electronic transitions of chromophores within the protein.

Phycoerythrin in vivo is assembled into so-called rods that serve as constituents of the light-harvesting PBsome. Within the rods of the cyanobacterium, Nostoc sp., grown in cool-white fluorescent light, PE exists as a stack of four disks each consisting of a PE trimer $(\alpha\beta)_3$. Each trimer contains 15 phycoerythrobilin chromophores. The four trimers comprising a dodecamer of PE are attached to a PC hexamer, also in the form of a double disk. The assembly is governed and facilitated by linker proteins that determine the final structure of the PBsome [12]. The spectral properties of isolated PE aggregates closely resemble those they manifest within the PBsome [10], thus allowing the study of electronic transitions in a simpler system.

MATERIALS AND METHODS

Phycoerythrin in the trimeric and hexameric forms were isolated from PBsomes of the cyanobacterium <u>Nostoc</u> sp. as described earlier [12]. Subsequent to dialysis of the PE hexamer against 0.75M K-phosphate, pH 7.0, the PE dodecamer was obtained by sedimentation in linear gradients of sucrose (0.15M-0.8M) containing 0.75M K-phosphate, pH 7.0, for 16h at 40,000 rpm in a SW40 Beckman rotor. The PE trimer and hexamer each contained approximately 5% PC, and the dodecamer contained approximately 7% PC; however, the PC was not physically associated with PE as demonstrated by the absence of fluorescence emission from PC following PE excitation. The curve-fitting analysis included a component to account for the contribution of the contaminating PC.

Aggregate size and polypeptide composition were determined respectively by sedimentation in linear gradients of sucrose and by sodium dodecyl sulfate polyacrylamide gel electrophoresis as described earlier [12]. Absorption and circular dichroism (CD) were measured as previously described [13] with a Cary 17D and a Cary 60 spectrometer, respectively. The circular dichroism and absorption spectra were resolved into Gaussian components using a curve-fitting program written for the Apple II computer [1] with the blue edge of the spectra approximated by a wide bandwidth chimeric component attributed to sensitizing chromophores.

RESULTS AND DISCUSSION

The absorption and CD spectra of PE trimers, hexamers and dodecamers were resolved into Gaussian components representing individual electronic transitions of phycoerythrobilin chromophores within these protein aggregates. Figure 1 (left panel) shows the computer-generated fits of both

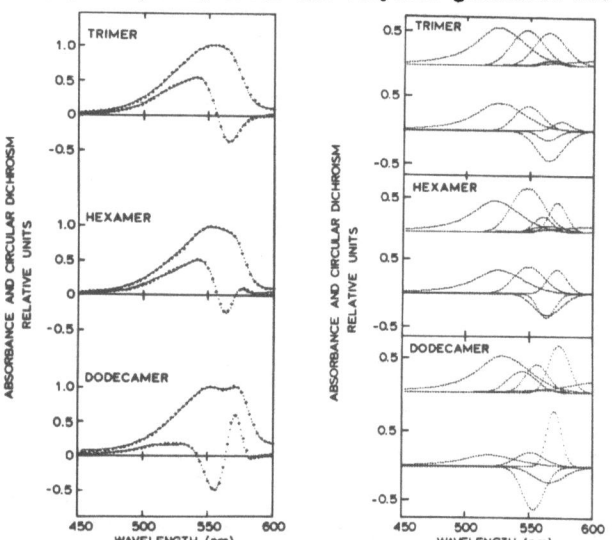

FIGURE 1. Left panel. Absorption and CD spectra of PE aggregates. The experimental and fitted curves are plotted with dots and crosses respectively. Only every third point in the fitted curve is shown for clarity. The absorption spectra were normalized for unit area.
FIGURE 1. Right panel. The resolved components of the absorption and CD spectra of the three PE aggregates.

the absorption and CD spectra for each of the aggregates. Only the experimental curve and the sum of the calculated components are shown for clarity and to demonstrate the error of the fit.

The evolution of a band in both the absorption and CD spectra at 570 nm is seen. Also, the negative band in the CD spectrum is shifted toward the blue as a function of aggregate size. These changes have previously been interpreted in terms of chromophore-chromophore interactions [10]. A reasonably close match was found between the absorption and CD components with regard to the essential parameters of peak wavelength and half-bandwidth (see Table I).

Figure 1 (right panel) shows the individual component bands and allows a quantitative treatment of these changes. The resolved components are plotted on the same scale as the experimental and fitted curves in Fig. 1 (left). The corresponding spectral parameters are given below.

TABLE 1. Spectral parameters of the resolved components of the phycoerythrin absorption and circular dichroism spectra.

Trimer	Abs	L	524	546	564	563	568	600	E=2.5%
		W	45	29	28	17	16	66	
		A	.50	.25	.21	.02	.02		
	CD	L	525	547	564	563	574		E=6.1%
		W	43	25	25	17	17		
		A	.48	.19	.23	.05	.05		
Hexamer	Abs	L	526	547	562	558	570	593	E=1.8%
		W	45	28	28	17	16	66	
		A	.51	.27	.05	.08	.12		
	CD	L	524	548	563	562	570		E=7.8%
		W	42	26	25	16	17		
		A	.38	.23	.17	.11	.12		
Dodecamer	Abs	L	527	544	563	554	572	601	E=2.6%
		W	49	26	26	21	18	58	
		A	.51	.14	.02	.14	.20		
	CD	L	517	548	564	553	568		E=10.7%
		W	49	26	28	20	14		
		A	.19	.13	.16	.26	.25		

L=wavelength maximum, W=half-bandwidth, A=area under the component curve as the percentage of total area, and E=error of the fit. All numerical data are rounded to the last digit shown. The component at ~600nm is attributed to contaminating PC.

Normalizing the absorption spectra with respect to the area under the spectral curve allows their comparison based on changes in the relative areas under each component curve, a quantity proportional to the oscillator strength of a given transition. All the absorption spectra feature a broad 45 nm wide band in the high energy region. This component, centered around 525 nm, remains constant in area, comprising about 50% of the total absorbance. The corresponding band in the CD spectra has the same features with the exception that in going from the trimer to the dodecamer it decreases in intensity and is blue-shifted to 517 nm. In fact, fitting two wide chimeric components at 513 and 532 nm, instead of the one at 525 nm, better accounted for the CD features in the dodecamer spectrum but did not change the total area ascribed to the high energy band nor did it significantly influence the parameters or the discussion which follows for the two exciton band pairs. Applying Occam's razor, we restricted the resolutions to one component in this region although two wide bands representing sensitizing chromophores may be present.

Although the results discussed in this paper pertain solely to PE aggregates, we consider the conclusions relevant to the native PBsome since spectroscopically the exciton features are essentially the same for the dodecamer of PE, the isolated PBsome rod which is comprised of a PE dodecamer associated with a PC hexamer, and the isolated PBsome. The only difference seen in comparing the PE dodecamer with the PBsome or the PBsome rod is the diminished high energy CD band attributed to sensitizing PE chromophores. In the PBsome or in the rod, this CD band remains at the level observed for the isolated PE trimer. There may be a slightly different conformation of the sensitizing chromophores in the isolated PE dodecamer which affects its optical activity but not its absorption.

The trimer spectrum also envelops a pair of bands centered around 555 nm with maxima at 546 and 564 nm and bandwidths of 27 ± 2 nm. Barely distinguishable in the fit of the trimer absorption spectrum is another set of bands with peaks at 563 and 568 nm. Their total contribution to the trimer absorption spectrum is almost within the error of the fit. On the other hand, the hexamer absorption and CD spectra show that the contribution from this low energy pair of transitions is significant. It continues to increase as aggregation proceeds; the dodecamer absorption and CD spectra are dominated by this long-wavelength pair. At the same time, the gradual and concomitant decline in the contribution from the pair of bands centered around 555 nm is evident from the resolutions.

Whereas the bisignate bands in the CD spectra are close to being conservative, it is interesting to note that the two low energy components, representing a chromophore dimer in the long-wavelength region of the spectra (centered around 563-564 nm), increase in intensity primarily at the expense of the low energy (563 nm) transition of the chromophore dimer whose absorption is centered around 555 nm. In terms of rod assembly, this implies that the same chromophores take part in both interactions, the long-wavelength dimer emerging as a result of interaction either at the interface of two proximal disks or due to the change in the local concentration and orientation of chromophores within the protein.

It is readily seen that while the contribution from the broad band is constant, that of the short-wavelength pair of bands centered about 555 nm decreases. Concurrently, the long-wavelength pair of chromophore transitions increases dramatically in intensity.

Figure 2 shows a representation of the change in area under the broad band as well as under the short- and long-wavelength pair of bands as a function of aggregate size, i.e., the number of phycoerythrobilin chromophores in each aggregate. A linear regression analysis for each of the sets of three data points, representing the change in area under the absorption bands attributed to the short- and long-wavelength chromophore dimer, gives slopes of -0.66 and 0.63, respectively, substantiating the complementary relationship between the two chromophore interactions. In a comparable presentation of the CD data, the trend is similar with that for absorption. However, due to the diminished optical activity of the sensitizing chromophore band in the dodecamer, the relative magnitude of the exciton bands is exagerrated and consequently the absorption and CD data quantitatively do not coincide.

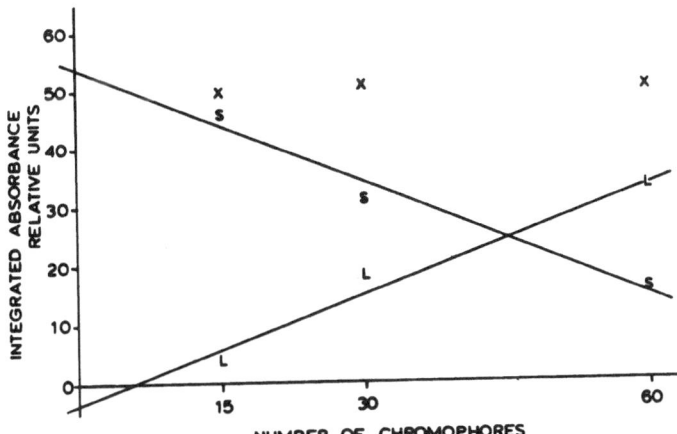

Figure 2.   Relative areas of the absorption band at 525 nm (x), the short-
wavelength pair of bands of 27±2 nm half-width (s), and the long-
wavelength pair of bands of 17±2 nm half-width (L), plotted as a function
of the number of chromophores in the three PE aggregates.

A curious feature of the complementary behavior of the two pairs of
exciton interactions is the change in the spectral half-bandwidth parameter. The broad, presumably non-interacting, band is approximately 45 nm
in width, whereas the half-widths of the dimer bands are 27±2 nm and 17±3
nm respectively. This series of values is close to that which is expected
upon chromophore-chromophore interaction where the half-widths decrease by
a factor of $1/\sqrt{M}$, M being the number of chromophores taking part in the
interaction [4].

The fact that this relationship holds for the chromophore-chromophore
interactions in PE means that the chromophore configuration responsible
for the long-wavelength split may either be a quatromer or that the disap-
pearing 564 nm band of the short-wavelength dimer interaction acquired in
part properties of a monomer. This would imply that the excited state is
not shared equally between the interacting chromophores. Such an idea has
been invoked as a possibility to explain the behavior of the 870 nm band
in bacteriochlorophyll [5] based on differences in the "environmental
interactions" which "reduce the amount of mixing of the excited states
localized on single molecules". The 545 band retains a considerable
portion of its intensity; therefore, it may well be less affected by the
attachment of another disk of PE.

In terms of transfer of electronic excitation energy, a cascade of
energy levels among chromophores of precise spatial organization ensures a
directional flow toward ever longer wavelength chromophores. The chromophores with short-wavelength absorption bands sensitize the fluorescence
of those absorbing at longer wavelengths, and thus, the terms "sensitizing" (`s´) and "fluorescent" (`f´) chromophores were introduced by
Teale and Dale [8]. The wide, 525 nm band absorption in PE may therefore
be attributed to sensitizing chromophores. Since its contribution to the
total absorption is constant regardless of the state of aggregation, the
amount of light energy harvested increases linearly with the number of
disks in the PE rod constituents of the PBsome.

The larger the rod, however, the further away the sensitizing chromophores will be from PC (the next pigment in the energy transfer sequence). The same is true for the fluorescent chromophores with transitions at longer wavelengths. Upon assembly of two trimers into a hexamer, the low energy exciton split becomes conspicuous. Were this the result of exciton interaction at the interface of the two trimer disks assembled into a hexamer, there would be little evolutionary advantage in assembling even larger aggregates because the intradisk transfer of electronic excitation energy quickly ends up on the lowest energy exciton band via resonance transfer and internal conversion from the higher energy exciton band. Subsequent interdisk transfer would necessitate uphill transfer to the next layer of sensitizing chromophores, in effect abolishing any directional flow of excitation energy toward the core of the PBsome.

However, if the lowest energy level is distributed within the aggregate or spans across the stack of disks, then the disadvantage stemming from increased distances within the light-harvesting apparatus is balanced by an increase in the spectral overlap between the fluorescence from the lowest, 572 nm exciton level at 580 nm, and the absorption band of the sensitizing phycocyanobilin chromophore at 586-600 nm. In fact, in the dodecamer, at least 35% of the total absorption of PE is due to the chromophores participating in the long-wavelength exciton interaction centered around 561 nm; that is, about 20-24 chromophores are able to create a conduit for delocalized excitation. The two-fold narrowing of the spectral bands already indicates that the excitation energy is delocalized, albeit only partially, over at least four chromophores.

In PBsomes of Synechocystis 6701, it was shown [2] that compared to a fast excitation transfer between sensitizing and fluorescent chromophores within the disk, disk-to-disk transfer is the rate-limiting step in PBsome rods.

In the scheme of chromophore-chromophore interactions discussed above, symmetry considerations would warrant that if at least four chromophores interact as the disks are assembled into a hexamer and subsequently into a rod and if there is no total mixing of excited states, internal conversion to the lowest level will not always coincide with the vectorial transfer toward one particular end of the rod. In an in vivo assembly involving dodecameric PE, this would result in a slowdown of exciton transfer across the disk interfaces. This slowdown may not be critical since recently [5] time-resolved fluorescence spectra of chromatically adapted Tolypothrix tenuis showed that excitation energy is transferred faster in PE-rich, PE-excited PBsomes than in PBsomes whose rods contained only PC [6]. The difference was attributed to slower transfer among the `f´ chromophores of PC in the PE-less system [6], and we suggest that it may be due to the fact that the delocalized excitation on interacting chromophores in PE is able to span greater distances with fewer transfer steps than would be involved in transfer via fluorescent non-interacting chromophores as is the case for PC.

With the advent of tunable dye-lasers in picosecond spectroscopy, these ideas may be tested since individual transitions can be excited and the fate of excitation monitored on the timescale of chromophore-chromophore transfers (see, for example, reference [9]). Internal conversion between exciton components should give faster fluorescence rise-times than `s´ to `f´ transfer. In addition, x-ray crystallographic analysis has already yielded information concerning possible chromophore arrangements in PC [7], and work on the other biliproteins is sure to follow.

ACKNOWLEDGMENTS
    New Jersey Agricultural Experiment Station, Publication No. D-01104-
3-85, supported in part by State funds and by the United States Hatch Act.
This work was also supported by the Science and Education Administration
of the United States Department of Agriculture under Grant 85-CRCR-1-1562
from the Competitive Research Grants Office. We are grateful to Dr. Peter
Kahn and Dr. Jozef Grabowski for helpful discussions.

REFERENCES
1. Csatorday K, MacColl R, Csizmadia V, Grabowski J and Bagyinka C
    (1984) Biochemistry 23: 6466-6470
2. Glazer AN, Yeh SW, Webb SP and Clark JH (1985) Science 227: 419-423
3. Glick RE and Zilinskas BA (1982) Plant Physiol 69: 991-997
4. Hemenger RP (1977) J Chem Phys 67: 262-264
5. Mar T and Gingras G (1984) Biochim Biophys Acta 765: 125-132
6. Mimuro M, Yamazaki I, Yamazaki T and Fujita Y (1985) Photochem
    Photobiol 41: 597-603
7. Schirmer T, Bode W, Huber R, Sidler W and Zuber H (1985) J Mol Biol
    184: 257-277
8. Teale FWJ and Dale RE (1970) Biochem J 116: 161-169
9. Wehrmeyer W, Wendler J and Holzwarth AR (1985) Eur J Cell Biol 36:
    17-23
10. Zilinskas BA, Campbell S and Grabowski J (1984) In: Advances in
    Photosynthesis Research, (Sybesma,C,ed.) Vol. II, pp. 687-690,
    Martinus Nijhoff Dr. W. Junk Publishers, the Hague
11. Zilinskas BA and Greenwald LS (1985) Photosynthesis Res: in press
12. Zilinskas BA and Howell DA (1983) Plant Physiol 71: 379-387

Photosynthesis Research 10: 217–222 (1986)
© *Martinus Nijhoff Publishers, Dordrecht*   217

THE LIGHT-HARVESTING CHLOROPHYLL A/B PROTEIN ACTS AS A TORQUE ALIGNING CHLOROPLASTS IN A MAGNETIC FIELD

J.G. KISS,[*] GY.I. GARAB, ZS.M. TÓTH and Á. FALUDI-DÁNIEL[**]

Department of Plant Physiology, Biological Research Center, Hungarian Academy of Sciences, Szeged, P.O.Box 521, 6701 Hungary

Key words: light-harvesting chlorophyll a/b protein, liquid crystal, magnetic susceptibility, polarized fluorescence

Abstract. Displacement of particles from the purified light-harvesting chlorophyll a/b protein aggregate (LHC) was studied in magnetic fields of various strengths (0 to 1.6 T) by polarized fluorescence measurements. Macromolecular aggregates of LHC have a considerable magnetic suscepti-bility which enables the particles to rotate and align with their nematic axes parallel with $\vec{H}$. As LHC is embedded in a transmembrane direction thylakoids should align perpendicular to $\vec{H}$, the mode of alignment experimentally observed in thy-lakoids. The value of the magnetic susceptibility could be estimated by relating it to the integral susceptibility of the chlorophyll molecules in LHC. The fitting of this value with the field strength dependency of the fluorescence polar-ization ratio (FP) revealed a relationship between the LHC content of various photosynthetic membranes and their capac-ity for alignment, which suggested that LHC might be the torque ordering chloroplasts in a magnetic field.

INTRODUCTION

LHC, the pigment-protein complex which controls the dis-tribution of quanta between the two photosystems might have a dynamic structure enabling the complex to change its con-nectivity with the individual photosystems depending on the physiological demands of the plant [11]. Such mobility is expected in a system where polypeptides with the attached pigments can, under certain conditions, transmute to liquid crystal organizations. Such a structure was suggested by a giant CD signal of LHC [5], a characteristic of pigments embedded in liquid crystals [14]. A further attribute of

[*]Clinic of ENT, Medical University, Szeged, P.O. Box 422, 6701 Hungary
[**]To whom correspondence should be sent.
Abbreviations: LHC: light-harvesting chlorophyll a/b pro-tein; FP: fluorescence polarization ratio, I_z/I_y;

liquid crystals is their diamagnetic susceptibility and their orientation by electric fields [15]. The former can be easily demonstrated by the rotation of LHC macromolecules resulting in an alignment where the nematic axis in the particles is parallel to the direction of the magnetic field. Supposing that thylakoids are aligned in the magnetic field by LHC as a torque one would expect that the success and speed of alignment depends not only on environmental conditions (temperature and viscosity of the medium) but on the amount of LHC contained in the membranes.

In this paper experiments are reported which investigate the magnetic susceptibility of LHC in order to determine whether LHC in vivo can bring about the alignment of thylakoids in a magnetic field.

MATERIALS AND METHODS

Purified LHC was prepared [4] from chloroplasts isolated from spinach. The isolation medium contained 0.35 M sucrose, 0.015 M phosphate, 0.01 M KCl, pH 7.2 [1]. Mesophyll and bundle sheath chloroplasts of maize were prepared as in [3], chloroplasts grown under intermittent light were obtained by the method described in [2].

Magnetic alignment of the purified LHC and of various types of chloroplasts was carried out in an electromagnet (Type Phylatex 1316, DDR) equipped with a sample holder with a cuvette which could be illuminated by the 488 nm line of an Argon laser ILA 120 Zeiss Jena. The magnetic field

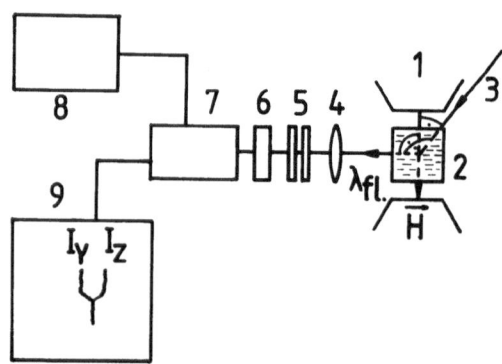

FIGURE 1. Block-scheme of the apparatus for measuring the alignment of thylakoids in a magnetic field. 1: magnet, 2: samples with membranes aligned with their plane perpendicular to the field, 3: light exciting fluorescence, 4: lens, 5: polaroid filters alternating the direction of observation between the two fluorescence components emitted parallel and perpendicular to the plane of the aligned membranes (I_z and I_y, respectively), 6: interference filter transmission at 680 $\pm$ 10 nm, 7: photomultiplier, 8: power supply, 9: recorder

strength was regulated from 0 to 1.6 T and measured with a
probe inserted into the cuvette by a Gaussmeter (RFL In-
dustries Inc. Boston New Jersey USA).

Fluorescence was excited and observed perpendicular to
the magnetic field direction. Polaroid filters separated the
fluorescence components emitted parallel and perpendicular
to the exciting beam [6]. The system is represented diagram-
matically in Fig. 1. Measurements were made at room temperature.

LHC particles of different sizes were obtained from large
aggregates of purified LHC by incubation with 2% v/v Triton
X-100.

RESULTS

In these studies we postulated that in chloroplasts of various LHC content, as in isolated LHC particles and their
fragments, fluorescence at 680 nm originates from Q_y dipoles
of chlorophyll a making the same angle of orientation with
respect to the normal of the membrane plane. If so, then
observing the alignment of various thylakoids in a steady

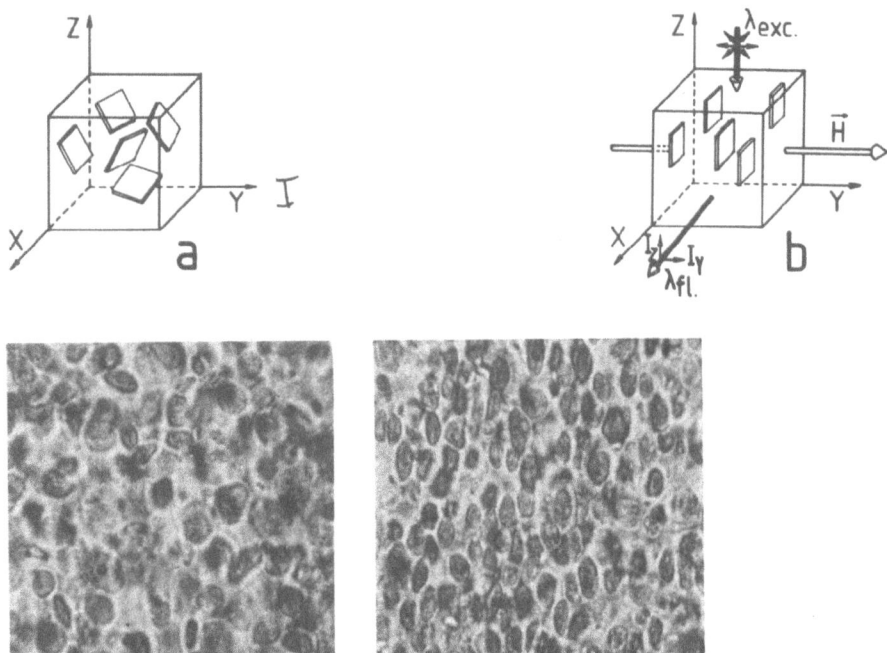

FIGURE 2. The geometry of thylakoids and chloroplasts suspensions prior to (a) and during (b) magnetic alignment.
(Chloroplasts suspensions in the light microscope - x 1500 -
at 0 (a) and 1.2 T (b) magnetic field strength.)

state should result in saturation FP values of the same
magnitude. The geometry of thylakoid alignment is shown in
Fig. 2. According to data obtained for the field strength
dependency of FP (Fig. 3) our presumption seemed to be correct: the orientability of the particles was determined by
the magnetic susceptibility which induced the particles to
reach the same level of FP. Calculations relating FP to the
magnetic field strength and the magnetic susceptibility were
performed using the theory developed in [10] and [13]. Regarding aligned particles as discs

$$p = \frac{I_z}{2I_z + I_y} = \frac{(1+\cos^2\gamma)}{4} + \frac{(1-3\cos^2\gamma)}{8}\left(\frac{e^{h^2}}{hD(h)} - \frac{1}{h^2}\right) \qquad (1)$$

$$\text{where } D(h) = \int_0^h e^{x^2} dx, \text{ and } h = \sqrt{\frac{\Delta X}{2kT}} \cdot \vec{|H|}$$

$\vec{H}$ is the field strength, ΔX is the magnetic anisotropy expressed as $X_y - X_z$, γ denotes the orientation angle of the Q_y
emission dipole with respect to the normal of the membrane
plane.
     FP used in our curve fitting procedure could be obtained
as

$$FP = \frac{1}{\frac{1}{p} - 2} , \text{ where } p = Eq (1) \qquad (2)$$

     The angle γ used in these calculations was throughout
58.6° (with respect to the normal of the membrane plane)
which was found by us earlier to be characteristic for the
fluorescence band localized at 680 nm [6].
     As seen from Fig. 3 the calculated values of FP obtained
from Eqs. (1) and (2) showed a reasonable fit with the experimental points of FP.
     In order to know whether the diamagnetic moment is in-
fluenced by the size of the aligning particles allowing
coupling between magnetic dipoles, we studied FP with LHC
aggregates of different sizes. Size determinations were
carried out on the basis of the speed of relaxation of
alignment (after switching off the magnet, till FP has
dropped to 1.0).
     Presuming that fragmentation did not change the form of
the particles.

$$r_1 = r_0 \sqrt[3]{\frac{\tau_1}{\tau_0}} \ [9] \qquad (3)$$

r_1 = radius of the fragment
r_0 = radius of the non-fragmented aggregate
τ_1 = relaxation time of the fragment
τ_0 = relaxation time of the non-fragmented aggregate

     The measurements show a straight line of direct relation-
ship between the particle size and FP indicating that no

special coupling between the magnetic dipoles is represented
by individual chlorophyll molecules.

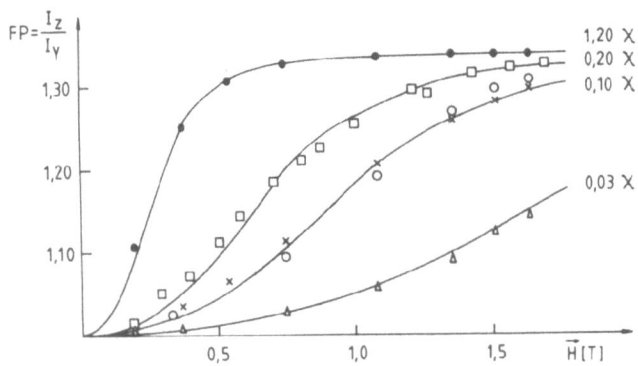

FIGURE 3. Field strength dependence of FP with thylakoids of
different LHC contents (experimental points). Curves calculated according to Eqs. (1) and (2) with γ = 58.6° showing
various susceptibilities fitted to the measured points.
•, mesophyll chloroplast of maize (Chl a/b = 2.8); □, macromolecular aggregate formed from purified light-harvesting
chlorophyll a/b protein (Chl a/b = 1.16); x, bundle sheath
chloroplast of maize (Chl a/b = 4.6);Δ , bean chloroplasts
developed at a light region alternating periods of 98 min
dark and 2 min light (Chl a/b=5.2); o, bean chloroplasts developed under weak continuous light for 24 h (Chl a/b=3.1).
The suspension traversed by light was adjusted to OD: 0.15 at
680 nm which corresponds to an approx. chlorophyll concentration of $3 \cdot 10^{-6}$M. Temp. 25°C, $X = 3.4 \cdot 10^{-20}$ cm^3 [13] for
chloroplasts.

DISCUSSION

The orientability of thylakoids has been recognized for many
years [7] but the physical basis has not been elucidated.
Here we could show that the magnetic susceptibility of LHC
is commensurable with that of the chloroplasts. This suggests that LHC may act as a torque which aligns the chloroplasts in a magnetic field.
     The value of the magnetic susceptibility estimated for
chloroplasts [13] was $X = 3.4 \cdot 10^{-20}$ cm^3. For mesophyll chloroplasts the best fit (see Fig. 3) was found with the curve
calculated at 1.2X, but for the LHC only 0.2X was applicable. This relatively low magnetic susceptibility of the
purified LHC can be explained by the difference in structure

between the artificially aggregated macromolecule and that
of the LHC in vivo [8]; the latter is suggested to be more
asymmetric [12].

Functional importance of the liquid crystal like structure
of LHC, mobile under the effect of magnetic and electric
fields [15] can be the signalization between Photosystem II
and LHC, hence high local electric fields around the reaction
center [16] may induce LHC to separate from Photosystem II by
moving away.

REFERENCES

1. Anderson JM and Boardman NK (1966) Biochim Biophys Acta
   112, 403-421
2. Argyroudi-Akoyunoglou J and Akoyunoglou G (1973) Photochem Photobiol 18, 219-223
3. Bialek GE, Horváth G, Garab GyI, Mustárdy LA and Faludi-
   -Dániel Á (1977) Proc Natl Acad Sci USA 74, 1455-1457
4. Burke JJ, Ditto CL and Arntzen ChJ (1978) Arch Biochem
   Biophys 187, 252-263
5. Faludi-Dániel Á and Mustárdy LA (1983) Plant Physiol
   52, 54-56
6. Garab GyI, Kiss JG, Mustárdy LA and Michel-Villaz M
   (1981) Biophys J 34, 423-437
7. Geacintov NE, Van Nostrand F, Becker FF and Tinkel JB
   (1972) Biochim Biophys Acta 267, 65-79
8. Gregory RPF, Demeter S and Faludi-Dániel Á (1980) Biochim Biophys Acta 591, 356-360
9. Keszthelyi L (1980) Biochim Biophys Acta 598, 429-436
10. Knox PS and Davidovich MA (1978) Biophys J 24, 689-712
11. Kyle DJ, Ting-Yun Kuang, Watson JL and Arntzen ChJ
    (1984) Biochim Biophys Acta 765, 89-96
12. Li J (1985) Proc Natl Acad Sci USA 82, 386-390
13. Papp E and Meszena G (1982) Biophys J 39, 1-5
14. Saeva FD (1979) In: Liquid Crystals (Saeva FD ed) Marcel
    Dekker Inc New York pp 249-273
15. Williams R (1974) In: Liquid Crystals and Plastic
    Crystals Vol. 2 (Gray CW and Winsor PA eds) J Wiley
    New York, London, Sidney, Toronto pp 110-122
16. Zimányi L and Garab GyI (1982) J Theor Biol 95, 811-821

Photosynthesis Research 10: 223 – 229 (1986)
© *Martinus Nijhoff Publishers, Dordrecht*

RELATIVE SENSITIVITY OF VARIOUS SPECTRAL FORMS OF PHOTOSYNTHETIC PIGMENTS
TO LEAF SENESCENCE IN WHEAT (TRITICUM AESTIVUM L.)

A. GROVER,S.C. SABAT and P. MOHANTY
Bioenergetics Laboratory, School of Life Sciences, Jawaharlal Nehru
University, New Delhi-110067, India

Key words : Chlorophyll forms, Chloroplast absorption spectrum, leaf
senescence, Wheat

ABSTRACT

The change in the characteristics of the absorption spectrum of chloroplasts which were isolated from the mature and senescing primary wheat
leaves, was examined at various wavelengths in which the photosynthetic
pigments mostly absorb. Chlorophyll (Chl) a was observed to be relatively
more sensitive to leaf senescence than Chl b and carotenoids. Furthermore,
the various spectral in vivo forms of Chl a, did not degrade to a similar
extent; the far red absorbing forms of Chl a including species that absorb
maximally at 692 nm (Chl a–692), 700 nm (Chl a–700) and 708 nm (Chl
a 708) were found to be extremely sensitive to senescence induced losses.
Both attached and detached senscing primary wheat leaves exhibited nearly
similar pattern in the loss of photosynthetic pigments which suggests that
the loss in long wavelength absorbing forms of Chl a is a selective indicator
of leaf senescence.

INTRODUCTION

Excepting some recent reports on the mutants of Festuca pratensis
where it has been shown that the loss of Chl and other metabolic changes
during leaf senescence are not stringently related (20,21), the decline in
Chl content in leaves has often been regarded as an indicator of leaf senescence (15,17). In fact, quantification of the Chl loss is generally employed
as a criterion for monitoring the influences of various physiological and
environmental factors on the aging process (8,11).

Besides Chl b, Chl a constitutes the major photosynthetic pigments
and it (Chl a), in vivo, exists in various spectroscopic forms (5). Biswal
and Mohanty(3) showed that the alteration of photochemical activities of
chloroplasts during leaf senescence takes place in a specific and sequential
manner. In this context, it is of interest to know how the senescence
induced loss in specific photosynthetic pigment forms is correlated with
the sequential changes in the photochemical activities of chloroplasts.
The study on this aspect is also likely to reveal the relative sensitivity
of Chl a, Chl b and carotenoids to the aging process.

With this aim in view, we have monitored the magnitude of change
in absorbance of isolated wheat chloroplasts, as a function of wavelength,
in mature and senescing primary wheat leaves. Senescence-associated changes
in absorbance in attached and detached leaves, incubated in light and dark,
have been analysed.

MATERIALS AND METHODS

Wheat (Triticum aestivum L. cv. Kalyansona) seedlings were raised
in petri plates on moist filter paper, at $25\pm1^{\circ}C$ under continuous light
(20 watts m^{-2}). Water in each petri plate was changed after every 24
h.

Under the present experimental conditions, our preliminary data showed that in primary wheat leaves, the Chl content, on g fr wt basis, and the photosystem II and photosystem I mediated electron transport activities, on mg Chl basis, peaked on 14th day of seedling growth, after which both these parameters steadily declined (data not shown). Thus from d14 onwards, the primary leaves of our seedlings were in the senescing phase of their development.

For senescence analysis of both attached and detached leaves, 16 d old seedlings were used. One set of these seedlings was transferred to dark and the other set was maintained under continuous white light (20 watts m^{-2}). The measurement of the spectral changes in chloroplalsts isolated from the primary leaves of these seedlings was made after 6 d of incubation in each treatment (attached leaves senescence). For examining the senescence in detached leaves, primary leaves of 16 d old seedlings were excised and were floated on distilled water randomly (around 50 leaves floated on 200 ml of distilled water per petri plate). One set of detached leaves was maintained in dark and the other set was kept under continuous white light (20 watts m^{-2}). The analysis of detached leaves was done after 3 d of incubation in each treatment (detached leaf senescence). The incubation temperature, for both attached and detached leaves senescence, was maintained at $25°±1°C$.

Total leaf Chl content was estimated following Arnon(1). Chloroplasts were isolated acording to Saha and Good (13), with the only modification that Tricine – NaOH was replaced by Hepes–NaOH buffer, both in isolation and suspension media. Although no special procedure was followed to isolate the chloroplasts from senescing leaves we have used a home made chloroplast chopper, specially designed to chop the leaves gently in such a way that the chloroplasts were not damaged during isolation. Chloroplast absorption spectra were recorded on Hitachi 557 spectrophoto-meter. The number of chloroplasts for a given value of absorbance at 750 nm (which was 0.085 in the present study) were almost the same in light and dark incubated leaves as shown by measuring absorbance of different volumes of chloroplast suspension (in 3 ml of suspension buffer) and by counting chloroplasts using haemocytometer (see Fig.1).

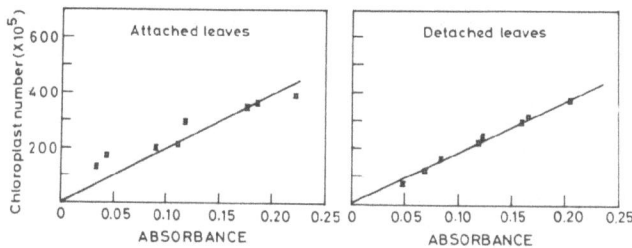

Fig. 1. Increase in absorbance(OD) at 750 nm as a function of number of chloroplasts, isolated from attached and detached senescing leaves. The measurements were done after 2 d of light/dark incubation of seedlings.O, light incubated leaves; ●, dark incubated leaves. The adjustment of absorbance of samples suggest that approximately equal number of chloroplasts were taken for comparison.

Fig.1 indicates for attached and detached leaves that the light scattering at 750 nm increased linearly with the number of chloroplasts. For the measurements of the absorption spectrum the baseline was carefully corrected and each spectrum was scanned 3 times. The opal surface of the quartz cuvettes was kept in the path of light for minimizing the scattering (3,4,14). The absorbance values were read at 4 nm intervals after correcting for scattering, and the absorption spectra were replotted.

RESULTS AND DISCUSSION

Senescence-associated loss of total Chl (on g fr wt basis) was appreciably faster in (a) detached leaves than in attached leaves, and (b) in dark incubated leaves than in light incubated ones (Table 1). This observation is in conformity with the results obtained by other workers (18,19). The Chl a to Chl b ratio remained nearly constant throughout the incubation period in all treatments(Table 1). Likewise, the ratio of absorbance at 680 nm (mostly absorbed by Chl a) and 652 nm (mostly absorbed by Chl b) declined only marginally during the senescence period (Table 1).

Table 1: Changes in total Chl content, Chl a Chl b ratio and OD 680 to OD 652 ratio in senescing attached and detached primary wheat leaves. Each value represents mean of three independent determinations ± SE.

Treatment	Chl content mg (g fr wt)$^{-1}$	Chl a /Chl b ratio	OD 680/OD652 ratio
Day 0*	1.46±0.1	2.74±0.29	2.26
Attached leaves, 6 d dark incubation	0.65±0.05	2.67±0.08	2.12
Attached leaves, 6 d light incubation	1.20±0.13	2.53±0.15	1.91
Detached leaves 3 d dark incubation	0.50±0.23	3.11±0.13	2.00
Detached leaves 3 d light incubation	0.83±0.75	2.54±0.70	1.92

* Leaves taken from 16 d old seedlings (prior to light/dark incubation treatment).

Both in attached and detached leaves, accompanying the overall loss of Chl content due to leaf senescence,the absorbance at all the wavelengths of chloroplast absorption spectrum declined (Fig 2A, 3A). As expected, the per se loss of absorbance was more (a) in detached leaves than in attached leaves, and (b) in dark incubated leaves as com pared to light incubated leaves. This data was obtained while maintaining nearly the same number of chloroplasts in each treartment (see Fig. 1)

Fig 2B and Fig 3B represent the percentage of decline in absorbance at various wavelengths. These values were calculated by considering the respective absorbance values observed with chloroplasts of 16 d old (control) primary leaves. In all treatments, the magnitude

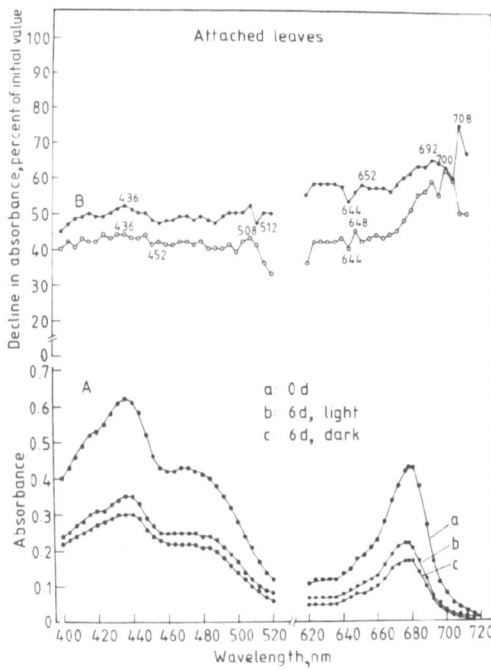

Fig. 2. Wave length dependent changes in the characteristics of the absorption spectra of chloroplasts isolated from light/ dark incubated attached senescing primary wheat leaves. A. Absorption spectra of isolated chloroplasts replotted after correcting for scattering a: 0 d.i.e. leaves taken from 16 d old seedlings prior to light/ dark incubation; b: attached leaves after 6 d of light incubation of 16 d old seedlings; c: attached leaves after 6 d of dark incubation of 16 d old seedlings. B. The percent decline in absorbance as a function of wavelength. The per cent decline in absorbance was calculated assuming absorbance value obtained with the leaves taken from 16 d old seedlings as 100. 0, attached leaves after 6 d light incubation; ● attached leaves after 6 d dark incubation.

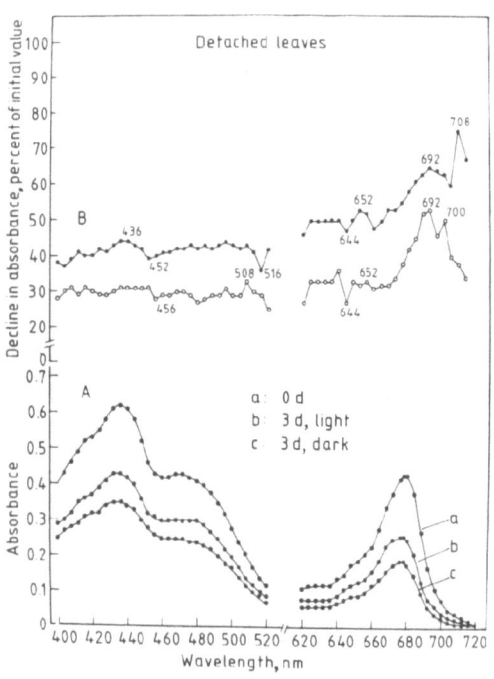

Fig. 3. Wavelength dependent changes in characteristics of the absorption spectra of chloroplasts isolated from light/dark incubated detached primary senescing wheat leaves. A. Absorption spectra of (a) control, 0 day, (b) 3 day , light (c) 3 d, dark incubated leaf chloroplast. B. The extent of loss shown as per cent of control, as function of wavelength. Other details as in Fig.2B.

of decline in absorbance varied at different wavelengths of the absorption spectrum; the extent of decline in absorbance was relatively higher in 660–710 nm range which is mostly absorbed by Chl a (5). The blue region of the chloroplast absorption spectrum, in all treatments showed markedly less sensitivity than the red region. In particular, Chl a–692 was maximally affected due to senescence. In an analogous study on in vitro chloroplast aging, relatively higher sensitivity for aging induced loss of Chl a 678 and Chl a 708 than other forms of Chl a, has been shown (4). In the present study on in vivo chloroplast senescence, besides the 692 peak one distinct additional peak of decline in absorbance towards the far red end of the spectrum was observed in all treatments. This additional peak was marked at 708 nm in dark incubated leaves, but at 700 nm in light incubated ones (Fig.2B, 3B). French et al.(7) have suggested that absorption at 708 nm is due to the aggregated forms of Chl a. It is likely therefore, that in vivo chloroplast aging involves some type of changes in membrane structure in such a way that spectroscopic properties of Chl a altered in a specific manner.

The results shown in Fig. 2 and 3 reveal that the Chl b which have the absorption maxima at about 645–652 nm, are less sensitive to aging than the Chl a.

During in vitro aging of chloroplast carotenoids absorbing at 510 nm have been shown to be highly sensitive to the aging stress (4). It has been suggested that decline in absorbance in 510–512 nm range reflects a decrease in electric potential and/or leakage of H^+ ions across the thylakoid membrane during aging (4,22). However, in the present study the maximum percentage of decline in absorbance in the range of 500–516 nm region of the absorption spectrum only 7 per cent. In both attached and detached leaves senescing in light or dark, the percentage of decline in carotenoids was relatively lesser than for either Chl a or Chl b. This suggests that unlike the case of in vitro chloroplasts, carotenoids are not very affected during in vivo aging. Similar observations have earlier been documented in other senescing sytems(16). In repeated experimentation, the percentage of loss in absorbance of both Chl and carotenoids varied with the degree of leaf senescence qualitatively; however, similar profile of changes was always observed.

The aging process is known to induce chemical modification of the lipid and protein molecules. This suggestion holds true for aging of chloroplasts under in vitro conditions as well as for intact leaves (2,6,9). Presumably, the chemical changes in lipid molecules affect the membrane potential and the index of refraction and thus induce spectral changes of Chl (4). In view of the observations that the physical state of membrane lipids influences the absorbance and fluorescence of Chl a (10), the differential sensitivity of the various spectralforms of Chl a to leaf senescence shown in the present work is probably related to their relative alterations in the lipid protein association in thylakoid membranes. Further, aging–induced alterations in lipid protein association in attached and detached wheat leaves was probably similar, since the change in spectra in the entire chloroplast absorption spectrum broadly followed a similar pattern in these two conditions. The significance of the spectral change at 708 nm region seen in light incubated leaves,but not in dark incubated leaves is not known however, this change reflect extensive disaggregation of aggregated forms of Chl a in light. Indeed, leaf–senescence associated alterations in chloroplasts under light and dark conditions (in vivo) have been shown to differ in several respects(12).

It is unlikely that the spectral changes observed in the present investigation originated from molecular modifications that result in spectral shift because the major peak positions of the absorption spectrum (i.e. 436 nm, 472 nm,652 nm, 678 nm) remained essentially unchanged. In a similar type of study on chloroplasts aging it has earlier been indicated that the spectral shifts are not the major factors in aging and that the spectral changes do not originate from phenophytinization of chlorophylls (4), but may be associated with the change in membrane lipid protein interaction.

In conclusion, in the present study, we have shown a distinct wavelength dependent pattern in the loss of photosynthetic pigments during senescence of attached and detached wheat leaves. It is suggested that the loss in long wavelength absorbing forms of chlorophyll a is the selective indicator of leaf senescence. It will be of interest to compare the kinetics of loss of specific spectral forms of Chl a with decline in photochemical activities catalyzed by two photosystems during senescence such that temperal sequence of loss in structure and function can be correlated.

ACKNOWLEDGEMENTS

This work was supported by ICAR-USDA (PL 480) research grant (FG-IN-575; IN-SEA-170) to PM.

REFERENCES

1. Arnon DI (1949) Plant Physiol 24:1-15.
2. Barber J, Chow WS, Scbufflaire C and Lannoye R (1980) Biochim Biophys Acta 591:92-103.
3. Biswal UC and Mohanty P (1976) Plant Cell Physiol 17: 323-331.
4. Brody SS (1983) Photochem Photobiol 37:585-586.
5. Brown JS (1972) Ann Rev Plant Physiol 23: 73-86.
6. Dhindsa RS, Dhindsa P and Thorpe TA (1981) J Exp Bot 32:93-101.
7. French CS, Brown JS and Lawrence MC (1972) Plant Physiol 49:421-429.
8. Grover Anil and Sinha SK(1985) Physiol Plant (in press).
9. Hoshina S, Kaji T and Nishida K (1975) Plant Cell Physiol 16: 465-474.
10. Hoshina S, Mohanty P and Fork DC (1983) Photosyn Res 5:347-360.
11. Lindoo SJ and Nooden LD (1977) Plant Physiol 59:1136-1140.
12. Mae TN, Kai N, Makino A and Ohira R (1984) Plant Cell Physiol 25: 333-336.
13. Saha S and Good NE (1970) J Biol Chem 245:5017-5021.
14. Shibata K (1973) Biochim Biophys Acta 304:249-259.
15. Stoddart JL and Thomas H (1982) In Boulter D and Parthier B eds. Enclycopedia of Plant Physiology, vol 14A, pp 592-636. Spriner-verlag, Berling Heidelberg New York.
16. Thimann KV (1978) Bot Mag (Tokyo) Special Issue 1:19-43.
17. Thimann KV (1980) In Thimann KV ed. Senescence in Plants, pp 85-115. CRC Press, Florida.
18. Thimann KV, Tetley RR and Thanh TV (1974) Plant Physiol 54:859-862.

19. Thimann KV, Tetley RM and Krivak BM (1977) Plant Physiol 59:448454.

20. Thomas H (1977) Planta 137:53–60.
21. Thomas H and Stoddart JL(1975) Plant Physiol 56: 438–441.
22. Witt HT(1979) Biochim Biophys Acta 505:355–427.

II.   Excitation and Energy Migration; Regulation of Energy
           Transfer; State Transition; and Variable Chlorophyll $\underline{a}$
           Fluorescence

Figure 4. Warren L. Butler, George Hoch, and L.N.M. Duysens .... early discussions.

Figure 5. Warren and Lila Butler .... at home in California.

Photosynthesis Research 10: 233–242 (1986)
© *Martinus Nijhoff Publishers, Dordrecht*

ENERGY MIGRATION AND EXCITON TRAPPING IN GREEN PLANT PHOTOSYNTHESIS

NICHOLAS E. GEACINTOV[a], JACQUES BRETON[b] and ROBERT S. KNOX[c]

[a]Chemistry Department, New York University, New York, NY 10003, [b]Service de Biophysique, Departement de Biologie, C.E.N. Saclay, 91191 Gif sur Yvette, cedex, France, and [c]Department of Physics and Astronomy, The University of Rochester, Rochester, NY 14627

Key words: chloroplasts, fluorescence decay components, reaction centers, charge recombination and fluorescence

Abstract. The possible origins of the different fluorescence decay components in green plants are discussed in terms of a random walk and Butler's bipartite model. The interaction of the excitations with the photosystem II reaction centers and, specifically, the regeneration of theses excitations by charge recombination within the reaction centers, are considered. Based on comparisons between fluorescence decay profiles, time-dependent exciton annihilation and photoelectric phenomena, it appears that the fast 200 ps decay component corresponds to primary energy transport from the antenna to the reaction centers and is dominant in filling the photosystem II reaction centers.

Introduction

In spite of much theoretical and experimental research, the details of exciton migration and trapping in photosynthetic systems are still not completely understood. In this overview we summarize recent advances in this field, and attempt to integrate the results obtained by different experimental approaches to this problem.

The central question is: how long does it take for an exciton to reach a reaction center and to give rise to the primary act of charge separation at that reaction center? Different experimental approaches have been utilized in attempts to answer this question; these include (1) time-resolved fluorescence decay profile measurements, (2) direct measurements of the kinetics of the appearance of ionic species in the reaction centers by differential absorption spectroscopy, and (3) rapid kinetic measurements of the transmembrane electric field which arises because of the separation of electric charges within the reaction centers. The availability of picosecond pulse lasers and other technological advances have provided an important impetus to this field. Recently, most of the activity in this area of photosynthesis research has centered on accurate measurements of fluorescence decay profiles in green plants and the interpretation of the multi-exponential decays in terms of exciton migration and the known mechanisms of fluorescence.

Butler's Models of Energy Transfer and Fluorescence

In the 1970's, using steady-state measurements of fluorescence parameters, Butler and his co-workers developed a model of the photosynthetic apparatus and provided important insights into the mechanisms of fluorescence (reviewed in [1]).

In Butler's bipartite model, the light harvesting chlorophyll a/b protein complexes and the PS II core chlorophyll a antenna molecules are considered to be tightly coupled by excitation transfer. These two types of chlorophyll - protein complexes thus represent, approximately, a single species of fluorescence - emitting antenna molecules. Energy transfer from the antenna to the reaction centers (RC) occurs with rate constant k_T, where energy can either be returned to the antenna (rate constant k_t) or undergo charge separation (photochemical conversion) with a rate constant k_p, according to the following scheme:

$$\text{fluorescence} \quad \xleftarrow{ k_A } \quad A \underset{k_t}{\overset{k_T}{\rightleftharpoons}} RC \xrightarrow{ k_p } \text{charge separation} \qquad (1)$$

where A represents the antenna, and k_A the rate constant for decay of excitons within the antenna. The bipartite model provides a reasonable explanation of steady-state fluorescence data [1]. Nevertheless, this model is known to represent an oversimplification, particularly since the coupling between the light-harvesting antenna molecules and PS II core antenna chlorophyll a molecules may be less than 100% efficient. This factor is taken into account in Butler's tripartite model [1]. Evidence has since been provided about an additional type of heterogeneity of the photosynthetic apparatus of green plants, namely the existence of α and β subunits [2]. Each type of PS II unit may be characterized by different bipartite fluorescence kinetics, thus further increasing the complexity of the fluorescence [3].

Multi-exponential fluorescence decay profiles

The heterogeneity of the fluorescence manifests itself in terms of multi-exponential decay profiles at low temperatures [4,5] and at room temperature [6,7,8,9]. The latter work, utilizing mode-locked, synchronously pumped dye lasers and time-correlated single photon counting methods, provides the most detailed description of the time dependent fluorescence decay profiles F(t) at ambient temperatures. These decay profiles can be described by a sum of 3 (or more) exponentials:

$$F(t) = \sum A_i \exp(-t/\tau_i) \qquad (2)$$

where the terms A_i (i=1,2,3) correspond to the initial amplitudes of the individual components, the decays of which are characterized by the lifetimes τ_i. When the sums of the amplitudes are normalized such that $\sum A_i = 1.0$, these amplitudes are proportional to the fraction of pigment molecules of each type excited at t = 0. The time integrated fluorescence yield of each component is given by the product $A_i\tau_i$, and the relative contribution of each component to the steady state fluorescence is $A_i\tau_i(\sum A_i\tau_i)^{-1}$.

In chloroplasts and green algae, lifetimes of about 100-200 ps, 400-600 ps, and 1.2 - 2.5 ns are generally found. These decay times are termed the fast, intermediate or middle, and slow components, respectively. The fast component has been shown to be comprised of a 50-80 ps component and a 150 - 200 ps component [6,9]. Relating these different components to the various excitation energy distribution models of the photosynthetic apparatus has proven to be quite difficult. Before discussing the various interpretations of the kinetic fluorescence data, we consider two theoretical approaches, the random walk and the kinetic model.

The random walk model

The migration of excitons is assumed to occur by a series of transfers from molecule to molecule by the "hopping" Förster energy transfer mechanism [10]. In a simple model, the different fluorescence components can be assigned to a particular pigment pool, e.g. PSI and PSII and their associated antenna pigment systems. In the simplest model of this type, N antenna molecules and one reaction center are considered to form a regular lattice such that [N]>>[RC]. The excitons, created with uniform probability anywhere within this array, perform a random walk until they are captured by the RC; the excited RC undergoes photochemical conversion (rate constant k_p), or an exciton is returned to the antenna. The details of such exciton motion have been described in detail by Pearlstein [11-13]. Under certain restrictive, but physically realistic conditions [13], the decay time τ of the excitons is mono-exponential and is given by:

$$\tau = \alpha\, N\tau_j + Nk_p^{-1} \qquad\qquad (3)$$

where α is a term which depends on the lattice parameters. The first term is the "first passage time", τ_{FPT}, the time required for an exciton to reach the RC for the first time; the second term is the lifetime of the exciton which starts a random walk at the reaction center. Thus, if the trap does not photoconvert efficiently, the exciton has a longer lifetime because it has a finite probability to be returned to the antenna. If $\tau_{FPT} \gg N/k_p$, the time of photoconversion at the trap is much faster than the diffusion time of the exciton to the trap, and the kinetics of exciton diffusion and trapping are said to be "diffusion controlled". In the opposite case, the kinetics are said to be "trap-limited".

In the above treatment, it is assumed that there is only one exciton in the array of N antenna molecules. A more complex random walk approach including energy transfer, trapping, and bimolecular exciton-exciton annihilation processes which can occur at high excitation intensities, has been recently described [14,15].

The coupled kinetic rate equation model

In this approach the details of the random walk of the excitons, as well as all other microscopic parameters, are neglected. The energy transfer between different pigment systems, or between the array of antenna molecules and the reaction centers, is described by macroscopic kinetic rate constants. The instantaneous fluorescence of each subsystem can usually be taken as an indication of the population of excitons in that subsystem.

In order to illustrate this approach, we consider two coupled (by energy transfer), distinct pools of fluorescence-emitting pigment systems, containing N_1 and N_2 molecules, respectively. The time dependence of the excitation is described by the functions $S_1(t)$ and $S_2(t)$, each of which depends on the absorption cross section per molecule and on N_1 and N_2, respectively. Under low-intensity excitation conditions, the excited state populations N_1^* and N_2^* obey the kinetic equations:

$$dN_1^*/dt = S_1(t) - (K + \tau_1^{-1})N_1^* + K'N_2^* \qquad (4)$$

$$dN_2^*/dt = S_2(t) + KN_1^* - (K' + \tau_2^{-1})N_2^* \qquad (5)$$

K is the overall rate constant of excitation transfer from pool 1 to pool 2, and K' the overall rate constant for exciton transfer from pool 2 to pool 1. The rate constants τ_1^{-1} and τ_2^{-1} are equal to the reciprocal excitation lifetimes in each pool, respectively, in the limit of complete decoupling (K = K' = 0).

In the limit of narrow pulse excitation, and for times t such that $S_1(t)$ and $S_2(t) = 0$, the decay of the excitons in each pool is governed by the two equations:

$$N_1^*(t) = A_1 \exp(-\lambda_1 t) + A_2 \exp(-\lambda_2 t) \qquad (6)$$

$$N_2^*(t) = A_3 \exp(-\lambda_1 t) + A_4 \exp(-\lambda_2 t) \qquad (7)$$

The quantities λ_1 and λ_2 are eigenvalues of a kinetic matrix constructed from the coefficients of equations (4) and (5) [3], and are equal to:

$$\lambda_{1,2} = 1/2 \, [\, a \pm \sqrt{b} \,] \qquad (8)$$

where $a = K + K' + \tau_1^{-1} + \tau_2^{-1}$, and $b = (K + \tau_1^{-1} - K' - \tau_2^{-1})^2 + 4KK'$.

In general, it is evident that the fluorescence decay of each pigment pool is double-exponential. Furthermore, the experimentally observable decay parameters λ_1 and λ_2 are complicated functions of the kinetic parameters K, K', τ_1^{-1} and τ_2^{-1}. The amplitudes A_i are determined by the boundary conditions of the problem, e.g. the initial concentrations of each fluorescence-emitting species at t = 0.

We now discuss the issue of fluorescence yield rise times which can occur under certain conditions. If one of the emitters is largely non-absorbing, its population, just after a short excitation pulse, is virtually zero. The fluorescence signal due to this emitter is proportional to

$$F(t) \propto \exp(-\lambda_1 t) - \exp(-\lambda_2 t) \qquad (9)$$

a factor which is sometimes said to contain an "exponential rise time". An example of a fluorescent component fairly well described by this time dependence is the fluorescence emitted from PSI at 735 nm at low temperatures (F735). Equation (9) was used in early work to describe the time-dependent increase in F735 by Campillo et al [16], and more recently by others [4,5]. The time λ_2^{-1} is related to the transfer time between the two components, but is not a simple rise time (example: if λ_2^{-1} = 200 ps and λ_1^{-1} = 100 ps, F(t) in equation (9) reaches a maximum at 139 ps and has a 10-90% risetime of 78 ps). The way a factor, such as (9), arises in formal n- component kinetics is through the eigenvectors of a kinetic matrix [3]. In the two-component case, the eigenvectors are precisely such that the rising component will have a value of zero at t = 0 ($A_3 = -A_4$ in eq. 7). When experimental data with rise times are subjected to an n-exponential fit (fluorescence decay profiles at low temperatures), a negative component appears, but no attempt has been made to explain such negative pre-exponential amplitudes in terms of kinetic models [4].

As we have stressed earlier [5], it is both possible and desirable to proceed directly with a numerical integration of model kinetic equations of the type shown above (eqs. 4 and 5), and to determine realistic values of parameters by trial and error. The direct integration method solves the problems of risetimes, the interpretations of the experimentally observed parameters λ_1 and λ_2 in terms of kinetic model parameters, and eliminates explicit deconvolutions.

Analyses of experimental fluorescence decay profiles

Current interpretations of the experimental data are based on the following models: (1) each of the lifetime components is qualitatively associated with excitons in a particular pigment system, or pool, (2) coupled kinetic rate equations, and (3) random walk parameters. A detailed examination of the excitation and emission wavelength dependence of each of the amplitudes A_i (eq. 2) has been particularly helpful in attributing the origins of the different components to light harvesting antenna containing chlorophyll b and associated with PS II reaction centers, and to chlorophyll a antenna molecules associated with PS I and PS II [9].

The long-standing problem of whether, and how much, PS I contributes to the overall fluorescence of green plants at ambient temperatures appears to be solved. The fastest fluorescence decay component with lifetimes in the range of 50-80 ps [6,9] is attributed to PS I fluorescence. The fluorescence yield of this component at ambient temperatures is small, and its excitation and emission wavelength characteristics correspond to the longer wavelength forms of chlorophyll a [9]. Additional evidence pointing to the PS I origin of this fast fluorescence component stems from studies with mutants which lack PS I [6,17,18], and its lack of sensitivity on (1) the state of PS II reaction centers, (2) the pH, and (3) the presence of cations [8].

The origins of the other, longer lifetime components are more difficult to explain. The observed lifetimes depend on the state of the PS II reaction centers. Butler et al [3], as well as Karukstis and Sauer [19], initially suggested that the origins of the intermediate lifetime components can be explained in terms of the α and β heterogeneity of the photosynthetic units [2]. Holzwarth et al [9] have proposed the most explicit explanation of fluorescence decay kinetics based on this concept. A fast 180 ps component in the F_o state (PS II reaction centers open) is attributed to α units, and is converted to a 2.2-2.4 ns lifetime in the F_{max} state (reaction centers closed). A third 500-600 ps lifetime in the F_o state is attributed to the β units, and is transformed to a 1.2-1.4 ns component in the F_{max} state. While this model is self-consistent, there are some difficulties; the antenna size of α units is larger than that of β units [20] and thus, according to the random walk model (eq.3) a longer lifetime should be observed in the α units. However, the opposite is observed in the F_o state; therefore the model of Holzwarth et al [9] requires some additional assumptions about differences in exciton quenching efficiencies of PS II reaction centers in the α and β units (in the F_o state),

and about differences in the interunit excitation transfer efficiencies in
the F_{max} state.

Butler et al [3], Berens et al [21], as well as Schatz and Holzwarth [22],
account for the multiplicity of lifetimes by utilizing the coupled
differential equations kinetic approach (eqs. 5 and 6) based on Butler's
bipartite model (eq. 1). The observed fluorescence decay times are
identified with the parameters λ_1^{-1} and λ_2^{-1}, where:

$$\lambda_{1,2} = -1/2[\ k_A + k_p + k_T + k_t \pm \sqrt{(k_A + k_T - k_p - k_t)^2 + 4k_Tk_t}\ ] \quad (10)$$

where the + sign is associated with λ_1 and the – sign with λ_2. It is
important to note that this approach predicts a double-exponential decay,
even though the emission occurs from a single pigment pool, the antenna
pigment system in the bipartite model. On the other hand, in the random
walk model discussed earlier, a single decay component is predicted; the
reasons for these different predictions of the two models are not
immediately apparent and deserve further investigation.

The different lifetime components in the F_o and F_{max} states are attributed
by Berens et al [21] to either the fast or the long reciprocal eigenvalues
(eq. 10) of the kinetic matrix, originating either in α or β units of PS
II.   This heterogeneous bipartite model seems to provide a more adequate
description of the fluorescence decay characteristics than the more complex
tripartite model [21].

**Origin of the long fluorescence component, variable fluorescence, and the
Klimov recombination mechanism**

Klimov and his co-workers advanced the hypothesis that the variable PS II
fluorescence, which is observed when the reaction centers are closed , is
due to a recombination luminescence mechanism described as follows [23]:

$$Exciton + P680-I-Q^- \rightarrow P680^+-I^--Q^- \rightarrow P680^*-I-Q^- \rightarrow P680-I-Q^- + exciton \quad (11)$$

The acceptor is denoted by Q, as usual, while I is an intermediate electron
acceptor identified as a pheophytin molecule. Haehnel et al [24] first
attributed the longest fluorescence component to the charge recombination
mechanism; the yield of this component increases by a factor of 20 as the
PS II reaction centers are closed. Both the lifetime and amplitude of this
long component increase in the F_{max} state, the former presumably because of
interunit transfer of excitons. Subsequently [9,25] doubts were raised
about this interpretation based on the observation that the amplitude (and
yield) of the fast 180 ps fluorescence decay component, attributed to PS
II, is strongly decreased in the F_{max} state. It was reasoned that the
amplitude and lifetime of this component should be constant in the F_o and
F_{max} states since the primary trapping and photochemical conversion should
be the same in the P680-I-Q and P680-I-Q$^-$ states. Since this is not
observed, it is argued that the 180 ps component is transformed to a longer
lifetime in the F_{max} state because of a decreased excitation quenching rate
constant. However this alternative explanation in itself does not
constitute a strong argument against the Klimov mechanism. In the random
walk model, the Klimov recombination can be viewed as a reduced efficiency
of photochemical conversion at the trap (lower k_p in eq. 3); thus the 180
ps lifetime in the F_o state is transformed into a longer, but still mono-

exponential decay phase in the F_{max} state, due to the re-injection of an excitation from the RC into the antenna.

The fastest fluorescence decay phases of 50-80 ps attributed to PS I, and the 150-180 ps attributed to PS II, could well correspond to primary energy transfer and reaction center trapping events. However, some of the slower fluorescence decay components have also been attributed to the same phenomenon. It is evident that fluorescence techniques alone cannot provide unequivocal information about the rates of primary exciton trapping and conversion at the reaction centers. Additional information can, in principle, be obtained from absorption and photoelectric measurements related to the appearance of charged species in the reaction centers.

Kinetics of absorption changes due to charge separation in reaction centers

Because there is only one reaction center per 100 - 250 antenna molecules, it is difficult to measure the rate of appearance of oxidized P680 or P700 molecules in intact membranes. However, such measurements have been performed with reaction center-enriched sub-chloroplast particles. Fenton et al [26] found that charge separation occurs within 10 ps of excitation in PSI particles in which the CHla/P700 ratio is 30-40. Kamogawa et al [27] using highly enriched PSI reaction center particles (8-10 Chla/P700), found a $t_{1/e}$ risetime of P700$^+$ of about 25 ps. The fluorescence of these particles, under low intensity excitation such that exciton-exciton annihilation processes were excluded, displayed similar decay kinetics [28]. Il'ina et al [29] found a risetime of 15-30 ps for P700$^+$ in PSI-enriched particles with Chla/P700 of ~ 60.

In intact chloroplasts a fluorescence decay phase of 50-80 ps [6,9] was attributed to the decay and capture of excitons in PSI. This lifetime is about 2-3 times longer than the photoconversion time measured by absorption techniques in PSI-enriched particles. This difference is consistent with larger antenna sizes in intact chloroplasts, giving rise to a larger number of exciton transfer steps before trapping, and thus to a longer lifetime according to eq. 3. The interpretation of the fast 50-80 ps fluorescence lifetime in terms of primary excitation trapping in PS I is thus entirely reasonable. Using the random walk model (eq. 3) and a trapping time of 53 ps, Gulotty et al [6] have estimated a molecule to molecule hopping rate $\tau_j = 0.1 - 0.7$ ps.

Photoelectric measurements of primary charge separation in reaction centers

Still another extremely fruitful and newly developed approach involves the measurement of the rate of appearance of photovoltages on subnanosecond time scales in chloroplasts following excitation with picosecond laser flashes [30]. The photoelectric signal, which is due to PSI and PSII photosynthetic transmembrane charge separation, occurs within less than 175 ps after excitation with a 30 ps laser pulse. Trissl and Kunze concluded that the mean exciton trapping time in both photosystems leading to charge separation is thus equal to, or shorter than 175 ps. There was no evidence for an intermediate rising phase of the photovoltage in the time range of 400 ps to 4 ns.

Fluorescence decay kinetics provide information on all types of excitons, including those decaying within a time interval comparable to the first passage time, excitons which have visited the traps without initiating photochemistry, and excitons which are generated by Klimov-type recombination processes and which have been returned to the antenna. As Trissl and Kunze point out, the power of the photoelectric method lies in the fact that only those exciton phases are detected which give rise to charge separation in the reaction centers. Their results thus suggest that only the first fluorescence decay phase (the < 200 ps components associated with PSI and PSII antenna fluorescence) should be attributed to primary exciton trapping events. The longer-lived fluorescence decay components may arise from excitons which have already visited the reaction centers and whose lifetimes are somehow prolonged by their interactions with the reaction centers. In open PS II reaction centers near the F_0 level, one such mechanism could involve ultrafast charge recombination as proposed by Breton [31] and by Klimov [23]:

$$P680^+-I^--Q \longrightarrow P680^*-I-Q \longrightarrow P680-I-Q + \text{exciton} \qquad (12)$$

In this model, the lifetime of the exciton could be limited by the $P680^+-I^-$ $-Q \longrightarrow P680^+-I-Q^-$ charge stabilization on the acceptor Q. The time for such a process is estimated as ~ 200 ps. If this model and the associated time scale of ~ 200 ps are correct, then it becomes more difficult to rationalize the experimentally observed longer (> 200ps) fluorescence decay phases. However, it appears intuitively that an expanded and more complex bipartite kinetic scheme taking such processes into account explicitly, could account for the experimental observations in which the observed decay times are complex functions of the model parameters as in eq. 10.

The fast fluorescence decay phase, and competition between exciton—exciton annihilation and trapping by reaction centers

In an effort to correlate the different fluorescence decay phases with capture of excitons by PS II reaction centers, high intensity picosecond laser excitation double pulse experiments were devised [32]. In these experiments it was found that only the initial 150 ps exciton decay phase produced by the first pulse was active in quenching excitons produced by the second pulse (F_0 state). The intermediate second phase of the first pulse, even though it dominates the overall fluorescence yield [24,25], is not active in annihilating excitons generated by the second pulse; the intermediate fluorescence decay phase in the F_0 state was thus attributed to relatively immobile excitons, in contrast to the short-lived (< 200 ps) excitons [32].

A detailed comparison of the closing of PSII reaction centers induced by single picosecond laser pulses of variable intensities, with the decrease in the overall fluorescence yield produced by the same laser pulses, lead to the startling conclusion that exciton—exciton annihilation processes are not accompanied by decreases in the efficiencies of exciton trapping by PS II reaction centers [33]. It was concluded that the fast exciton decay component displays a relatively low sensitivity towards bimolecular anni-hilation processes because of rapid trapping by PS II reaction centers. This observation is consistent with the results of the double-pulse experiments, and the fast photovoltage kinetics [30]. The intermediate and longer fluorescence decay phases appear to be due to photophysical events other than primary exciton trapping, and may be due to excitons which have already visited the RC, or have been re-injected from the RC into the antenna; because of their longer lifetimes, these excitons are susceptible to bimolecular exciton-exciton annihilations, thus giving rise to a decrease in the integrated fluorescence yield as the laser pulse excitation energy is increased (for a review of exciton annihilation, see [34]).

Finally, it is interesting to point out that in purple photosynthetic bacteria, competition between exciton-exciton annihilation and exciton trapping by reaction centers is indeed effective; because of this competition the efficiency of trapping decreases as the intensity of picosecond lasers is increased [35]. Thus, the topological details of the reaction centers embedded within the array of antenna complexes is probably quite different in green plants than in photosynthetic bacteria.

Conclusions

Recent, accurate measurements of fluorescence decay profiles in green plants have provided a new impetus to studies of primary energy transfer processes in green plants. The decay profiles are usually represented by a sum of 3-4 exponentials with positive amplitudes. These different decay components are attributed to exciton decay in the light harvesting antenna complexes containing chlorophyll b associated with PS II, or chlorophyll a antenna systems associated with PS I and PS II. Heterogeneity of photosynthetic units (α and β units) has been invoked in order to account for the complexities of the decay profiles utilizing either random walk or kinetic coupled differential equations based on Butler's heterogeneous bipartite model. Photoelectric and exciton-exciton annihilation experiments suggest that only the fast fluorescence decay components (< 200 ps) are directly involved in primary energy transfer involving the migration of excitons from the antenna to the reaction centers, followed by charge separation at the reaction centers.

Acknowledgements

This work was supported by National Science Foundation grants PCM-83-08190 to N.E.G. at New York University, and U.S. Department of Agriculture grant 82-CRCR-1-1129 to R.S.K. at the University of Rochester.

References

1. Butler, W.L. (1979) Ann. Rev. Plant Physiol., 29, 345-378.
2. Melis, A., and Homann, P.H. (1978) Arch. Biochem. Biophys., 190, 523-530.
3. Butler, W.L., Magde, D., and Berens, S.J. (1983) Proc. Natl. Acad. Sci. (USA) 80, 7510-7514.
4. Reisberg, P., Nairn, J.A., and Sauer, K. (1982) Photochem. Photobiol. 36, 657-661.
5. Wittmershaus, B., Nordlund, T.M., Knox, W.H., Knox R.S., Geacintov, N.E., and Breton, J. (1985) Biochem. Biophys. Acta 806, 93-106.
6. Gulotty, R.J., Mets, L., Alberte, R.S., and Fleming, G.R. (1985) Photochem. Photobiol. 41, 487-496.
7. Berens, S.J., Scheele, J., Butler, W.L., and Magde, D. (1985) Photochem. Photobiol. 42, 51-57.
8. Karukstis, K.K., and Sauer, K. (1985) Biochim. Biophys. Acta 806, 374-

388.

9. Holzwarth, A.R., Wendler, J., and Haehnel, W. (1985) Biochim. Biophys. Acta 807, 155-167.
10. Knox, R.S. (1977) in Primary Processes of Photosynthesis, Barber, J., editor, Elsevier, Amsterdam, 55-97.
11. Pearlstein, R.M. (1982) in Photosynthesis: Energy Conversion, Vol. 1, Govindjee, editor, Academic Press, New York, 293-330.
12. Pearlstein, R.M. (1982) Photochem. Photobiol. 35, 833-844.
13. Pearlstein, R.M. (1984) in Advances in Photosynthesis Research, Vol.1, Sybesma, C., editor, M. Nijhoff/Dr. W. Junk Publishers, The Hague, 13-20.
14. Den Hollander, W.T.F., Bakker, J.G.C., and Van Grondelle, R. (1983) Biochim. Biophys. Acta 725, 492-507..
15. Van Grondelle, R. (1985) Biochim. Biophys. Acta 811, 147-195.
16. Campillo, A.J., Shapiro, S.L., Geacintov, N.E. and Swenberg, C.E. (1977) FEBS Lett. 83, 316-320.
17. Karukstis, K.K., and Sauer, K. (1984) Biochim. Biophys. Acta 766, 148-155.
18. Green, B., Karukstis, K.K., and Sauer, K. (1984) Biochim. Biophys. Acta 767, 574-581.
19. Karukstis, K.K., and Sauer, K. (1983) Biochim. Biophys. Acta 725, 384-393.
20. Thielen, A.P.G.M., Van Gorkom, H.J., and Rijgersberg, C.P. (1981) 635, 121-131.
21. Berens, S.J., Scheele, J., Butler, W.L., and Magde, D. (1985) Photochem. Photobiol. 42, 59-68.
22. Schatz, G.H. and Holzwarth, A.R. (1986), this volume.
23. Klimov, V.V. (1984) in Advances in Photosynthesis Research, Vol. 1, Sybesma, C., editor, M. Nijhoff/Dr. W. Junk Publishers, The Hague, 131-138.
24. Haehnel, W., Nairn, J.A., Reisberg, P., and Sauer, K. (1982) Biochim. Biophys. Acta 680, 161-173.
25. Haehnel, W., Holzwarth, A.R., and Wendler, J. (1983) Photochem. Photobiol. 37, 435-443.
26. Fenton, J.M., Pellin, M.J., Govindjee, and Kaufmann, K.J. (1979) FEBS Lett. 100, 1-4.
27. Kamogawa, K., Namiki, A, Nakashima, N., Yoshihara, K., and Ikegami, I. (1981) Photochem. Photobiol. 34, 511-516.
28. Kamogawa, K., Morris, J.M., Takagi, Y., Nakashima, N., Yoshihara, K., and Ikegami, I. (1983) Photochem. Photobiol. 37, 207-213.
29. Il'ina, M.D., Krasauskas, V.V., Rotomskis, R.J., and Borisov, A.Yu. (1984) Biochim. Biophys. Acta 767, 501-506.
30. Trissl, H.W., and Kunze, U. (1985) Biochim. Biophys. Acta 806, 136-144.
31. Breton, J. (1983) FEBS Lett. 159, 1-5.
32. Dobek, A., Deprez, J., Geacintov, N.E., Paillotin, G., and Breton, J. (1985) Biochim. Biophys. Acta 806, 81-92.
33. Geacintov, N.E., Paillotin, G., Deprez, J. Dobek, A., and Breton, J. (1984) in: Advances in Photosynthesis Research, Vol. 1, Sybesma, C., editor, M. Nijhoff/Dr. W. Junk Publishers, The Hague, 37-40.
34. Breton, J., and Geacintov, N.E. (1980) Biochim. Biophys. Acta 595, 1-32.
35. Bakker, J.G.C., Van Grondelle, R., and Den Hollander, W.T.F. (1983) Biochim. Biophys. Acta 725, 508-518.

Photosynthesis Research 10: 243–253 (1986)
© *Martinus Nijhoff Publishers, Dordrecht*

REGULATION OF ENERGY TRANSFER BY CATIONS AND PROTEIN PHOSPHORYLATION
IN RELATION TO THYLAKOID MEMBRANE ORGANISATION

J. BARBER
Imperial College of Science and Technology, Department of Pure and Applied
Biology, Prince Consort Road, London, SW7 2BB, United Kingdom

ABSTRACT

A brief review is given of the state of knowledge which indicates that
the State I-State II transition in higher plants and green algae is due to
the reversible phosphorylation of the chlorophyll a/b light harvesting
complex. The importance of membrane reorganisational changes in this
process is discussed in terms of changes in electrostatic parameters as
emphasised by the interplay of the effect of phosphorylation and the
background levels of cations surrounding the membrane. It is argued that
recognition of this interplay is vital when using the bipartite or
tripartite models of Butler to obtain quantitative information of energy
transfer between the various pigment complexes.

1. INTRODUCTION

Without doubt, the bipartite and tripartite models developed by Warren
Butler (25) to explain changes in chlorophyll fluorescence yields under
different conditions form a firm basis for investigating energy
distribution between the different pigment systems of photosynthetic
organisms. The models are of particular value in attempting to account
for the remarkable constancy of the steady-state quantum yield of
photosynthesis over a broad spectral region where photosystem two (PSII)
and photosystem one (PSI) have quite different absorption properties
(71). This phenomenon has been termed 'the quantum yield anomaly' and
its explanation did not become clear until the discovery of the State I-
State II transition (24). This transitions represent a short term
adaptation mechanism for finely tuning the distribution of light to PSII
and PSI in order to achieve optimal rates of photosynthesis under the
particular lighting conditions which prevail. Warren Butler set himself
the task of formulating theoretical expressions which could be used to
give quantitative information about energy distribution between PSII and
PSI over a range of experimental conditions. For example, not only did
he, in collaboration with A.C. Ley, confirm the existence of State I-
State II transitions in the red alga Porphyridium cruentum, but was also
able to obtain for this organism estimates of changes in energy transfer
between PSII and PSI (known as spillover) which correlated quantitatively with changes in quantum yields of photochemistry and oxygen
evolution (25). This outstanding work was complemented by related
studies using isolated thylakoid membranes of higher plant chloroplasts.
It had been shown by Murata (70), from chlorophyll fluorescence
measurements, that the distribution of light between PSII and PSI could
be altered by changing the level of Mg^{2+} in a suspension of isolated
thylakoids. This effect was considered to be reminiscent of the
in vivo State changes and therefore was an ideal experimental system

for Butler to test the analytical expressions derived from his kinetic models (25). By monitoring chlorophyll fluorescence, and by using the bipartite model he found that for wavelengths ranging from 540nm to 675nm, the distribution of energy between PSII and PSI was approximately 73% and 27%, respectively, when 6mM Mg^{2+} was present. With the same ionic conditions the degree of energy transfer from PSII to PSI was 7% when the PSII traps were open but increased to 23% when fully closed. If Mg^{2+} was absent the distribution was 68% to PSII and 32% to PSI, with spillover changing from 12% to 28% on closing the PSII traps. Of importance in this analysis is that Butler had shown, contrary to Murata's conclusions, that the effect of Mg^{2+} was not only to change the extent of spillover but also to alter the absorption cross-sections of PSII and PSI. Later, when the tripartite model was adopted, the effect of Mg^{2+} could be further interpreted in terms of changes in energy transfer between the light harvesting chlorophyll a/b complexes (LHC-II) and the PSII and PSI complexes (see ref. 25).

Clearly the models and equations developed by Butler and the experimental approaches he adopted to exploit them provide an important framework for studying the regulation of energy distribution in quantitative terms. Indeed, today Butler's models find considerable usefulness in a range of different studies, including stress physiology (72,74). However, Warren Butler made no serious attempt to link his kinetic models with the structural organisation of pigment-protein complexes and with the organisation of the membrane in which they are located. This is surprising because, as indicated above, he made extensive use of Mg^{2+} to perturb fluorescence yields and alter the rate constants and yield parameters which occur in his mathematical expressions. But the addition of Mg^{2+} to isolated thylakoids causes extensive conformational changes involving the lateral movements of intramembrane particles (77) and restacking into granal and stromal lamellae (53). A thorough study of the relationships between conformational changes and chlorophyll fluorescence gave rise to a model in which changes in energy distribution could be understood in terms of alterations in the spatial relationships between LHC-II, PSII and PSI complexes (9). Moreover, it was clearly demonstrated that these changes could be induced by a range of different cations and were controlled by the interplay of electrostatic and electrodynamic forces (10). Coupled with the emergence of a structural model based on surface electrical properties came the finding that protein phosphorylation, rather than changes in cation levels, underlie the mechanism by which higher plants and green algae regulate energy distribution between PSII and PSI as manifested in the State transitions (20,48). In this review paper I present, as concisely as possible, the evidence that it is protein phosphorylation which gives rise to the 'quantum yield anomaly' in higher plants and algae. The subject has been reviewed previously (2,3,13,14,19,40,45,78), but in this case I want to further emphasise the concept of a dynamic membrane system in which the lateral movements of various chlorophyll protein complexes is carefully regulated by the manipulation of their surface charges so as to alter the balance between attractive and repulsive forces. Consideration of these lateral movements and resulting changes in energy transfer between pigment complexes is in the spirit of the fluid-mosaic model of membrane structure but can, in principle, also be used for extrapolation of the parameters which appear in Butler's models.

2. PROTEIN PHOSPHORYLATION

In 1977, Bennett (15) reported that when isolated intact chloroplasts were incubated in the light with ^{32}P-orthophosphate, several thylakoid proteins became phosphorylated. The most conspicuous were the polypeptides of LHC-II which have apparent molecular weights of approximately 27kDa and 25kDa although other phosphoproteins with molecular weights of about 9kDa, 33-35kDa and 45kDa were also detected. Since then considerable knowledge of the phosphorylation properties of these polypeptides, especially LHC-II, has accumulated and is summarised below.

2.1 (a) LHC-II

(i) The kinase responsible for phosphorylating LHC-II is membrane bound (17), is removable by a detergent mixture of cholate and octylglucoside (1,62) and may have a molecular weight of 50kDa (31).

(ii) When in the thylakoid membrane the kinase is activated by light below 700nm (82) or in darkness in the presence of reducing agents such as ferredoxin (17), dithionite (4) or duroquinol (5).

(iii) The kinase can be fully activated in the presence of uncouplers but can be partially inhibited by the formation of a pH gradient across the thylakoid membrane (33).

(iv) The kinase requires Mg^{2+} for its activation with a $C_{\frac{1}{2}}$ for maximum activity of 0.3mM for isolated pea thylakoids stacked in the presence of 10mM lysyl-lysine (81).

(v) The kinase has a K_m for ATP of about 90uM (20) although its activity is modified by the relative levels of other adenylates (8).

(vi) The kinase probably has thiol groups at its active site (68).

(vii) The phosphorylation site is at the threonine residues of the surface exposed N-terminal segment of the LHC-II polypeptides and mild trypsin treatment removes a 2kDa fragment containing the phosphorylatable threonines (16,69).

(viii) The ability of the kinase to phosphorylate LHC-II seems to be enhanced when the thylakoids are stacked, as compared with the unstacked conformation (81).

(ix) The kinase activity can be stimulated by low levels of Zn^{2+} (64).

(x) With spinach LHC-II it has been found that there is a preferential phosphorylation of the 25kDa relative to the 27kDa polypeptide (59).

(xi) In contrast to the kinase, the phosphatase, which brings about dephosphorylation of LHC-II is insensitive to illumination conditions or to redox potentials, although it is inhibited by molybdate and fluoride (18).

(xii) Like the LHC-II kinase, the phosphatase is membrane bound (18).

(xiii) The kinase(s) responsible for the phosphorylation of other membrane polypeptides differ to that acting on LHC-II in having a different concentration requirement for Mg^{2+} (81) and ATP (22), sensitivity to Zn^{2+} (64) and being less sensitive to the nucleotide affinity inhibitor, 5'-p-fluorosulphonyl-benzyl-adenosine (31). Recently, two different kinases have been isolated. These kinase did not phosphorylate LHC-II but a crude isolate did (61,62).

(xiv) As with the LHC-II kinase, the additional kinase activity is stimulated by light and reducing conditions in the dark (16,17).

2.2. Other polypeptides

The identity of the other phosphoproteins is still a matter

of investigation. It seems likely that the 9kDa polypeptide is neither the DCCD-binding subunit of the coupling factor (80) nor the apoprotein of cytochrome b-559 (88). However, like the other main phosphoproteins, this low molecular weight polypeptide is probably a component of PSII. The protein(s) in the region of 33kDa has been suggested to be the Q_B herbicide-binding protein often referred to as the psbA gene product or the D1 protein (76,87). However, very recently new evidence has emerged that the 33kDa phosphoprotein is the product of the psbD gene, known as the D2 protein (29,89). To my knowledge there have been no attempts to identify the origin of the higher molecular weight phosphoproteins in the region of 45kDa, although they do seem to form a part of the PSII core complex (67).

3. FUNCTIONAL ROLE FOR THE PHOSPHORYLATION OF LHC-II

There are a number of key observations which indicate that the phosphorylation of LHC-II is the underlying mechanism for controlling energy distribution between PSII and PSI during the State transitions.

(i)   The light induced phosphorylation of LHC-II is inhibited by DCMU but not by low concentrations of DBMIB (4)

(ii)   The phosphorylation of LHC-II occurs in light which is preferentially absorbed by PSII and is inhibited by far-red PSI absorbing light (48,82).

(iii) The redox potential for the dark phosphorylation of LHC-II has a midpoint potential of 0mV and titrates with a requirement of $2e/2H^+$ (47,48).

These observations led to the conclusion that the phosphorylation of LHC-II only occurs when the plastoquinone pool is in a reduced state (see Fig. 1). Further observations listed below link the LHC-II phosphorylation process with changes in the quantum yields of PSI and PSII activities, indicative of an in vivo regulatory process.

(i)   LHC-II phosphorylation reduces the quantum yield of PSII mediated electron flow (32,52,79) and the chlorophyll fluorescence with maxima below 700nm at both room temperature (20,31,48,49) and 77K (20,54).

(ii)   LHC-II phosphorylation increases the quantum yield of PSI mediated electron flow (32,50), primary charge separation in PSI (83) and 77K chlorophyll fluorescence above 700nm (20,54).

(iii) After LHC-II phosphorylation there is an increase in the ability of chlorophyll-b to excite PSI fluorescence at room temperature (56) and 77K (57).

(iv)   The effect of LHC-II phosphorylation is also to reduce overall connectivity between LHC-II-PSII complexes (55,73,86).

All these findings support a scheme in which a population of LHC-II, which on undergoing phosphorylation due to the over-reduction of the plastoquinone pool, can redirect excess excitation energy from PSII to PSI (4,20,48). In this way the redox poise of the plastoquinone pool (or another intersystem component with a similar redox potential) can delicately regulate the distribution of quanta to PSII and PSI and therefore maintain a maximum quantum yield of electron transport for any particular lighting conditions. As suggested in refs. (4) and (48) the optimisation of energy distribution by the LHC-II phosphorylation/dephosphorylation process could be the basis of the in vivo State I-State II phenomenon. Indeed, studies with intact leaves (26,28) and with the green unicellular alga Chlorella (75) support this relationship. However, to-date the links between the in vitro and in vivo mechanisms have been qualitative and, indeed, some have claimed that LHC-II phosphorylation does not bring about any increase in the quantum yield of PSI photochemistry (41) as would be required for a State I to State II transition to occur.

4. STRUCTURAL REORGANISATION IN RESPONSE TO LHC-II PHOSPHORYLATION

Accepting that the phosphorylation/dephosphorylation of LHC-II is the basis of the State transitions in higher plants and green algae, how can this be understood in terms of our modern concepts of thylakoid membrane structure? The answer to this question comes from considering three important features of the chloroplast thylakoid:

(i) Under non-phosphorylating conditions (dark or far-red light) the majority of LHC-II, together with the PSII core complexes, are located in the appressed lamellae of the grana while PSI is predominantly found in the non-appressed membranes which constitute the stromal and end-granal lamellae (6,10,). The LHC-II-PSII inter-connecting domains within the appressed regions probably give rise to PSII -units while the low levels of separate LHC-II-PSII complexes in the non-appressed regions can be equated with PSII- units (65,66).

(ii) The lateral separation of complexes, as described above, is due mainly to differences in the electrical charge properties of their outward facing surfaces (9,10). Arguments have already been presented based on the principles of electrostatic and electrodynamic interactions (11,12) that those components in the appressed regions carry a low net surface charge density while those in the non-appressed regions are more electrically charged.

(iii) The thylakoid is a dynamic system having a lipid matrix of high fluidity (35) such that perturbation of the electrical characteristics of the exposed surfaces of the intrinsic protein complexes will induce changes in their spatial relationships in response to alterations in coulombic forces (12). Indeed, the fatty acids of the polar lipids are extremely unsaturated (36) and variation in coulombic forces due to changes in ionic levels do bring about lateral movements of protein complexes (9,77).

If these three features are taken at face value then it is reasonable to speculate that the introduction of additional negative charges onto the surface of LHC-II complexes by phosphorylation will destabilize their aggregation within the appressed regions and force them to migrate laterally into the more fluid non-appressed membranes where they will statistically interact with PSI via energy transfer mechanisms (10,13,14). There may be no requirement for them to become physically attached to PSI and there is no evidence for this. They could, however, transfer energy to PSI via PSII -units. The removal of a fraction of the LHC-II population from the appressed regions would be expected to decrease energy transfer between LHC-II-PSII as is observed (55,73). Dephosphorylation of LHC-II would lead to a return to its original surface charge properties and thus to its re-establishment as a part of the LHC-II-PSII domain in the appressed regions. There are a number of observations which give strong support to this lateral mobility model and its dependence on surface charge properties and membrane fluidity.

(i) Titration of the yield of chlorophyll fluorescence with mono-, di- and tri-valent cations in the presence of DCMU indicates an overall increase in the surface charge density after LHC-II phosphorylation (81, 85).

(ii) After LHC-II phosphorylation the chlorophyll a/b ratio decreases in the non-appressed lamellae (28) due to an increase in the LHC-II level in this membrane region (57).

(iii) Radiolabelling with ^{32}P indicates higher specific activities of LHC-II in the non-appressed membranes (7) which are time dependent in a pulse-chase experiment (57,59) and which occur mainly on the 25kDa apopolypeptide (59)

(iv) Freeze-fracture analyses show that LHC-II phosphorylation is accompanied by changes in the distribution of particles in fracture faces indicative of a proportion of LHC-II complexes migrating from appressed to non-appressed membranes (58).

(v) The phosphorylation induced lateral movement of LHC-II and not the phosphorylation process itself, is dependent on the fluidity of the thylakoid membrane (37).

(vi) The extent of reorganisation of the membrane, and therefore degree of interactions between different chlorophyll complexes, is dependent on the background cation levels as would be expected for a process under electrostatic control. At sub-optimal background levels of cations (e.g. 1-2mM Mg^{2+}), the effect of LHC-II phosphorylation is to cause considerable membrane unstacking (84) and intermixing of PSI, PSII and LHC-II complexes as indicated by changes in both absorption cross-sections and spillover, as monitored by chlorophyll fluorescence (50, 51,84). At higher levels of cations (e.g. above 5mM) the degree of membrane unstacking is much less, about 10% (28,39,58,85) and it seems that the absorption cross-section changes predominate, indicative of only LHC-II movement (43,50,84)

Figure 1 presents a diagrammatic representation of the structural changes which seem to occur when LHC-II is phosphorylated in the presence of a high level of cations which effectively screen surface electrical charges (e.g. above 5mM). Under these conditions a pool of LHC-II acts as a mobile antenna system able to regulate the absorption cross-section of PSII and PSI. Although absorption cross-section changes dominate there is also evidence for some changes in the extent of spillover (27,39,55,86). The pool of LHC-II involved in this process is probably only weakly associated with PSII in the appressed membranes and contains a high level of the 25kDa apoprotein compared with the 27kDa (59). In contrast, the more tightly held LHC-II is enriched in the 27kDa apoprotein (59). At lower cation levels, when electrostatic screening is poorer (e.g. below 5mM Mg^{2+}), the effect of LHC-II phosphorylation is more dramatic leading to a considerable degree of unstacking and inter-mixing of complexes (85). Under these conditions both absorption cross-section and spillover changes occur (38). All evidence to-date, whether obtained with higher plants (27,63) or with green algae (42) indicate that the cation level of chloroplasts is sufficiently high that the in vivo State transitions involve mainly changes in absorption cross-sections due to LHC-II movement. The only exception to this is with chloroplasts isolated from developing leaves where the LHC-II phosphorylation seems to induce spillover, as well as absorption cross-section changes, even at high Mg^{2+} levels (23).

Although the model given in Figure 1 indicates the control of LHC-II phosphorylation/dephosphorylation by the redox state of plastoquinone, it does not take into account the possible role of chloroplast adenylate levels, transmembrane pH gradients and NADPH/ATP requirements, all of which effect kinase activity and may also be important in controlling excitation distribution to PSII and PSI (3,46). The model also does not

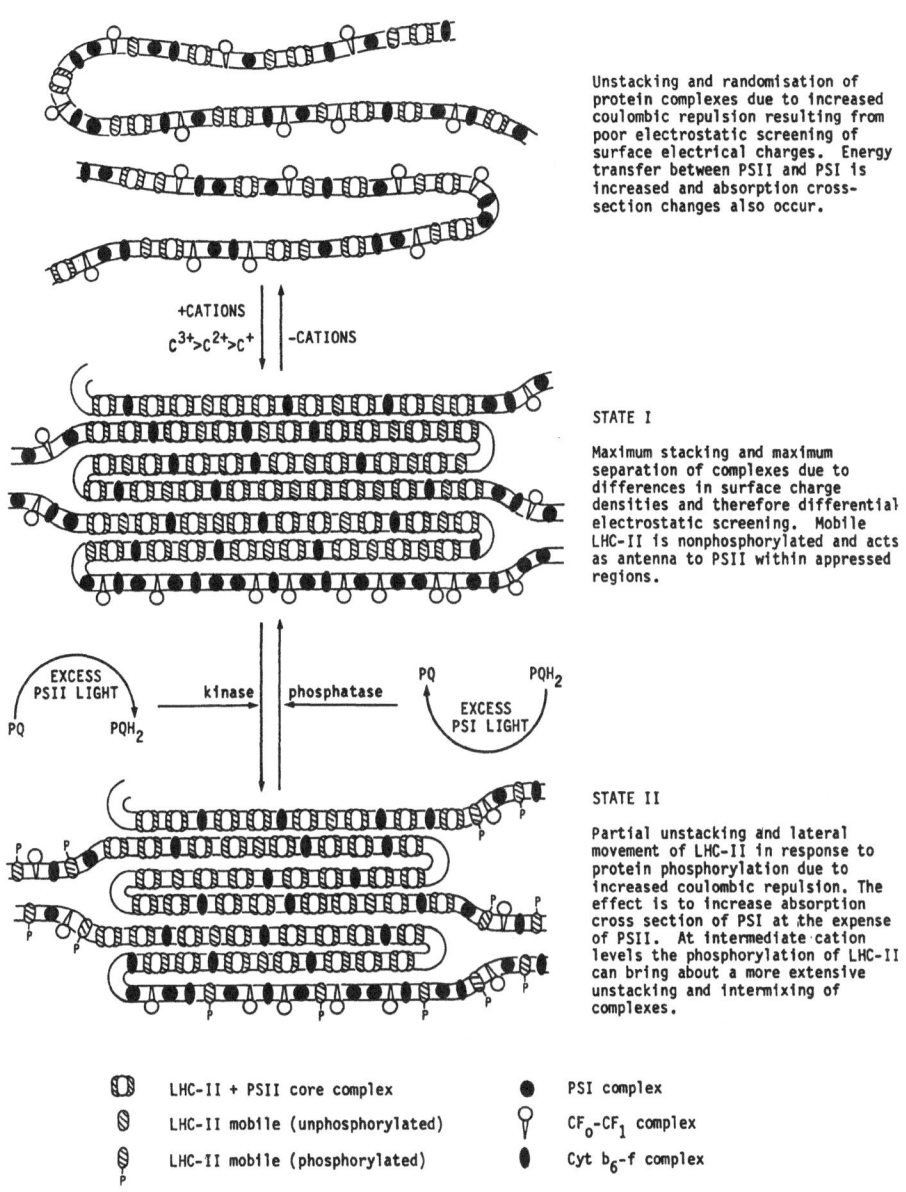

Unstacking and randomisation of
protein complexes due to increased
coulombic repulsion resulting from
poor electrostatic screening of
surface electrical charges. Energy
transfer between PSII and PSI is
increased and absorption cross-
section changes also occur.

+CATIONS

$c^{3+}>c^{2+}>c^+$ -CATIONS

STATE I

Maximum stacking and maximum
separation of complexes due to
differences in surface charge
densities and therefore differential
electrostatic screening. Mobile
LHC-II is nonphosphorylated and acts
as antenna to PSII within appressed
regions.

EXCESS
PSII LIGHT   kinase   phosphatase   PQ   PQH$_2$

PQ   PQH$_2$ EXCESS
                                                   PSI LIGHT

STATE II

Partial unstacking and lateral
movement of LHC-II in response to
protein phosphorylation due to
increased coulombic repulsion. The
effect is to increase absorption
cross section of PSI at the expense
of PSII. At intermediate cation
levels the phosphorylation of LHC-II
can bring about a more extensive
unstacking and intermixing of
complexes.

LHC-II + PSII core complex	●	PSI complex
LHC-II mobile (unphosphorylated)		CF_0-CF_1 complex
LHC-II mobile (phosphorylated)		Cyt b_6-f complex

FIGURE 1. Diagrammatic representation of changes in thylakoid membrane
organisation in response to protein phosphorylation and modification of
cation levels.

include the possible regulatory function of the other phosphoproteins or the membrane locations of the kinase(s) and phosphatase(s) involved. Work to understand these latter considerations has started (34,44,46,52,87) but no clear picture has emerged which can match the clarity of the relatively simple concepts presented in Figure 1.

5. CONCLUSIONS

The segregation of LHC-II PSII and PSI into the appressed and non-appressed regions of the thylakoids of higher plants and green algal chloroplasts allows these photosynthetic organisms to regulate excitation distribution in such a way as to obtain optimal rates of electron flow under limiting light conditions. The process of changing surface charge densities by protein phosphorylation/dephosphorylation and therefore inducing reshuffling of LHC-II between PSII and PSI is elegant, not only in its conceptional simplicity, but also because it can be described in physico-chemical terms. However, what is sadly lacking at present is a rigorous quantitative analysis of energy transfer changes which occur between the various pigment beds. It is in this context that the bipartite and tripartite models, devised by Warren Butler, have such a lot to offer. Analyses using Butler's models of the LHC-II phosphorylation phenomenon has been restricted to a few studies (23,38). Further work is needed, but one word of caution is that before undertaking such a study it is important to recognise that the interactions between the pigment complexes is an interplay of the increase in surface charge density by LHC phosphorylation and the background cation levels screening these surface electrical charges. Therefore both parameters must be given consideration. Also because the thylakoid membrane structure is dynamic and under the control of electrostatic forces great care should be taken in isolating and manipulating chloroplasts for experiments. Leaves should be pretreated in order to standardise the phosphorylation and organisation of their thylakoids (e.g. preillumination with far-red light). If the thylakoids are then isolated into media containing 5mM Mg^{2+} they are likely to maintain their in vivo organisation. However, if they are allowed to unstack in low salt containing media and then restacked by adding 5mM Mg^{2+} they may not precisely regain the configuration of the in vivo state.

Unfortunately as yet the 'phosphorylation-mobile antenna' properties of LHC-II cannot be extrapolated to those organisms which show State transitions but do not contain LHC-II. Studies by Biggins and colleagues (21) concluded that no phosphorylation processes can be correlated with State I-State II transitions in Porphyridium cruentum and previous to this Ley and Butler (60) had shown that in this organism the State transitions involved large changes in extent of spillover rather than alterations in absorption cross-sections. Unfortunately then, the 'quantum yield anomaly' is not understood in structural terms for the phycobilin containing organisms having no obvious differentiation into appressed and non-appressed membranes. Nevertheless, Warren Butler's thorough studies with Porphyridium established without doubt that this organism can regulate energy transfer between PSII and PSI and that this regulation can be described by his models. We can look forward to using this knowledge as a basis for obtaining explanations about the conformational changes which occur and comparing the similarities and differences with higher plant and green algal systems.

6. ACKNOWLEDGEMENTS

I wish to thank both the Science and Engineering Research Council (SERC) and the Agricultural and Food Research Council (AFRC) for financial support. I would also like to thank my colleagues for their support during the course of the writing of this review, particularly Alison Telfer and Julian Whitelegge.

REFERENCES

1. Alfonzo R, Nelson N, Racker E (1980) Plant Physiol. 65:730-734
2. Allen JF (1983) Trends in Biochem. Sci. 8: 369-373
3. Allen JF (1983) Crit. Revs. Plant Sci 1: 1-22
4. Allen JF, Bennett J, Steinback KE, Arntzen CJ (1981) Nature 291:21-25
5. Allen JF, Horton P (1981) Biochim. Biophys. Acta 638:290-295
6. Anderson JM (1981) FEBS Lett 124: 1-10
7. Andersson B, Akerlund H-E, Jergil B, Larsson C (1982) FEBS Lett. 149: 181-185
8. Baker NR, Markwell JP, Thornber JP (1982) Photobiochem. Photobiophys. 4:211-217
9. Barber J (1980) FEBS Lett. 118:1-10
10. Barber J (1982) Ann. Rev. Plant Physiol. 33:261-295
11. Barber J (1980) Biochim. Biophys. Acta 594: 253-308
12. Barber J (1982) BioSci. Rep. 2: 1-13
13. Barber J (1983) Photobiochem. Photobiophys 5: 181-190
14. Barber J (1985) Photosynthetic Membranes (Staehelin LA, Arntzen CJ, eds.) Encycl. Plant Physiol., Springer Verlag, Heidelberg, in press
15. Bennett J (1977) Nature 269:344-346
16. Bennett J (1979) Eur. J. Biochem. 99:133-137
17. Bennett J (1979) FEBS Lett. 103:342-344
18. Bennett J (1980) Eur. J. Biochem. 104:85-89
19. Bennett J (1983) Biochem J. 212:1-13
20. Bennett J, Steinback KE, Arntzen CJ (1980) Proc. Natl. Acad. Sci. (USA) 77: 5253-5257
21. Biggins J, Campbell CJ, Bruce D (1984) Biochim. Biophys. Acta 767: 138-144
22. Black MT, Foyer CH, Horton P (1984) Biochim. Biophys. Acta 767: 557-562
23. Black MT, Horton P (1983) Biochim. Biophys. Acta 767: 568-573
24. Bonaventura C, Myers T (1969) Biochim. Biophys. Acta 189:366-383
25. Butler WL (1978) Ann. Rev. Plant Physiol. 29:345-378
26. Canaani O, Barber J, Malkin S (1984) Proc. Natl. Acad. Sci. (USA) 81: 1614-1618
27. Canaani O, Malkin S (1984) Biochim. Biophys. Acta 766: 513-524
28. Chow WS, Telfer A, Chapman DJ, Barber J (1981) Biochim. Biophys. Acta 638: 60-68
29. Delepelaire P (1984) EMBO J. 3: 701-706
30. Dominy P, Williams WP (1985) FEBS Lett. 179:321-324
31. Farchaus J, Dilley RA, Cramer WA (1985) Biochim. Biophys. Acta 809:17-26
32. Farchaus JW, Widger WR, Cramer WA, Dilley RA (1982) Arch. Biochem. Biophys. 217: 362-367
33. Fernyhough P, Foyer CH, Horton P (1984) FEBS Lett. 176:351-353
34. Fernyhough P, Horton P (1983) Biochim.Biophys. Acta 725: 155-161
35. Ford RC, Barber J (1983) Biochim. Biophys. Acta 722: 341-348
36. Gounaris K, Barber J (1983) Trends in Biochem. Sci. 8: 378-381

37. Haworth P (1983) Arch. Biochem. Biophys. 226: 145-154
38. Haworth P, Kyle D, Arntzen CJ (1982) Biochim. Biophys. Acta 680:
    343-351
39. Haworth P, Kyle DJ, Arntzen CJ (1982) Arch. Biochem. Biophys. 218:
    199-206
40. Haworth P, Kyle DJ, Horton P, Arntzen CJ (1982) Photochem. Photobiol.
    36: 743-758
41. Haworth P, Melis A (1983) FEBS Lett. 160: 277-280
42. Hodges M, Barber J (1983) Plant Physiol. 72: 119-1122
43. Hodges M, Barber J (1985) Biochim. Biophys. Acta 767: 102-107
44. Hodges M, Packham NK, Barber J (1985) FEBS Lett. 181: 83-87
45. Horton P (1983) FEBS Lett. 152: 47-52
46. Horton P (1985) Photosynthetic mechanisms and the environment, Vol.6
    Topics in Photosynthesis (Barber J, Baker NR, eds.) pp135-187
    Elsevier Amsterdam
47. Horton P, Allen JF, Black MT, Bennett J (1981) FEBS Lett. 125:
    193-196
48. Horton P, Black MT (1980) FEBS Lett. 119:141-144
49. Horton P, Black MT (1981) Biochim. Biophys. Acta 635: 53-62
50. Horton P, Black MT (1982) Biochim. Biophys. Acta 680: 22-27
51. Horton P, Black MT (1983) Biochim. Biophys. Acta 722: 214-218
52. Horton P, Lee P (1984) Biochim. Biophys. Acta 767: 563-567
53. Izawa S, Good NE (1966) Plant Physiol. 41: 544-552
54. Krause GH, Behrend U (1983) Biochim. Biophys. Acta 723: 176-181
55. Kyle DJ, Haworth P, Arntzen CJ (1982) Biochim. Biophys. Acta 680:
    336-342
56. Kyle DJ, Baker NR, Arntzen CJ (1983) Photobiochem. Photobiophys.
    5:79-85
57. Kyle DJ, Kuang T-Y, Watson JL, Arntzen CJ (1984) Biochim. Biophys.
    Acta 765: 89-96
58. Kyle DJ, Staehelin LA, Arntzen CJ (1983) Arch. Biochem. Biophys. 222:
    527-541
59. Larsson UK, An   ?son B (1985) Biochim. Biophys. Acta 809:396-402
60. Ley AC, Butler W. (1980) Plant Physiol. 65: 714-722
61. Lin ZF, Lucero HA, Racker E (1982) J. Biol. Chem. 257:12153-12156
62. Lucero HA, Lin ZF, Racker E (1982) J. Biol. Chem. 257: 12157-12160
63. Malkin S, Telfer A, Barber J (1986) Biochim. Biophys. Acta 848: 42-48
64. Markwell JP, Baker NR, Thornber JP (1983) Photobiochem. Photobiophys.
    5:201-207
65. Melis A and Anderson JM (1983) Biochem. Biophys. Acta 724: 473-484.
66. Melis A, Homann PH, (1978) Arch. Biochem. Biophys. 190: 523-530
67. Millner PA, Barber J (1985) Physiol. Vegetal. in press
68. Millner PA, Widger WR, Abbott MS, Cramer WA (1982) J. Biol. Chem.
    257:1763-1742
69. Mullet JF (1983) J. Biol. Chem. 258:9941-9948
70. Murata N (1969) Biochim. Biophys. Acta 189:171-181
71. Myers J (1971) Ann. Rev. Plant Physiol. 22:289-312
72. Orgen E, Oquist G (1984) Physiol. Plant. 62: 193-200
73. Percival MP, Webber AN, Markwell JP and Baker NR (1985) Biochim.
    Biophys. Acta. in Press.
74. Powles SB, Bjorkman O (1982) Planta 156: 97-107
75. Saito K, Williams WP, Allen JF, Bennett J (1983) Biochim. Biophys.
    Acta 724: 94-103
76. Shochat S, Owens GC, Hubert P, Ohad I (1982) Biochim. Biophys. Acta
    681:21-31
77. Staehelin LA (1976) J. Cell Biol. 71:136-158

78. Staehelin LA, Arntzen CJ (1983) J. Cell Biol. 97: 1327-1337
79. Steinback KE, Bose S, Kyle DJ (1982) Arch. Biochem. Biophys. 216: 356-361
80. Suss, KH (1980) FEBS Lett. 112:255-259
81. Telfer A (1986) Biochem Soc. Trans. 14:52-53
82. Telfer A, Allen JF, Barber J, Bennett J (1983) Biochim. Biophys. Acta 722:176-181
83. Telfer A, Bottin H, Barber J, Mathis P (1984) Biochim. Biophys. Acta 764: 324-330
84. Telfer A, Hodges M, Barber J (1983) Biochim. Biophys. Acta 724: 167-175
85. Telfer A, Hodges M, Millner PA, Barber J (1984) Biochim Biophys Acta 766: 554-562
86. Torti F, Gerola PD, Jennings RC (1984) Biochim. Biophys. Acta 767: 321-325
87. Vermaas WFJ, Steinback KE, Arntzen CJ (1984) Arch. Biochem. Biophys. 231: 226-237
88. Widger WR, Farchaus JW, Cramer WA, Dilley RA (1984) Arch. Biochem. Biophys. 233:72-79
89. Wildner GF, Dmoch R, Fiebig C, Dedner N (1985) Quinones in Photosynthetic Membranes. Abstract of meeting held in Saint Lambert des Bois.

Photosynthesis Research 10: 255–276 (1986)
© *Martinus Nijhoff Publishers, Dordrecht*   255

MONO- BI- TRI- AND POLYPARTITE MODELS IN PHOTOSYNTHESIS

RETO J. STRASSER

SUMMARY: It is shown how energy fluxes in mono-bi-tri- and polypartite photosystems can be described. The derivation of the energy distribution term α and the probability of spill over p_{21} as proposed by W.L. BUTLER are reviewed.

1. INTRODUCTION

1.1 Models

A model can be considered as a tool to give us a better understanding of the problem being investigated. The proposed model pursues two distinct goals:

- to connect and explain the known data within a common framework,
- to make predictions on the behaviour of a sample under new experimental conditions.

A tool is a very modest construction compared to a piece of art an artist can create with it. In analogy, a model is a very crude representation of the biological system a scientist tries to describe. Therefore, a model is a valuable tool for unifying and explaining data, as well as making predictions. These predictions encourage the development of new measuring techniques which enable us to control the validity of the model under new experimental conditions. Even if a model has to be modified or rejected due to new information, it is still considered to be an intellectual creation.

1.2 The presentation of a model

Every model can be presented as:

- a pure verbal description
- an analogical graphic representation
- an analytical and mathematical formulation.

These three forms of presentation carry identical information. That means e.g. a tripartite mathematical formulation and a tripartite verbal description both belong to a tripartite graphic presentation of a model. However, this is not strictly followed in the literature and may lead to confusion.

[109]

1.3   The theoretical information of a model

A paint brush can be used for drawing or for other purposes depending on the technique applied. The technique of handling a model, determines how it can be associated with the biological system under consideration. These abstract handling techniques of a model are **theories.** The verbal description, graphic representation and mathematical formulation of a model should be unambigously linked to well defined theories. Different theories for the same model may lead to similar, different or the same conclusions. Therefore, the reader can only understand what a model means if he knows which theories are being referred to by the author pertaining to a model.

1.4   The practical information of a model

There is no limit as far as formulation of fantastic and very sophisticated models are concerned. However, the terms needed for the description of a model increase exponentially with its complexity. A model of practical usefulness should be measurable, which means that a correlation between the experimental signal and each theoretical term is needed. The inability to associate an experimental signal with each term (variable or constant) in a model forces us to formulate an assumption of its value like zero, infinite constant or negligible. Therefore, the information supplied by a model depends on the amount of experimental information which can be associated with it. There is a constant battle to find an optimum compromise between theoretical and practical information which a model supplies. The more complex the model, the better it helps to understand biological systems. More speculations and assumptions however, are needed to associate it with the experimental data. The simpler the model, the more rigid it is and the less it represents the biological system. But it can be strictly associated with the experimental data. W.L. Butler and his colleagues have been always aware of the highly **multipartite** structure of any photosystem. But the availability of independent experimental data forced one to formulate the photosynthetic apparatus as a **bipartite, tripartite** or **polypartite** model. All these formulations are in fact, consequent extensions of the formulation of a **monopartite** photosystem.

2.   THE MONOPARTITE PHOTOSYSTEM

The model of a monopartite photosystem developed by W.L. Butler (1) and M. Katajima supports a concept which is able to explain principal activities such as photochemistry and fluorescence emission of PSII. It uses the biochemical terms of pigment concentration and first order rate constants for the de-excitation events of excited pigments. The <u>static</u> concept of the model is based on the established opinions that a pigment pool acts as an antenna to absorb photons and that some of the absorbed photons are trapped by the reaction centers. W.L. Butler introduces a <u>dynamic</u> concept in his model as follows: The energy which flows from the excited antenna pigment pool to the reaction center has several options:

1st   **The excited reaction center transforms its excitation energy into photochemistry by reducing an electron acceptor while the reaction center gets oxidized.**

2nd   **The excited reaction center is unable to perform photochemistry. The excitation energy of such a closed reaction center migrates back to the antenna pool.**

The rate constants of the de-excitation of the excited reaction center of PSII are k_F, k_D, k_T (fluorescence, heat dissipation or energy transfer). The rate constants of the de-excitation of the excited reaction center are k_p, k_d, k_t (photochemistry, heat dissipation, energy transfer from the reaction center back to the antenna pool). A reaction center which is able to perform photochemistry upon excitation is named **open**, otherwise **closed**. The energy migrates back and forth from a closed reaction center to an antenna pool. The concept of **energy cycling** between the antenna pool and the reaction center was introduced this way. The equations and concept of energy cycling between the antenna and the reaction center are identical to the equation for the energy cycling between the antenna pools of neighbouring photosynthetic units described much earlier by A. Joliot and P. Joliot (2).

3.   ENERGY CYCLING DETERMINES THE COMPLEXITY OF A MODEL

Energy cycling is nothing else but a flux of energy which repeatedly moves back and forth from one location to another. Each exciton has the probability to migrate in one direction. The energy cycling degenerates into an irreversible one way energy migration if the probability of energy transfer back to its origin is zero. Therefore, real energy cycles or one way energy transfers (like spill over) can be formulated using the same equations. The complexity of a photosynthetic model is determined firstly, by the number of pigment pools and reaction centers assumed and secondly, by the energy cycles one attributes to the model. The following different types of energy cycle can be defined. Their distinctions are crucial inasmuch as each type influences the experimental data differently:

1. Trapping is the energy migration from the antenna pool of a photosystem to its reaction center (e.g. trapping in PSI RC or trapping in PSII RC).

2. Coupling is the energy migration from one antenna pool to another antenna pool of the same photosystem (e.g. energy cycling between the core antenna and the light harvesting complex of PSII).

3. Spill over is the energy migration from one antenna pool of a reaction center to an antenna pool of another reaction center (e.g. core antenna of PSII to antenna of PSI or LHC of PSII to core antenna of PSI etc.)

4. Grouping is the energy migration from one antenna pool of a photosystem to an antenna pool of a neighbouring photosynthetic unit (e.g. LHC of PSII to the LHC of a PSII of a neighbouring unit).

A monopartite model includes only trapping.
A bipartite model includes trapping and spill over in its simplest form, or trapping, spill over and grouping if unit-unit energy transfer is considered.
A tripartite model includes trapping, spill over, coupling with or without grouping.
A polypartite model is the general description of a model which includes all biochemically known pigments.
Every model can be arranged as a separate pack model (no grouping) or as a grouped pack model (allowing grouping to occur).

The models used and proposed by different authors can all be classified according to the above expressions. Each author, however, uses different terminologies to describe the energy migration in his model so that it is often difficult to compare one model with another. Nevertheless, it is possible to formulate all models using the same terminology provided that the following rules (generally accepted in Biochemistry and Photochemistry) are taken into consideration:

1) To each pigment pool an index number is given. e.g. 1 for core antenna of PSI, 2 for core antenna of PSII, 3 for the light harvesting Chl a/b complex etc.

2) Each reaction <u>center</u> is labeled with an index <u>letter</u> indicating that it is a single molecule and not a pigment pool. e.g. for the reaction center of PSI index a and for the reaction center of PSII index b.

3) The non-defined location where the dissipated energy goes are labeled with an index e.g. F for fluorescence and D for heat dissipation.

4) The absorption fluxes of each pigment pool are represented by J_1, J_2, J_3 etc, the excitation rates by E_1, E_2, E_3 etc and the energy fluxes by E_{21} (spill over from pigment pool of PSII to pigment pool of PSI) or by E_{2b} (energy flux from pigment pool 2 to the reaction center of PSII) or by E_{2F} (the total fluorescence emission of the pigment pool 2). The measured fluorescence signal is labeled F_2 and it is proportional to the total flux E_{2F}. The same is valid for F_1 and E_{1F} etc. (All energy fluxes or rates are determined by the amount of photons or excitons moving per time).

5) The rate constants are designated by the term k_{ij}, the probability that an exciton goes from the location i to the location j is p_{ij} and the lifetime of an excited pigment P_i^* is τ_i etc.

6) All other terms can be derived by applying these rules. e.g. the quantum yield of photochemistry of PSII is $\psi_{2b} = E_{2b}/J_2$. In the case of PSI, the quantum yield of photochemistry is $\psi_{1a} = E_{1a}/J_1$.

More details are supplied by the energy flux theory in bio-membranes (3).

4.   EXPERIMENTAL BASIS OF THE MODELS

The monopartite model of Butler explains the two extreme points of a fluorescence induction curve at room temperature or low temperature in the presence of DCMU. The two extreme points are the initial fluorescence F_o and the maximal fluorescence F_M. The maximal variable fluorescence is defined as $F_v = F_M - F_o$. The fluorescence rise from F_o to F_M can be attributed to the energy cycling between the antenna and the reaction center of PSII (trapping). The shape of the fluorescence induction kinetics at room temperature is either exponential or sigmoidal. The unit-unit transfer model of Joliot and Joliot associates the sigmoidal form of the curve with the existence of energy cycling between several photosynthetic units (grouping). At low temperature the shape of the fluorescence curve is never sigmoidal. A model has been developed by Strasser which allows the conversion of low temperature induction curves to room temperature curves and vice versa (4).

The low temperature (-196°C) emission spectra of green chloroplasts show two to three emission peaks. Their position vary from organism to organism. The long wavelength peak (in higher plants at 735 nm) is attributed to the emission of PSI, the middle peak (nearly in all plants at 695 nm) is attributed to the emission of core antenna or the phaeophytin of PSII. The short wavelength peak (between 680 and 695 nm) can be mostly attributed to the core chlorophyll (CP 43)(CP 47) and to the light harvesting chl a/b complex. However, some antenna of PSI and some early chlorophyll forms emit in this region as well. At all wavelengths where fluorescence emission occurs, an intact photosynthetic apparatus exhibits variable fluorescence at low, as well as at room temperature due to the redox state of the reaction center of PSII. In most cases, the observed variable fluorescence does not depend on the redox state of P700. However, some PSI particles and some algae do show some variable fluorescence which is dependent on the redox state of P700.

W.L. Butler concluded (at least in the case of higher plants and most green algae) that the variable fluorescence observed on the emission band of PSI (735 nm) is entirely due to energy transfer from PSII to PSI. This statement predicted that the excitation spectrum of the variable fluorescence at 735 nm and 695 nm would be proportional to one another. This prediction was confirmed experimentally (5). A second prediction of the energy transfer concept by Butler was that the excitation spectra for the variable fluorescence of the PSII would be proportional to the excitation spectra for initial and maximal fluorescence of PSII. A set-up which allows simultaneous measurements of the fluorescence induction kinetics at two or three different wavelengths (e.g. at 685, 695 and 735 nm) was used to confirm this prediction (6). (This experimental set-up was developed by the author in 1970 in the Photobiology Laboratory of the University of Liège, Belgium, using multibranched fibreoptics and a HeNe-laser for excitation) (7).

It is an experimental fact that at 77K the plot of the time dependent fluorescence rise signal of PSI (signal F_1) versus the time dependent fluorescence signal of PSII (signal F_2) is a straight line.

Therefore, we can correlate the two signals empirically as follows (see Fig. 2):

$$F_{1(t)} = \text{Intercept on } F_1 \text{ axis} + \text{slope} \cdot F_{2(t)}$$

In cases where the redox state of P700 influences the variable fluorescence at low temperature, it is necessary to pre-illuminate the sample with far red light in order to oxidize the reaction center of PSI.

All models (bi-tri-poly-partite, with or without grouping, separate pack or grouping pack formulation) which include the energy cycle of trapping and spill over from PSII to PSI in any form, predict a straight line plot of F_1 versus F_2.

The empirical description of this plot with its experimental terms as indicated in Fig. 2 is:

$$F_{1(t)} = \underbrace{\left(F_{1(M)} - F_{2(M)} \cdot F_{1(v)}/F_{2(v)}\right)}_{\text{Intercept on } F_1\text{-axis}} + \underbrace{\frac{F_{1(v)}}{F_{2(v)}}}_{\text{slope}} \cdot F_{2(t)}$$

The above equation of a straight line can be plotted if two points are given. The first point has the coordinates $F_{2(o)}$, $F_{1(o)}$ while the second point has the coordinates $F_{2(M)}$, $F_{1(M)}$. The maximal variable fluorescence are $F_{2(v)}$ and $F_{1(v)}$. They are calculated as $F_{2(v)} = F_{2(M)} - F_{2(o)}$ and $F_{1(v)} = F_{1(M)} - F_{(o)}$. The biophysical meaning of the two constants (intercept and slope) depends on the model applied.

Chl*
P.A*
k_F
k_D
k_T
k_t
k_d
P+A−
Chl
P.A

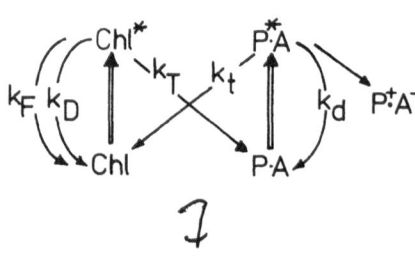

Figure 1:
Energy fluxes as proposed by W.L.Butler for a monopartite photosystem as PSII. The term Chl symbolizes the antenna pigment of PSII and P.A. represents the reaction center of PSII. This model is identical with the model of Joliot and Joliot for energy cycling between photosynthetic units if the term P.A is replaced by Chl. For details see text.

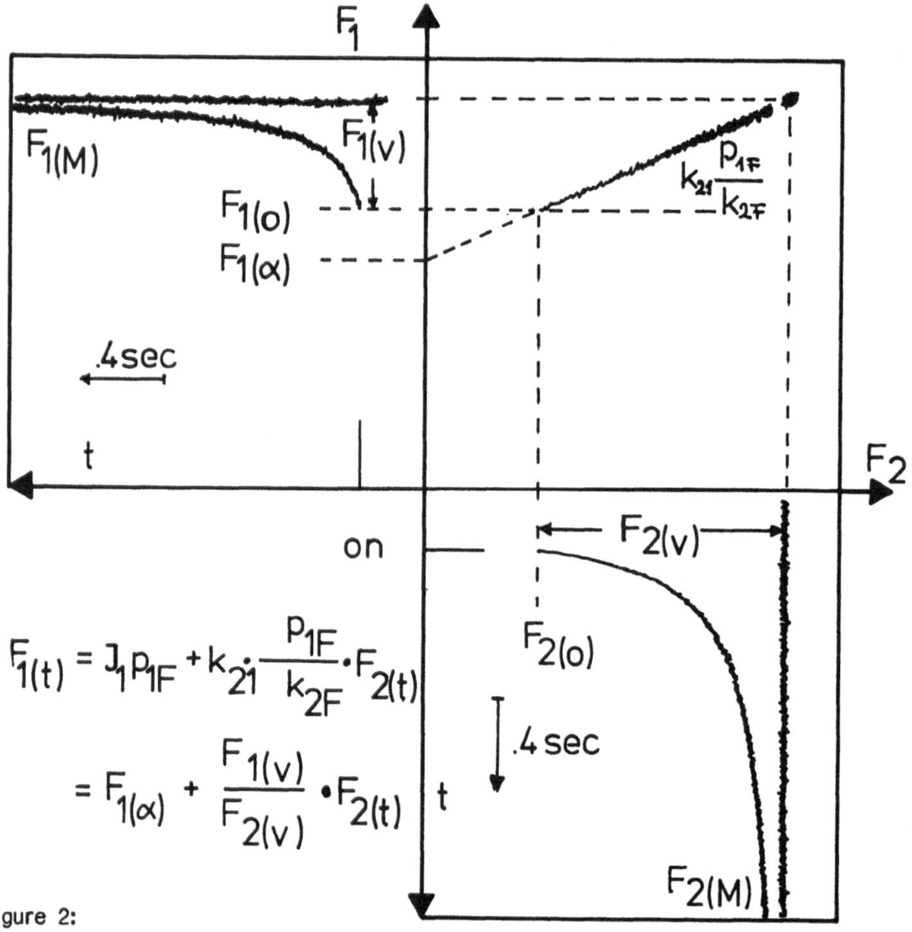

F_1
$F_{1(M)}$
$F_{1(v)}$
$F_{1(o)}$
$F_{1(\alpha)}$
$k_{21}\frac{p_{1F}}{k_{2F}}$
.4sec
t
F_2
on
$F_{2(v)}$
$F_{2(o)}$
$F_{1(t)} = J_1 p_{1F} + k_{21} \cdot \frac{p_{1F}}{k_{2F}} \cdot F_{2(t)}$
$= F_{1(\alpha)} + \frac{F_{1(v)}}{F_{2(v)}} \cdot F_{2(t)}$
.4 sec
t
$F_{2(M)}$

Figure 2:
The experimental traces of the fluorescence induction curves measured simultaneously at 735 nm for F_1 and 695 nm for F_2 at low temperature using chloroplasts of higher plants or leaves. The three dimensional square with the axis F_1, F_2 and time shows fluorescence induction at 77K. The fluorescence signals F_1, F_2 are given in relative intensities and the time is indicated. ON means the moment when light was switched on. Both kinetics F_1 vs time, F_2 vs time were measured simultaneously and stored in the computer.

5. BIOPHYSICAL SIGNIFICANCE OF EXPERIMENTAL EXPRESSIONS

5.1 The ratio $F_{2(v)}/F_{2(M)}$

Each fluorescence induction curve measured at an emission wavelength of PSII (at low temperature or room temperature, in the presence of DCMU) exhibits the two distinct signals, $F_{2(o)}$ and $F_{2(M)}$. Therefore, $F_{2(v)}=F_{2(M)}-F_{2(o)}$.

The fluorescence of a PSII with open reaction centers (no LHC and no grouping) is equal to the absorption rate (J_2) of PSII times the probability that an absorbed photon gets emitted (p_{2F}). As soon as a light harvesting complex or energy transfer from unit to unit (grouping) is taken into account, then the excitation rate of the antenna of PSII is equal to the sum of all three energy fluxes which have reached the antenna pool of PSII after the occurrence of absorption, grouping and coupling of the LHC. The fluorescence of a PSII with closed reaction centers is equal to the product of the photon flux absorbed by PSII and the gain factor (due to the energy cycling between the antenna and the reaction center of PSII) times the probability p_{2F}. The fluorescence of PSII with open or closed reaction centers of any model is equal to the excitation rate E_2^{op}(open centers) and E_2^{cl}(closed centers) of the antenna times the probability p_{2F} (that an exciton is dissipated as fluorescence).

$$F_2 = E_2 \cdot p_{2F} \text{ or } F_2^{op} \cdot p_{2F} \text{ or } F_2^{cl} = E_2^{cl} \cdot p_{2F}$$

op and cl refer to open and closed reaction centers. If all reaction centers are open, the fluorescence signal corresponds to $F_{2(o)}$ and if all reaction centers are closed, the fluorescence signal corresponds to $F_{2(M)}$. The ratio therefore is:

$$F_{2(v)}/F_{2(M)} = (F_{2(M)}-F_{2(o)})/F_{2(M)} = (E_2^{cl}-E_2^{op})/E_2^{cl}$$

In the case of a bipartite model $E_2^{op} = J_2$ and $E_2^{cl}=J_2 \cdot (1-T)^{-1}$ J_2 is the absorption flux of the antenna of PSII and T is the trapping product of $p_{2b} \cdot p_{b2}$ (p_{2b} is the probability that an exciton from the antenna reaches the reaction center of PSII, p_{b2} is the probability that the exciton at the reaction center goes back to the antenna pool). The energy cycling between two types of antenna of one photosystem is described in analogy to the <u>trapping product T</u> and the <u>coupling product C</u>. An energy cycle between units is described by the overall <u>grouping product G</u>. In the literature, G is indicated as p_{22} or p_{2G} depending on the model discussed (2)(8). The biophysical meaning of the ratio $F_{2(v)}/F_{2(M)}$ of a real tripartite model with trapping T, coupling C and grouping G can be formulated in a general way ref (3)(8) namely:

1) $F_{2(v)}/F_{2(M)} = T/(1-C).(1-G)$ Tripartite model (grouped)

A tripartite model without grouping (G=O):

2) $F_{2(v)}/F_{2(M)} = T/(1-C)$ Tripartite model (separate pack)

A bipartite model with grouping (C=O):

3) $F_{2(v)}/F_{2(M)} = T/(1-G)$ Bipartite model (grouped)

A bipartite model without grouping (G=O and C=O):

4) $F_{2(v)}/F_{2(M)} = T$ Bipartite model (separate pack)

The last expression was derived from W.L.Butler (1). Based on the definition of $T=p_{2b} \cdot p_{b2}$ and on the assumption that the probability of energy transfer from a closed reaction center of PSII to the antenna is almost unity, ($p_{b2}=1$) one can say that the experimental ratio $F_{2(v)}/F_{2(M)}$ is **proportional** to the trapping probability p_{2b}. In all models however, the quantum yield $\varphi_{2b(o)}$ of photochemistry when all reaction centers are open can be expressed by the following ratio:

$$\varphi_{2b(o)} = \frac{\text{initial excitation flux to the RC II}}{\text{light absorption flux by PS II}} = \frac{E_{2b(o)}}{J_2} = \frac{F_{2(v)}}{F_{2(M)}}$$

This expression shows that a change in the ratio $F_{2(v)}/F_{2(M)}$ can be attributed to a change in trapping, only if we are certain that our sample has no unit-unit (grouping), which is rarely the case under natural conditions. The correlation made by W.L. Butler (assuming $p_{2b}=1$) that $\varphi_{2b(o)} = p_{2b} = F_{2(v)}/F_{2(M)}$ is only valid if the sample is ungrouped and has no LHC. This expression does not allow calculations of the probability of photochemistry (p_{2b}) in any tripartite model or bipartite grouped model. The above equation also shows that the quantum yield of photochemistry is not equal to the ratio of the rate constants $k_{2b}/\sum k_{2x}$ as long as $E_2^{op} \neq J_2$.

5.2   The intercept $F_{1(\alpha)}$ of the plot F_1 versus F_2 (fig.2)

The intercept of the plot F_1 versus F_2 has the experimental description of $F_{1(M)} \cdot F_{2(M)} \cdot F_{1(v)}/F_{2(v)}$ defined originally by Butler as a measure for the absorption energy distribution expression α (1). But later,[12] he defined it as a measure proportional to α. However, both statements are wrong. Butler defined α as the fraction of light which is absorbed by PSI.

$$\alpha = J_1/(J_1+J_2) = (1 + J_2/J_1)^{-1} \qquad \text{bipartite model}$$

where J_1 is the absorption flux of PSI and J_2 is the absorption flux of PSII. α **is a distribution term.** The shape of the absorption or excitation spectrum for α is a function of the <u>ratio</u> of the PSII and PSI absorption or excitation spectra.

The intercept $F_{1(\alpha)}$ is a fluorescence term for PSI.

This fluorescence term includes an excitation spectrum of PSI <u>only</u> and the unit: photons emitted by PSI per time and per cross-section. As a compromise, Butler agreed to call the intercept in the F_1 versus F_2 plot as $F_{1(\alpha)}$ which is that part of the PSI emission due only to photons absorbed by PSI. Therefore:

$$F_{1(\alpha)} = J_1 \cdot P_{1F}$$

The F_1 versus F_2 plot can now be written as: $F_{1(t)} = F_{1(\alpha)} + F_{1(\beta)}$
$F_{1(\beta)}$ is a function of the state of the reaction center of PSII and expressed as:

$$F_{1(\beta)} = F_{2(t)} \cdot F_{1(v)}/F_{2(v)}$$

$F_{1(\beta)}$ symbolizes an energy flux which was absorbed by PSII and partially spilled over (with the spill over energy transfer probability p_{21}) to PSI. The term $F_{1(\beta)}$ includes therefore, an <u>excitation spectrum</u> of PSII and an <u>emission spectrum</u> of PSI (due to spill over). $F_{2(o)}$ and $F_{2(M)}$ have the same excitation spectrum as $F_{1(\beta)}$ however, they have an emission spectrum of PSII. The three dimensional plot of fluorescence emission versus excitation wavelength and versus emission wavelength shows the terms F_2, $F_{1(\alpha)}$, $F_{1(\beta)}$ of a real biological bipartite system (in flashed bean leaves without LHC measured at 77K) (Fig.3).

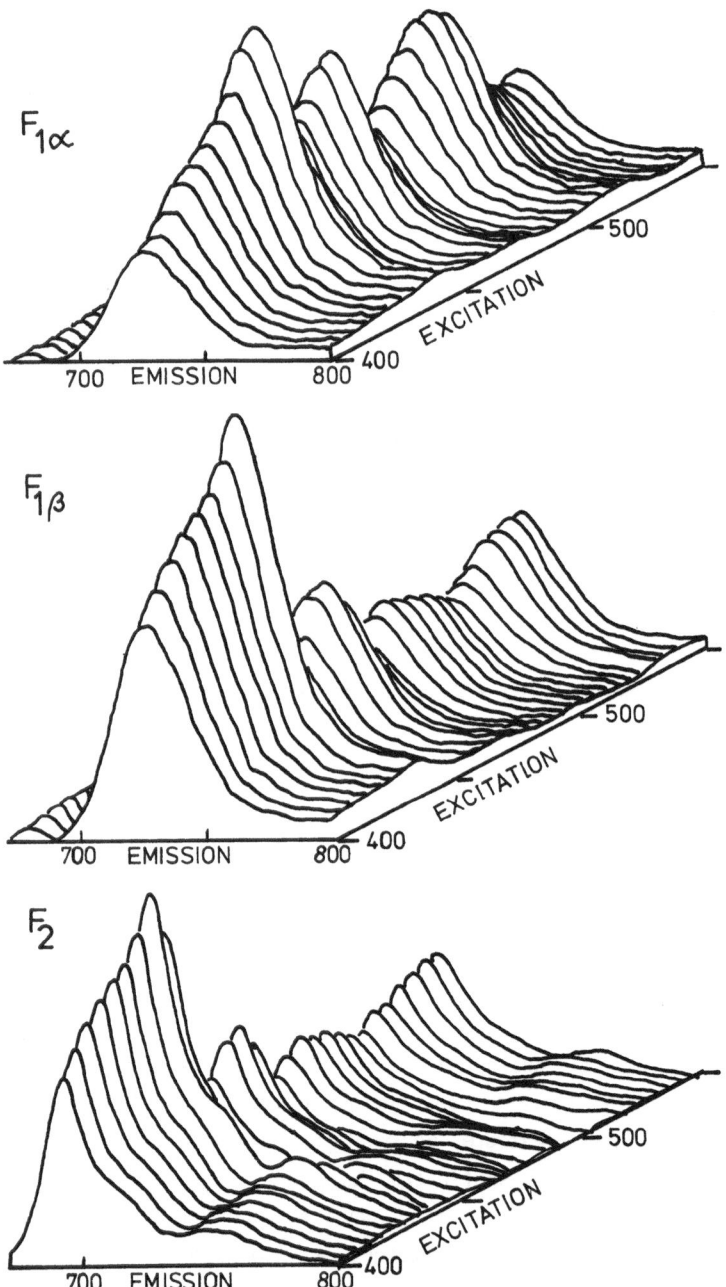

Figure 3:
Excitation and fluorescence emission spectra of "pure" PSI (indicated as
$F_{1(\alpha)}$) of "pure" PSII (indicated as F_2) and of the energy transfer flux from
PSII to PSI (indicated as $F_{1(\beta)}$ in flashed bean leaves at 77K. Excitation and
emission spectra were measured at the $F_{(O)}$ and at the $F_{(M)}$ levels, which
allow the calculation of $F_{1(\alpha)}$ and $F_{1(\beta)}$.

5.3   The slope of the plot F_1 versus F_2 (fig.2)

The slope of the plot F_1 versus F_2 can be written empirically as:

$$\text{slope} = F_{1(v)}/F_{2(v)} = F_{1(\beta)(o)}/F_{2(o)} = F_{1(\beta)(M)}/F_{2(M)}$$

Its biophysical meaning is:

$$\text{slope} = p_{1F} \cdot p_{21}/p_{2F} = p_{1F} \cdot k_{21}/k_{2F} = F_{1(v)}/F_{2(v)}$$

where k_{21} and k_{2F} are the rate constants of spill over or fluorescence emission of PSII respectively and $p_{ij} = k_{ij}/\sum_x k_{ix}$.

The experimentally measurable slope of F_1 versus F_2 is proportional to the **rate constant** of spill over k_{21}. The terms for the probabilities p_{ij}, the rate constants k_{ij} and the lifetime τ_i of the excited pigment complex are determined by the conformation of a photosynthetic system. Hence, they are referred to as conformation terms.

5.4   The biophysical significance of the plot F_1 versus F_2

The absorption terms (J_1, J_2, α) conformation terms $(p_{21}, k_{21}, \tau_1, \tau_2)$ and emission terms $(F_1, F_2, F_{1(\alpha)}, F_{1(\beta)})$ describe the energy fluxes flowing through a photosynthetic apparatus. All these terms can be linked together to the biophysical equation of a bipartite model as follows: (Bear in mind the definitions: $\alpha = J_1/(J_1+J_2)$; $p_{21} = k_{21} \cdot \tau_2$; $F_1 = F_{1(\alpha)} + F_{1(\beta)}$

$$\underbrace{\frac{\alpha}{(1-\alpha)}}_{\substack{\textit{Energy distri-}\\ \textit{bution term}}} = \underbrace{\frac{J_1}{J_2}}_{\substack{\textit{Absorption}\\ \textit{term}}} = \underbrace{p_{21}}_{\substack{\textit{Conformation}\\ \textit{term}}} \cdot \underbrace{F_{1(\alpha)} / F_{1(\beta)(o)}}_{\substack{\textit{Terms for}\\ \textit{fluorescence of PS I}}}$$

==

This equation carries the same information as equation 1 below. Furthermore, it shows that it is impossible to calculate the value of α with one set of data of $F_{1(\alpha)}$ and $F_{1(\beta)}$. The terms α and p_{21} are unknown. A second independent signal is needed to solve the above equation for α. This second information can be found in excitation _or_ fluorescence lifetime measurements. Both experiments lead to the same conclusion.

6. THE DETERMINATION OF ABSORPTION ENERGY DISTRIBUTION

The above equation for energy distribution in a bipartite system can be re-arranged and presented as :

Equation 1
(ref.9)

$$\alpha = \frac{F_{1(\alpha)}}{F_{1(\alpha)} + \frac{1}{p_{21}} \cdot F_{1(\beta)(o)}}$$

Replacing $F_{1(\beta)(o)} = F_{2(o)} \cdot \frac{p_{21}}{p_{2F}} \cdot p_{1F}$

leads to

$$\alpha = \frac{F_{1(\alpha)}}{F_{1(\alpha)} + \frac{p_{1F}}{p_{2F}} \cdot F_{2(o)}}$$

Replacing $p_{1F} = k_{1F} \cdot \tau_{1(o)}$ and
$p_{2F} = k_{2F} \cdot \tau_{2(o)}$ and
$k_{1F} = k_{2F}$

leads to

$$\alpha = \frac{F_{1(\alpha)}}{F_{1(\alpha)} + \frac{\tau_{1(o)}}{\tau_{2(o)}} \cdot F_{2(o)}}$$

Assuming: $F_{2(o)}/\tau_{2(o)} = F_{2(M)}/\tau_{2(M)}$ and $\tau_{1(o)} = \tau_{1(M)}$ leads to

Equation 2
(ref.10)

$$\alpha = \frac{F_{1(\alpha)}}{F_{1(\alpha)} + \frac{\tau_{1(M)}}{\tau_{2(M)}} \cdot F_{2(M)}}$$

This equation depends on the ratio of the experimental values $F_{1(\alpha)}/F_{2(M)}$. Therefore, it should be corrected to the shape of the total emission spectrum of F_1 and F_2.

The first equation (by Strasser and Butler, ref. 9) has been used in a combination of excitation and fluorescence data. The second equation (used by Wong and Govindjee and Merkelo, ref. 10) combines fluorescence lifetime measurements with fluorescence data. The first equation can be applied to a bipartite system but the combination of excitation and fluorescence data is technically very difficult to do. The second equation contains several assumptions. The signals for $F_{1(\alpha)}$ and $F_{2(M)}$ should be corrected to denote signals of the same relative area of the whole $F_{1(\alpha)}$ and $F_{2(M)}$ emission spectra. This correction is not necessary in equation 1 since it is cancelled by the ratio $F_{1(\alpha)}/F_{1(\beta)}$. Nevertheless, both equations and measuring techniques supply reasonable values for α and for the spill over probability p_{21}. It has to be emphasized here, that none of the equations consider either energy coupling between LHC and core antenna of PSII or grouping. However, as soon as the sample is placed in a high salt condition in the presence of DCMU, the fluorescence induction curve of chloroplasts is typically sigmoid at room temperature indicating that grouping occurs. The danger of all these equations is that, if the samples differ in their grouping or/and coupling constellations but have identical spill over constellations, then all these changes will appear in the calculations as changes in spill over properties **a priori**. So far, the **overestimation** of spill over can be avoided only when a theory is elaborated and when new simultaneous measuring techniques for trapping, coupling, grouping and spill over are developed. An analysis of the shape of the fluorescence induction curve can provide us with the necessary information about grouping. In a forthcoming paper, a new concept will be presented which allows measurement and calculation of a synergetic model including **trapping** of PSI and PSII, **spill over** from PSII to PSI, **grouping** between photosynthetic units, as well as absorption and dissipation fluxes of both photosystems. The data obtained from this concept show that when old experiments from the literature are analysed, the biggest conformational changes are always due mainly to changes in the grouping and slightly due to changes in the spill over constellation. The lack of mathematical freedom in the equations used by Butler led to the interpretation that every change in the energy distribution of the sample is a change in the spill over constellation (ref.15).

W.L. Butler proposed a method of calculating the absorption energy distribution α and the probability of spill over p_{21} by two sets of the four experimental values $F_{1(o)}$, $F_{1(M)}$, $F_{2(o)}$, $F_{2(M)}$ measured at low temperature. (Therefore, 8 independent experimental values are obtained). Many authors however, encountered some difficulties using this method. Many assumptions have to be made to solve the equations correctly but unfortunately, these assumptions are not in accordance with nature. The method uses two samples under different conditions (e.g. O mM Mg^{2+} and 10 mM Mg^{2+}) to solve the equations. If three samples under specific conditions were to be tested (e.g. O mM Mg^{2+}, 1 mM Mg^{2+} and 10 mM Mg^{2+}) then we could get three possible combinations each of which consists of a pair of two different samples (e.g. sample O and 1, sample O and 10, sample 1 and 10 mM Mg^{2+}).

Therefore, <u>two values</u> of α and p_{21} for each sample could be obtained. These two values should be <u>identical</u> if the behaviour of the sample corresponds to the statements assumed.
(see table 1 and 2)

The two sample or eight-point-method of calculating   energy distribution by W.L. Butler is stated in the appendix of this paper without comments. Each author has to decide for himself whether or not he wants to base his data on this method.

Mono-, bi-, tri- and polypartite models are trials to quantify the complexity of a photosynthetic apparatus. Every model is far from nature but-it serves as a stimulus to correlate biological experimental measurements to their biophysical meaning. All these models provide the basis for the analysis of the photosynthetic apparatus in terms of non-equilibrium thermodynamics. This line of investigation may tell us some day (in biological terms like structure stability, adaptation ability, trend of development etc.) why variables such as trapping, coupling, spill over and grouping tend to optimize the overall state of a photosynthetic system in its natural environment.

7.   APPENDIX: W.L. Butler's method of calculating the energy distribution term α and the probability p_{21} for energy transfer from PSII to PSI (ref. 11)

7.1   The experiment

Determination of the initial and maximal fluorescence at low temperature measured at an emission wavelength of PSI and simultaneously at an emission wavelength of PSII. The signals are: $F_{1(o)}$, $F_{1(M)}$, $F_{2(o)}$, $F_{2(M)}$. Two samples are measured under different conditions (e.g. high salt (+) and low salt (-) conditions).

7.2   Empirical definitions

1) $F_{i(M)} - F_{i(o)} = F_{i(v)}$ i can be substituted for either 1 or-2

2) $(F^-_{1(v)}/F^+_{1(v)}) \cdot (F^-_{2(v)}/F^+_{2(v)}) = R$

3) $F_{1(\alpha)} = F_{1(o)} - F_{2(o)} \cdot F_{1(v)}/F_{2(v)} = F_{1(o)} - F_{1(\alpha)(o)}$

7.3   Definitions of a bipartite model

1) Absorption energy distribution $\alpha = J_1/(J_1+J_2)$
2) Energy transfer probability from PSII to PSI

$$P_{21} = \frac{\text{Energy flux from PS II to PS I}}{\text{Total excitation rate of PS II}} = \frac{E_{21}}{E_2}$$

3) Intercept in the plot $F_{1(t)}$ versus $F_{2(t)}$ is $F_{1(\alpha)} = J_1 \cdot p_{1F}$
4) Slope of the plot $F_{1(t)}$ versus $F_{2(t)} = p_{1F} \cdot k_{21}/k_{2F}$

Table 1

Data from ref:	(11)		(14)		
mM Mg^{2+}	0	5	0	1	10
$F_{1(o)}$	76	60	80	73	55
$F_{1(M)}$	97	83	105.5	97.5	75.5
$F_{2(o)}$	28	32	34.5	39.5	43
$F_{2(M)}$	67	114	95	130	156.5
	.119	.066	****	.050	.034
p_{21}			.231	.161	****
(according to Butler)			.125	****	.058
	.324	.271	****	.226	.171
α			.510	.486	****
(according to Butler)			.361	****	.260

All calculations for α and p_{21} of the same sample should be identical.

Table 2

	pH 6.2			pH 8.8			pH 7.0		
	SD	Na^+	Mg^{2+}	SD	Na^+	Mg^{2+}	SD	Na^+	Mg^{2+}	
$F_{1(o)}$	237.2	277.0	216.8	232.6	204.9	174.4	266	347	219	
$F_{1(M)}$	300.8	339.2	257.8	265.9	225.0	186.3	323	426	276	
$F_{2(0)}$	22.3	29.5	36.6	19.9	24.0	47.7	59	56	59	
$F_{2(M)}$	65.1	62.1	100.0	47.7	36.4	63.6	100	100	133	
	-.368	-.527	****	.151	.193	****	.049	.062	****	calculations of Butler (12)
p_{21}	****	.184	.071	****	.557	.368	****	.040	.018	
	.510	****	.312	.364	****	.263	0	****	0	
	1.790	1.937	****	.569	.452	****	.099	.133	****	
α	****	.419	.366	****	.704	.588	****	.090	.063	
	.759	****	.718	.761	****	.506	0	****	0	
p_{21}	.14	.24	.09	.13	.23	.18	.25	.27	.13	of Wong, Govindjee, Merkelo (10)
α	.46	.48	.43	.54	.49	.42	.36	.40	.33	

7.4 Necessary assumptions to make

1) Total light absorption flux for all samples is constant

$$J_1^+ + J_2^+ = J_1^- + J_2^- \text{ therefore } \Delta J_2 = - \Delta J_1 \text{ !!}$$

2) The probability of fluorescence emission of PSI is constant

$$p_{1F}^+ = p_{1F}^-$$

3) The sum of the rate constants of fluorescence emission, heat dissipation and photochemistry of PSII is constant.

$$K^+ = k_{2F}^+ + k_{2D}^+ + k_{2b}^+ = k_{2F}^- + k_{2D}^- + k_{2b}^- = K^-$$

4) The rate constants of fluorescence emission of PSII are constant:

$$k_{2F}^+ = k_{2F}^- \text{ and } \Delta k_{2b} = - \Delta k_{2D} \text{ !!}$$

7.5 Correlation of data of the two samples

1) The ratio of the intercept of plot $F_{1(t)}$ versus $F_{2(t)}$

$$F_{1(\alpha)}^+ / F_{1(\alpha)}^- = J_1^+ \cdot p_{1F} / J_{1F}^- \cdot p_{1F} = J_1^+ / J_1^- = \alpha / \alpha^-$$

2) The ratio of the slope of plot $F_{1(t)}$ versus $F_{2(t)}$

$$R = \frac{p_{21}^-}{p_{21}^+} \cdot \frac{p_{2F}^+}{p_{2F}^-} \cdot \frac{p_{1F}^-}{p_{1F}^+} = \frac{k_{21}^-}{k_{21}^+}$$

3) The ratio of the energy transfer probability

$$p_{21}^- / p_{21}^+ = R \cdot \Sigma k_{2x}^+ / \Sigma k_{2x}^- \quad \text{this ratio can be written as}$$

$$= R \cdot \frac{\Sigma k_{2x}^+ / k_{21}^+}{(k_{21}^- + k_{2F}^- + k_{2D}^- + k_{2b}^- + (k_{21}^+ - k_{21}^+))/k_{21}^+}$$

this equation is identical with

$$= R \cdot \frac{1 / p_{21}^+}{R + \dfrac{1}{p_{21}^+} - 1}$$

7.6 The four necessary equations for α^+, α^-, p_{21}^+, p_{21}^-
are:

1) $\quad p_{21}^- \quad = \quad R \cdot p_{21}^+ / (1 + (R - 1) \cdot p_{21}^+)$

2) $\quad \alpha^+ / \alpha^- \quad = \quad F_{1(\alpha)}^+ / F_{1(\alpha)}^-$

3) $\quad \alpha^+ \quad = \quad F_{1(\alpha)}^+ / (F_{1(\alpha)}^+ + F_{1(\beta)(o)}^+ / p_{21}^+)$

4) $\quad \alpha^- \quad = \quad F_{1(\alpha)}^- / (F_{1(\alpha)}^- + F_{1(\beta)(o)}^- / p_{21}^-)$

7.7 The solution for the system of equations

The four equations can be used to solve for α^+, α^-, p_{21}^+, p_{21}^-. The probability of energy transfer from PSII to PSI and the absorption energy distribution term α can be calculated from the experimental data as follows:

$$p_{21}^+ = 1 / \left(1 - \frac{F_{2(v)}^+ \cdot (F_{1(o)}^+ - F_{1(o)}^-)}{F_{1(v)}^+ \cdot (F_{2(o)}^+ - F_{2(o)}^-)} \right)$$

$$\alpha = 1 / \left(1 - \frac{1 / p_{21}}{1 - F_{1(o)} \cdot F_{2(v)} / F_{1(v)} \cdot F_{2(o)}} \right)$$

The following table (see table 1) shows the calculations on energy distribution α and spill over probability p_{21} (from the literature) based on the experimental data obtained by W.L. Butler (11,14) using the two sample method. This method has an interesting theoretical approach, however, it lacks experimental consistency (a warning to those who intend to use the method).

A comparison of calculations on α (energy distribution) and p_{21} (spillover probability) based on the data obtained by Butler (two sample method) and the data obtained by Wong, Govindjee (fluorescence emission and lifetime method) is shown in table 2. SD (sucrose buffer), Na$^+$ (buffer with Na$^+$), Mg^{2+} (Buffer with Na^{2+} and Mg^{2+}).

Table 1 does not show any consistency at all since many completely different random values of p_{21} (spill over probability) and α (incident absorption energy distribution) are obtained from one and the same sample.

Table 2 shows that the method of Wong, Govindjee, Merkelo offers reasonable and generally acceptable values of p_{21} and α. The inconsistency found in Butler's two sample method is attributed to very strict and unbiological assumptions made by him e.g. $\Delta J_1 = -\Delta J_2$ or $\Delta k_{2b} = -\Delta k_{2D}$

They suggest that every absorption change in PSII should be compensated by an absorption change in PSI or/and every change in the rate constant of photochemistry should be parallel to an opposite change in the rate constant of heat dissipation. However, nature seems to vary these terms independently of one another. Both reported methods do not give any attention to grouping which is reflected in the lateral movements of protein complexes as revealed by electron microscopy. Furthermore, it is also reflected in the sigmoid shape of the fluorescence induction curve at room temperature in the presence of DCMU.

The message is: More experimental signals rather than initial/maximal fluorescence intensities and lifetime measurements are needed to describe energy distribution in a model. The model should include the four distinctly different types of energy transfer fluxes namely: trapping, spill-over, grouping and coupling.

REFERENCES

1. Butler WL and Kitajima M (1975) Biochim.Biophys.Acta 376:116-125
2. Joliot A and Joliot P (1964) C.R.Hebd.Séance Acad.Sci. 258:4617-4625
3. Strasser RJ (1978) Photosynthesis (Akoyunoglou G ed.) Elsevier/North Holland Biomed.Press: 513-524
4. Strasser RJ, Grepping H (1981) Photosynthesis (Akoyunoglou ed.) Balaban Intern.Science Serv. Philadelphia 3: 717-726
5. Strasser RJ, Butler WL (1977) Biochim.Biophys.Acta˙460: 230-238
6. Strasser RJ, Butler WL (1976) Biochim.Biophys.Acata 449: 412-419
7. Strasser RJ, Sironval C (1973) FEBS Letters 29: 286-288
8. Strasser RJ (1981) Photosynthesis (Akoyunoglou G ed.) Balaban Intern. Science Serv. Philadelphia 3: 727-737
9. Strasser RJ, Butler WL (1977) Biochim.Biophys.Acta 460: 230-238
10. Wong D, Govindjee, Merkelo H (1980) Biochim.Biophys.Acta 592:546-558
11. Butler WL, Kitajima M (1975) Biochim.Biophys.Acta 396: 72-85
12. Butler WL (1977) Encyclopedia of Plant Physiology (Trebst A ed.) Springer Verlag, Berlin 149-167
13. Wong D, Govindjee, Merkelo H (1981) Photochem.Photobiol.33: 97-101
14. Butler WL, Strasser RJ (1978) Photosynthesis (Hall DO et al eds) The Biochemical Society, London: 11-20
15. Strasser RJ, (1986) Proceedings of the international congress of Photosynthesis.

Photosynthesis Research 10: 277–282 (1986)
© *Martinus Nijhoff Publishers, Dordrecht*

INFLUENCE OF THYLAKOID PROTEIN PHOSPHORYLATION ON PHOTOSYNTHETIC ELECTRON TRANSPORT AND PHOTOPHOSPHORYLATION

GIORGIO FORTI and PAOLA M.G. GRUBAS

Centro CNR Biologia Cellulare e Molecolare delle Piante, Dipartimento di Biologia, Università di Milano, Via Celoria 26, 20133 Milano, Italy.

KEY WORDS: Thylakoids, protein phosphorylation, photosystem I, photophosphorylation.

ABSTRACT

Data are reported which show that thylakoid protein phosphorylation decreases photosystem II fluorescence yield and enhances the photosystem I dependent photophosphorylation catalyzed by phenazinemethosulphate in the presence of DCMU. The stimulation is larger at low light intensity, but is still observed at high intensity. These observations are interpreted to demonstrate that thylakoid protein phosphorylation causes a transfer of excitation energy from PS II to PS I, but may also have an independent stimulatory effect on PS I dependent photophosphorylation.

INTRODUCTION

High efficiency of photosynthetic electron transport (H_2O to NADP) requires an equal flux of light quanta to the two photosystems, PS I and PS II. However, the light harvesting chlorophyll-protein complex (LHCP) is largely associated with PS II in the grana whilst PS I is thought to be mainly located in the stromal membranes (1,2), so that the absorption cross section of PS II and PS I may be unequal. The mode of regulation of excitation energy distribution between the two photosystems, which can be discussed in terms of the model proposed by Butler and Kitajima (3), has therefore become a major problem in photosynthesis research.

The phosphorylation of LHCP by a thylakoid-bound, PQH_2-activated protein kinase has been proposed to lead to a decrease of excitation energy in PS II and an increase in that of PS I (4,5,6). Dephosphorylation by a phosphatase causes the reverse changes (4,5,6). On this basis phosphorylation-dephosphorylation of LHCP would regulate the distribution of excitons between the photosystems. It has indeed been shown that a fraction of LHCP becomes detached from the PS II-LHCP matrix in the phosphorylated membranes, and subsequently associates functionally with PS I (7,8). These results and their interpretation are based mainly on the measurement of fluorescence changes. The increase of PS I photochemical activity following thylakoid phosphorylation has been shown by Jennings and Zucchelli (9) as the increase of the ratio Q/Q^-. They also observed an increase of the rate of NADP reduction only in the presence of the uncoupler gramicidin, and their results could be accounted for by a kinetic model of electron transport which evaluates the ratio Q/Q^- in terms of the relative photochemical activity of PS II and PS I (9).

Inhibition of PS II and stimulation of PS I photochemistry presumably due to thylakoid protein phosphorylation has been reported (10). However, the interpretation of those results was complicated by the inhibition due to the preillumination treatment designed to activate the protein kinase (10). Contrasting observations on the effect of LHCP phosphorylation on P700 oxidation have been reported by other authors (11,12).

We have studied the influence of thylakoid protein phosphorylation on the PS I dependent cyclic photophosphorylation catalyzed by phenazinemethosulfate (PMS) in the presence of DCMU, and the non cyclic photophosphorylation coupled to NADP reduction. Whilst the latter is unaffected, the former is considerably enhanced. The enhancement is larger at low light intensity, as expected if the size of the PS I antenna were increased, but is also observed at high light intensity.

MATERIAL AND METHODS

Spinach leaves were collected and kept 1 hr in darkness before the chloroplasts were extracted as previously described (13), in the presence of 5 mM $MgCl_2$. Thylakoid protein phosphorylation was performed in the dark (to avoid changes due to preillumination) at 21°, in a medium containing sucrose 0.4 M, tricine-NaOH 30 mM, pH 8.0, NaCl 10 mM, $MgCl_2$ 5 mM, NaF 10 mM, ferredoxin 5 uM, NADPH 0.5 mM, 500 ug/ml of chlorophyll and ATP 0.5 mM. Control thylakoids were treated in the same manner except for the omission of ATP. At the indicated times the membranes were diluted 50 - fold in the tricine-NaCl-NaF-$MgCl_2$ buffer containing 0.2 M sucrose, which also served as the reaction medium with the appropriate additions of ADP 1 mM and orthophosphate 2.5 mM (labelled with ^{32}P) when photophosphorylation was measured. NADP reduction (14) and photophosphorylation (15) were measured as previously reported. Fluorescence was measured at 680 nm in the presence of 10 uM DCMU. The excitation light was filtered through a Corning 4-96 filter and a heat filter. The same light was used as actinic light in all the experiments. PS I dependent cyclic photophosphorylation was measured in the presence of 10 uM DCMU and 30 uM PMS (saturating concentration). No difference between phosphorylated and control thylakoids was observed with respect to activity as a function of PMS concentration.

RESULTS

We have investigated the influence of thylakoid protein phosphorylation on PS I photochemistry by measuring the PS I dependent cyclic photophosphorylation catalized by PMS. A large stimulation was observed (fig. 1).

No effect of ATP was observed in the absence of the reducing system (not shown), indicating that activation of the protein kinase was required as was the case for the fluorescence decline effect. The large stimulation observed at low light intensity decreased at higher light intensity (fig. 1, insert), as expected according to the notion that LHCP phosphorylation increases the size of PS I antenna (4,5,7). The rate of ATP formation was a linear function of light intensity in the phosphorylated thylakoids, whilst at the lowest light intensity (32 $W.m^{-2}$) the rate of control thylakoids showed a "light intensity lag" probably a consequence of the low proton motive force, due to low electron transport rate, insufficient to support ATP synthesis.

Fig. 1 shows that the enhancement of photophosphorylation due to thylakoid protein phosphorylation is still observed even at the high photon flux of 800 W.m^{-2} (not far from saturating under our conditions), and in fact reaches a plateau above 100 W.m^{-2}.

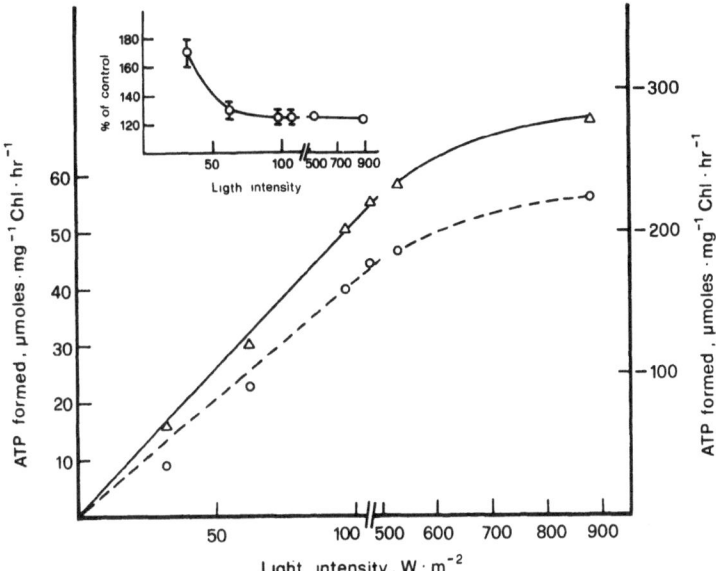

FIGURE 1. Effect of thylakoid protein phosphorylation on PMS cyclic photophosphorylation as a function of light intensity.
Membrane proteins were phosphorylated for 12 min (see Methods). Bars indicate the standard error of 6 photophosphorylation experiments with 6 different thylakoid preparations. PS II fluorescence decline was 14.2%, $\pm$ 1.6 (standard error). △—△ phosphorylated and non-phosphorylated o--o membranes. The ordinate scale at the left refers to abscissae values up to 100; the right ordinates are referred to abscissae values above 100 W. m^{-2}.

The time-course of changes of fluorescence and cyclic photophosphorylation during thylakoid protein phosphorylation was also studied (fig. 2).

The enhancement of cyclic photophosphorylation seems to be a faster response than fluorescence decline. The time-course of phosphorylation of LHCP and other thylakoid proteins appears to be almost linear up to 30 min, whilst the fluorescence decline is usually complete in a shorter time (16), although considerable variability is observed both in extent and kinetics.

Photophosphorylation coupled to NADP reduction was unaffected by thylakoid protein phosphorylation, whilst the usual pattern of fluorescence decline was observed (fig. 3)

The ratio ATP/NADPH was 1, indicating a good energy coupling and, as previously reported by Jennings and Zucchelli (9), no effect of LHCP phosphorylation on the rate of NADP reduction was observed under these conditions (fig. 3). This very reproducible observation was explained on the basis of the opposed effects on the photochemical activities of PS II and PS I of the phosphorylation-induced detachement of a fraction of LHCP from PS II and its association with PS I (9).

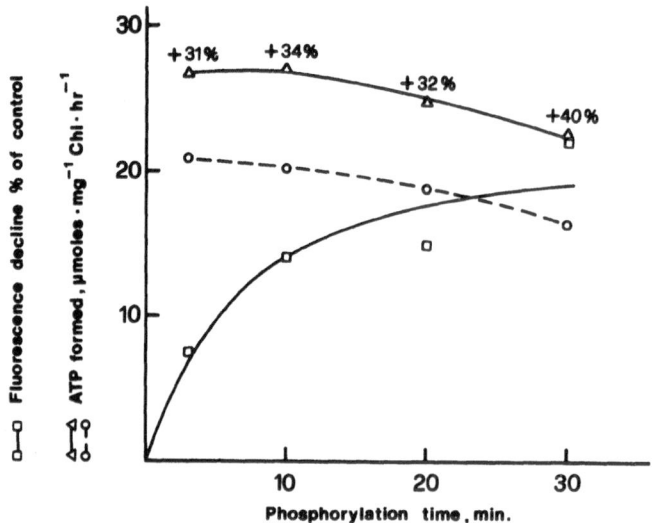

FIGURE 2. The time-course of protein phosphorylation-dependent changes in fluorescence and PMS cyclic photophosphorylation.
The membranes were preincubated in the presence (△———△) or absence (o— — — —o) of ATP (see Methods) and at the indicated times PMS cyclic photophosphorylation was determined using a light intensity of 62 W.m^{-2}.
The fluorescence quenching (□————□) is shown as a % of the non phosphorylated control membranes.

DISCUSSION

The observations reported in figs. 1 and 2 show a large enhancement of PMS cyclic photophosphorylation in the presence of DCMU (a PS I dependent reaction) by thylakoid protein phosphorylation, uncomplicated by secondary effects as those due to preillumination reported in (10), nor by the opposing effects of LHCP phosphorylation on PS I and PS II reported previously (9).

The extent of stimulation of PS I dependent photophosphorylation shown in fig. 1 (insert) at light intensities above 60 W m^{-2}, i.e. in the linear range of the activity versus light intensity, can be accounted for if one assumes that in unphosphorylated thylakoids light were absorbed about 60% by PS II and 40% by PS I.

In such a situation, the observed decrease of 14.2% of the PS II absorbed energy and its transfer to PS I would produce an increase of 21%

of the energy available to PS I.

A stimulation of 23% was indeed observed when the reaction rate was proportional to light intensity (106 W.m^{-2}, fig. 1). However, the fact that approximately the same level of stimulation persisted at light intensities close to saturation suggests that thylakoid protein phosphorylation might have an additional effect on PS I dependent photophosphorylation besides the increase of PS I antenna.

No stimulation of photophosphorylation coupled to NADP reduction was observed in well-coupled thylakoids (fig. 3) and the ATP/NADPH ratio of 1 was also unaffected by protein phosphorylation, demonstrating that the relative rates of NADP reduction, cyclic electron transport and phosphorylation-coupled Mehler reaction are not affected under these conditions by membrane protein phosphorylation.

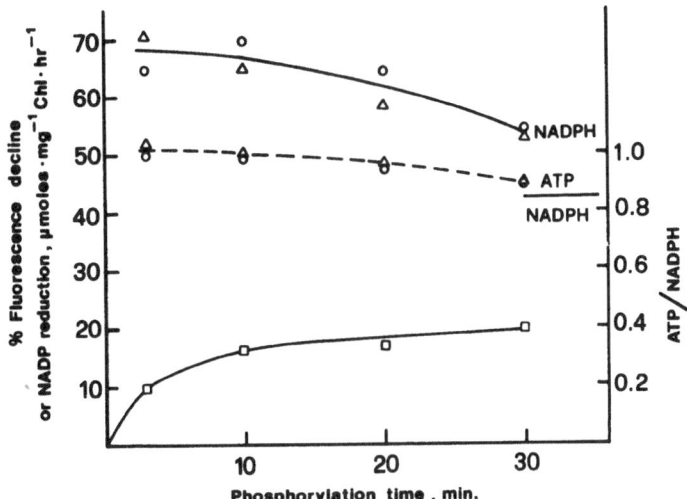

FIGURE 3. Effect of protein phosphorylation time on non cyclic photophosphorylation coupled to NADP reduction and on fluorescence decline.
△—△ phosphorylated thylakoids; O—O control; □—□ fluorescence decline, %.
Light intensity was 106 W.m^{-2}. Dashed line: ATP/NADPH ratio.

REFERENCES

1. Andersson B and Anderson J M (1980) Biochim. Biophys. Acta 593:427-440.
2. Anderson J M (1980) FEBS Letters 117:327-331.
3. Butler W L and Kitajima M (1975) Biochim. Biophys. Acta 396:72-85.
4. Bennett J (1983) Biochem. J. 212:1-13.

5. Horton P (1983) FEBS Letters 152:47–52.
6. Staehlin L A and Arntzen C J (1983) J. Cell Biol. 97:1327–1337.
7. Torti F, Gerola P D and Jennings R C (1984) Biochim. Biophys. Acta 767:321–325.
8. Kyle D J, Kuang T Y, Watson J L and Arntzen C J (1984) Biochim. Biophys. Acta 765:89–96.
9. Jennings R C and Zucchelli G (1986) Arch. Biochem. Biophys., in press.
10. Farchaus J W, Widger W R, Cramer W A and Dilley R A (1982) Arch. Biochem. Biophys. 217:362–367.
11. Haworth P and Melis A (1983) FEBS Letters 160:277–280.
12. Telfer A, Bottin H, Barber J and Mathis P (1984) Biochim. Biophys. Acta 764:324–330.
13. Jennings R C (1984) Biochim. Biophys. Acta 766:303–309.
14. Forti G and Grubas P M G (1985) FEBS Letters 186:149–152.
15. Lindberg O and Ernster L (1956) in "Methods of Biochemical Analysis" D. Glick Ed. Vol. III, p. 1 – Interscience Inc. N.Y.
16. Islam K and Jennings R C (1985) Biochim. Biophys. Acta 810:158–163.

Photosynthesis Research 10: 283–290 (1986)
© Martinus Nijhoff Publishers, Dordrecht

ENERGY DISTRIBUTION IN THE PHOTOCHEMICAL APPARATUS OF
PORPHYRIDIUM CRUENTUM: PICOSECOND FLUORESCENCE
SPECTROSCOPY OF CELLS IN STATE 1 AND STATE 2 AT 77 K

DOUG BRUCE, CHERYL A. HANZLIK*, LUCIA E. HANCOCK*,
JOHN BIGGINS and ROBERT S. KNOX*†

Division of Biology and Medicine, Brown University, Providence, R.I. 02912
and the *Department of Physics and Astronomy, and † Laboratory for Laser
Energetics, University of Rochester, Rochester, N.Y. 14627

Keywords: energy transfer, fluorescence kinetics, photosynthesis,
Porphyridium cruentum, spillover, state transitions.

Abstract. Excitation energy distribution in Porphyridium cruentum in
state 1 and state 2 was investigated by time resolved 77 K fluorescence
emission spectroscopy. The fluorescence rise times of phycoerythrin,
phycocyanin and allophycocyanin (in cells in state 1 and state 2) were
very similar in contrast to the emission from chlorophyll a (Chl a)
associated with the two photosystems. In state 2 photosystem II (PSII)
Chl a fluorescence emission rose faster than the PSI Chl a emission and
decayed more rapidly, and the converse was observed in state 1. These
kinetic data support the concept of increased energy transfer from PSII
Chl a to PSI Chl a in state 2 in P. cruentum.

Abbreviations: allophycocyanin, APC; chlorophyll a, Chl a;
photosystem II, PSII; phycocyanin, PC; phycoerythrin, PE.

Introduction

    In 1969 Murata [1] and Bonaventura and Myers [2]
independently discovered that the distribution of absorbed
excitation energy between PSII and PSI was variable in
photosynthesis. The concept that excitation energy absorbed
by Chl a closely associated with PSII could be transferred to
the Chl a of PSI ("spillover") had been introduced earlier by
Myers [3], and was suggested by Murata [1] to be the
mechanism controlling excitation energy distribution between
the two photosystems in the red algae Porphyridium cruentum.
Ley and Butler [4,5] estimated the initial distribution of
excitation energy between PSII and PSI, as well as the amount
of "spillover", in P. cruentum using the tripartite model
for photosynthesis developed earlier by Butler and Kitajima
[6]. They reported that the yield of energy transfer from
PSII to PSI varied from 50 % when PSII reaction center traps
were open, to 90 - 95 % when the traps were closed. Using
the same experimental technique and model, Ley and Butler [7]
also determined that the rate constant for energy transfer
from PSII to PSI doubled on transition from state 1 to
state 2 in P. cruentum.

More recently, the mechanism of the state transition in green plants was reported to involve primarily a direct change in the absorbance cross-section of PSII and PSI [8]. Biochemically, the process implicates a long range lateral migration of the light harvesting complex (LHC II) in the thylakoid membrane accompanied by, and perhaps driven by, a phosphorylation of LHC II. For a recent review see [9].

There is considerable evidence for a different mechanism in the case of phycobilisome-containing organisms. For example, no reversible protein phosphorylation event accompanying the state transition has been detected in these organisms [10], and the kinetics [11] and response to ionophores [12] are very different to those observed in green plants. As suggested by Murata [1], these data are consistent with a small reversible change in thylakoid membrane conformation that regulates the efficiency of excitation energy transfer from PSII Chl a to PSI Chl a.

In an earlier investigation of excitation energy transfer with time-resolved fluorescence spectroscopy (using a single photon-timing technique), the decay kinetics of PSII Chl a fluorescence were found to be faster in state 2 than in state 1 in both P. cruentum and Anacystis nidulans [13]. Limitations in the time resolution of the detection apparatus precluded a detailed analysis of fluorescence rise times in that study and it was not determined whether the change in PSII Chl a fluorescence decay was accompanied by a change in the PSI Chl-a fluorescence rise kinetics as presupposed by the "spillover" model.

The objective of the present project was to directly investigate the pathway of energy transfer in P. cruentum in state 1 and state 2 using time-resolved fluorescence spectroscopy with faster time resolution (streak camera detection). The results support previous time-resolved fluorescence studies on P. cruentum on the pathway of excitation energy from phycoerythrin (PE), to phycocyanin (PC), to allophycocyanin (APC) to PSII Chl a [14,15]. The kinetics of fluorescence emission from PE, PC, and APC were very similar for cells in state 1 and state 2, but, the kinetics of Chl a emission associated with PSII and PSI were dependent on the state transition. In state 2 the PSII Chl a fluorescence rose faster and decayed more quickly than the PSI Chl a fluorescence. The converse was true in state 1. These data are consistent with increased "spillover" in state 2 and were described by a simple kinetic analysis.

Materials and methods

P. cruentum was grown autotrophically on enriched seawater [16] and cells were chemically fixed in light state 1 or light state 2 using glutaraldehyde as previously described [12]. After fixation, the cells were resuspended in 2.2 M sucrose to an absorbance of between 0.1 and 0.15 at 532 nm for a 1 mm pathlength. The 77 K fluorescence emission spectra of the fixed cells (not shown) were very similar to previously published spectra of cells frozen in state 1 or state 2 [7,10-13].

For all measurements, 500 µL samples were placed in a
1 mm optical pathlength cuvette and brought to 77 K in a
cryostat. Samples were excited at 532 nm with single 30 ps
pulses from a frequency doubled active-passive mode-locked
Nd^{3+}:YAG laser at a repetition rate of 0.5 Hz. The
excitation fluence was 10^{13} photons cm^{-2}. Time-resolved
fluorescence emission was determined with a streak camera
detection apparatus as described previously [17-19].
Fluorescence was spectrally resolved by placing 10 nm
bandwidth interference filters between the sample and
detection apparatus. The data presented are averages of 200
to 400 streak traces.

Results
1. Experimental Data
Figure 1 shows the kinetics of fluorescence emission
from PE, PC and APC for cells in state 1, along with the
measured excitation pulse of the laser. PE fluorescence rose
with the excitation pulse (to within the instrumental
response time) in both state 1 and state 2, i.e. half maximal
emission occurred at the peak of the excitation pulse. The
times to reach maximum fluorescence intensity, relative to

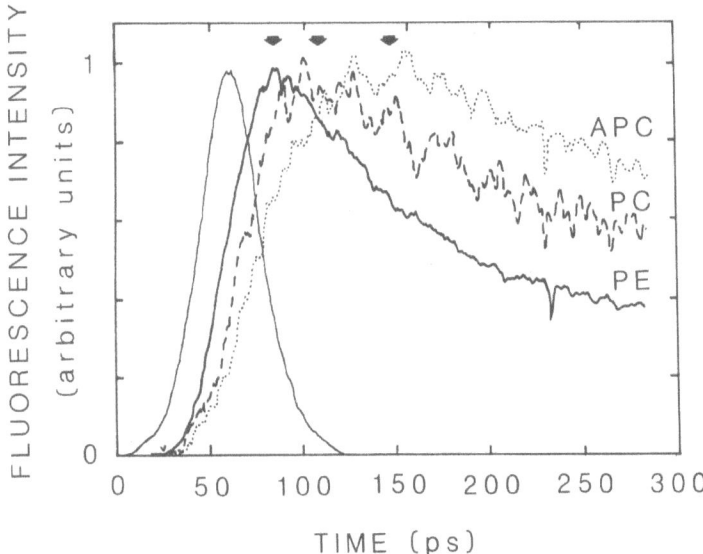

Figure 1. Fluorescence emission kinetics of phycoerythrin (PE, 570
nm), phycocyanin (PC, 640 nm), and allophycocyanin (APC, 660 nm). The
laser excitation pulse is shown by the light solid line. The data were
obtained using whole cells of P. cruentum in state 1. The arrows show
the approximate positions of the peaks of the fluorescence emission
curves. Fluorescence intensity is shown on a linear scale. All curves
were normalized to peak emission.

the PE peak, were 25 ± 8 ps for PC, and 60 ± 15 ps for APC in state 1. Corresponding times for cells in state 2 were 30 ± 8 ps for PC, and 55 ± 15 ps for APC. The fluorescence kinetics of PE, PC and APC were very similar for cells in state 2 (not shown).

In Figure 2 the kinetics of fluorescence emission from Chl a associated with PSII and PSI are shown for cells in state 1 and state 2. In contrast to the observations on energy transfer between components of the phycobilisome, a major change in the relative kinetic responses occurs with the state transition. In state 2 the PSII Chl a emission precedes PSI Chl a emission and decays more rapidly, whereas in state 1 PSI Chl a emission rises more rapidly and decays slightly faster than PSI Chl a emission.

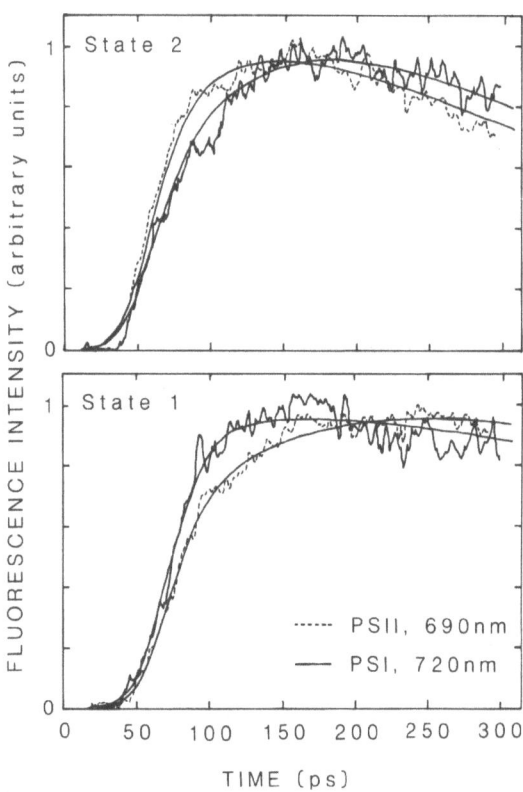

Figure 2. Fluorescence emission kinetics of PSII and PSI Chl a for whole cells of P. cruentum in state 2 and state 1. The smooth curves are fits to the data (see text for details). Fluorescence intensity is shown on a linear scale. All curves were normalized to peak emission.

2. Analysis of 690 nm and 720 nm Fluorescence Kinetics

The smooth curves in Figure 2 are computer generated fits to the experimental data. The fits were based on a simplified version of the three body kinetic model described by Wittmershaus et al. [20], in which the shape of the excitation pulse was considered explicitly in computing the time dependence of the fluorescence. The data were fitted by solving differential equations which allow for excitation of an emitter via an intermediate energy transferring species. The species absorbs the laser excitation pulse and then excites an emitter by way of transfer at a rate K_t. Our "rise time" was defined as the inverse of this rate.

The emissions at 690 nm and 720 nm were clearly biphasic in rise time, and were fitted with two components. It was not possible to fit any of the 690 nm or 720 nm data within the noise level using only one rise component. One component (fast component) had a fast rise (6-11 ps) and a relatively slow decay (640-740 ps). Only this component was seen at 700 nm and it appeared to overlap PSII and PSI emission at 690 nm and 720 nm resulting in the biphasic rises observed. The second component (slow component) of both photosystems had a rise time of about 200 ps.

Table I

Parameters used to fit the experimental data with the kinetic analysis described in the text. Times are in picoseconds. The rise and decay times of the slow components have large uncertainties because of our fast sweep time of 250 ps. (A, amplitude or % of initial excitation energy absorbed)

parameter	PSII Chl-a (690 nm emission)				PSI Chl-a (720 nm emission)			
	S1	A	S2	A	S1	A	S2	A
fast component								
"rise time"	10+1	33%	7+1	33%	10+1	67%	7+1	20%
"decay time"	740+120	33%	640+70	33%	740+120	67%	640+70	20%
slow component								
"rise time"	210+60	67%	190+70	67%	135+100	33%	190+50	80%
"decay time"	550+180	67%	100+60	67%	400+200	33%	150+50	80%

As shown in Table I, the change in 690 nm fluorescence emission induced by the state transition was modelled by assuming a faster decay time of the slow component in state 2, with no change in either the rise times of the fast and slow components, or the lifetime of the fast component. For the 720 nm data, the state transition was modelled by assuming an increase in the relative amplitude of the the slow component in state 2. In contrast to the 690 nm data, it was not possible to use the same relative amounts of fast and slow rise components in both states and fit the 720 nm data within the experimental noise.

Discussion   Energy transfer through the phycobilisome.
As reported previously for intact cells of P. cruentum
at room temperature [14,15], excitation of intact cells at
77 K with light absorbed primarily by PE yields successive
fluorescence emission from PE, PC and APC. The times to
reach maximal fluorescence emission of 0, 30 and 55 ps
(state 2) and, 0, 25 and 60 ps (state 1) for PE, PC and APC
respectively, were very similar to those recently determined
(0, 30 and 57 ps) for whole cells at room temperature by a
photon-timing technique with much lower intensity laser
pulses [14]. However, the times in the present study were
considerably slower than those (0, 12, and 24 ps) reported by
Porter et al. [15] using streak camera detection and higher
intensity laser pulses (10^{14}-10^{15} photons cm^{-2}). The
discrepancies most likely result from a lifetime shortening
effect caused by exciton annihilation events concomitant with
the higher intensity pulses. The agreement between the times
observed in this study and those reported in the
photon-timing study [14] substantiate that the energy
transfer kinetics PE --> PC --> APC in whole cells at room
temperature are similar to those of glutaraldehyde fixed
cells at 77 K, and also that the laser pulses used in this
study did not result in exciton annihilation effects. In
confirmation of an earlier report [13], the energy transfer
kinetics PE ---> PC ---> APC did not change during the state
transition.

Energy transfer from PSII Chl a to PSI Chl a.
In confirmation of a previous study [13] we observed the
largest change in fluorescence decay rate, induced by the
state transition, to be an acceleration of the PSII Chl a
fluorescence decay on transition to state 2. In addition,
the faster time resolution of the detection apparatus in the
present study showed that the rise kinetics of PSII and PSI
Chl a fluorescence were also dependent on the state
transition. In state 2 the fluorescence of PSII Chl a
preceded that of PSI Chl a and then decayed while PSI Chl a
fluorescence was still increasing. These kinetics are
consistent with energy transfer from PSII Chl a to PSI Chl a
in state 2. As the relative order of fluorescence emission
from PSII and PSI Chl a was reversed in state 1 our data are
consistent with the model for the state transition in P.
cruentum proposed by Biggins et al. [10] where the
distribution of excitation energy between PSII and PSI is
modulated by "spillover" which is maximal in state 2

Kinetic analysis of PSII and PSI Chl a fluorescence.
The intention behind the kinetic analysis conducted in
this study was to determine if model parameters used to fit
the data could be chosen which were consistent with the
concept of increased "spillover" in state 2. The rise
kinetics were fit satisfactorily in all cases with a sum of a
fast and a slow rise component. We believe the slow rise
component, on the order of 200 ps, seen in both PSII and PSI
reflected mainly the time of energy transfer through the

phycobilisome and into the photosystems. Although we cannot exclude a contribution by the "intrinsic" slow rise kinetics that have been observed in isolated PSII and PSI complexes [20,21]. We assume the fast rise component, on the order of 10 ps, reflects energy transfer from antenna Chl a to the fluorescing Chl a species and possibly fast energy transfer from APC to Chl a in PSII.

The state transition induced change in PSII fluorescence kinetics was modelled by assuming only a change in decay rate. Thus, it was possible to fit the PSII data without assuming a relative change in the amplitudes of the two rise components. In contrast, the PSI fluorescence rise could not be fit without assuming an increase in the relative amplitude of the slow rise component in state 2. This suggested an enhanced excitation of PSI in state 2 via energy initially absorbed in the phycobilisomes. Either direct energy transfer from the phycobilisomes to PSI or an increase in "spillover" would be consistent with this result. However, an increase in the association of the phycobilisome with PSI at the expense of PSII would be expected to decrease the slow rise component of PSII fluorescence and would not be expected to affect PSII fluorescence decay. Therefore, the data in this study were consistent with an increase in energy transfer from PSII Chl a to PSI Chl a in state 2 where the slow rise component of PSI Chl a emission reflects energy transfer through the phycobilisome to PSII Chl a and then to PSI Chl a.

We stress that although the data in this study support the concept of increased "spillover" in state 2 they do not prove its existence. The degree of experimental noise combined with the spectral overlap between the PSII Chl a emission and the PSI Chl a emission leave further interpretation to the results of a more exhaustive study of both the excitation and emission wavelength dependences of these energy transfer rate constants and decay times.

In conclusion, the data presented support our working model for the mechanism of the light state transition in P. cruentum, where a small change in the physical proximity between the two photosystems controls the "spillover" of excitation energy from PSII to PSI.

Acknowledgements

        This research was supported by The Science and Educational Administration of the United States Department of Agriculture under Grant Number 81-CRCR-1-0767 (J.B.) and Grant Number 82-CRCR-1-1128 (R.S.K.) from the Competitive Research Grants office; PCM-8302983 from the National Science Foundation (J.B.), and Sponsers of the Laser Fusion Feasibility Project of the Laboratory for Laser Energetics (R.S.K.). In addition D.B. gratefully acknowledges the support of a Natural Sciences and Engineering Research Council of Canada Post-doctoral Fellowship.

        We thank Drs. J.H. Sommer and B.P. Wittmershaus for assistance in experimental research and helpful discussions. We thank Dr. K. Sauer for suggestions with the manuscript.

References

1. Murata N (1969) Biochim Biophys Acta 172:242-251
2. Bonaventura C and Myers J (1969) Biochim Biophys Acta 189:366-383
3. Myers J (1963) In Photosynthetic Mechanisms of Green Plants. (B Kok and A Jagendorf eds) Washington D.C., Natl Acad Sci, Natl Res Council Publ no. 1145, pp 301-317
4. Ley AC and Butler WL (1976) Proc Natl Acad Sci USA 73: 3957-3960
5. Ley AC and Butler WL (1977) In Photosynthetic Organelles, special ed of Plant Cell Physiol (Miyachi S, Katoh S, Fujita Y and Shibata K eds), pp 33-46
6. Butler WL and Kitajima M (1974) In Proceedings of the Third International Congress on Photosynthesis, (Avron M ed) Elsevier Scientific Publishing Co, Amsterdam, pp 13-24
7. Ley AC and Butler WL (1980) Biochim Biophys Acta 592:349-363
8. Haworth P, Kyle DJ and Arntzen CJ (1982) Biochim Biophys Acta 680:343-351
9. Staehelin LA and Arntzen CJ (1983) J Cell Biol 97: 1327-1337
10. Biggins J, Campbell C and Bruce D (1984) Biochim Biophys Acta 767:138-144
11. Biggins J and Bruce D (1985) Biochim Biophys Acta 806:230-236
12. Biggins J (1983) Biochim Biophys Acta 724:111-117
13. Bruce D, Steiner T, Biggins J and Thewalt M (1985) Biochim Biophys Acta 806:237-246
14. Yamazaki I, Mimuro M, Murao T, Yamazaki T, Yoshihara K and Fujita Y (1984) Photochem Photobiol 39:233-240
15. Porter G, Tredwell CJ, Searle GFW and Barber J (1978) Biochim Biophys Acta 501:232-245
16. Kratz WA and Myers J (1955) Amer J Bot 43:282-287
17. Mourou G and Knox W (1980) Appl Phys Lett 36:623-631
18. Hanzlik CA, Hancock LE, Knox RS, Guard-Friar D and MacColl R (1985) J of Luminescence 34:99-106
19. Knox W and Forsley L (1983) ACS Symposium Series 236:221-231
20. Wittmershaus BP, Nordlund TM, Knox WH, Knox RS, Geacintov NE and Breton J (1985) Biochim Biophys Acta 806:93-106
21. Yamagishi A, Karukstis KK and Sauer,K (1986) submitted to Biochim Biophys Acta

Photosynthesis Research 10: 291–296 (1986)
© *Martinus Nijhoff Publishers, Dordrecht*

ANALYSIS OF EMERSON ENHANCEMENT UNDER CONDITIONS WHERE PHOTOSYSTEM II IS INHIBITED - ARE THE TWO PHOTOSYSTEMS INDEED SEPARATED ?

S. MALKIN, O. CANAANI and M. HAVAUX

ABSTRACT
Photoacoustic measurements of photosynthetic oxygen evolution and its enhancement by addition of background far-red light (Emerson enhancement) were made on both intact and inhibited or stressed leaves. The extent of enhancement increased with the dehydration treatment and decreased with a mild heat treatment or with the addition of DCMU. It nevertheless persisted even at a very high degree of inhibition - a result which indicates high population ratio of single pairs of combined photosystem II and photosystem I units, functional in whole chain electron transport. This implies a restriction on the separation between photosystems I and II, in contrast to existing concepts.

INTRODUCTION
There is a general view that PS I and PS II are located in separate topographical regions of the photosynthetic membrane (1-3). It is possible to separate membranal fractions assigned either to stroma lamellae, granal nonapressed zones and granal partitions. The last ones are recognized by forming inverted thylakoids (4,5). The various membranal fractions differ in their gross photochemical activities (5), protein chlorophyll complexes composition (6), and show different activity of the reaction centers of each of the two photosystems (7). This led to the conclusion that granal partitions are largely devoid of PS I, which mainly exists in the end granal zones and in the stroma lamellae, and that PS II resides mainly in the granal partitions. From the fluorescence kinetics it was deduced (7) that the granal partitions contain mainly PS II alpha units while the beta units reside mainly outside the apressed regions, but it seems (8) that the beta units may not be functional in electron transport. Electron transfer communication between PS I and PS II must thus occur over relatively large distances (in the order of 0.1 μm) and requires small mediator molecules (e.g. plastoquinones) to provide such a link by random diffusion (3).
In contradiction to the above, stands much of the latest work of Butler *et al*, who in their model of the bipartite or the tripartite arrangement of the photosynthetic apparatus (9,10) assumed close interaction between the two photosystems. Significant excitation energy transfer from PS II to PS I was indeed demonstrated in isolated chloroplasts, when PS II reaction centers were closed and under conditions where the granal membrane structure is preserved (11,12). Such experiments limit the extent of separation between the two photosystems to Forster type distances (i.e. < 100 Å).
Clearly one faces a dilemma in deciding between these two contradicting points of view. We have obtained independent experimental evidence which supports pairing between PS II and PS I centers, in a significant fraction of the photosystems. Our results are based on the measurement of the Emerson

enhancement effect, resulting from activity imbalance in favor of PS II (13). Such imbalance was still preserved even when PS II was inhibited severely - an incomprehensible result if electron equivalents are randomly transferred from PS II centers to distant PS I centers. This result, however, is easily explained if there is a link between a reaction center of PS II and a specific partner center in PS I - a possibility which requires close proximity between the two photosystems.

EXPERIMENTAL
Small pieces of leaves (usually 1 cm diameter discs) from tobacco *(Nicotiana tabacum L., var. Xanthi)* were cut, and put, after treatment, in a photoacoustic apparatus for measurements of photosynthetic activities. Dehydration was affected by exposure in air for several hours (14). Heat stress was carried out by incubation in water at $45^{O}C$ for several minutes in the dark. DCMU inhibition was achieved by 15 min. incubation in DCMU solution at various concentration.
The photoacoustic apparatus and method, and its use in detecting oxygen evolution and Emerson enhancement have been described before (13,15,16). The photoacoustic signal was excited by intensity modulated light at low frequency (19 Hz), of wavelength maximally distributed into PS II (650 nm - light-2) and in the light-limiting intensity range. Background non-modulated illumination at 701 nm, mainly absorbed in PS I (light-1), was applied at intervals to obtain an enhanced signal (the Emerson enhancement). Photosynthetic saturation was achieved by addition of background non-modulated high intensity light (400-680 nm), which served to fix a zero level for the modulated oxygen evolution signal. The relative quantum yield of oxygen evolution is given by the ratio of the "oxygen" signal amplitude to the amplitude of the photothermal signal (15, 16). Emerson enhancement was calculated as the ratio of the yields of oxygen evolution in presence and absence of background light-1 (13). All measurements were performed in state 1 - i.e. after a few minutes of adaptation to the combined lights 1 and 2 (13).
Auxillary measurement of PS I activity was performed by difference spectrophotometric determination of the in-vivo cytochrome-f photoxidation by lights 1 or 2, which opposes the dark reduction and PS II sentitized photoreduction (17).

RESULTS AND DISCUSSION
The most striking result is the preservation of the Emerson enhancement phenomenon for treated samples that exhibit a large degree of inhibition. Our first experiments were done on water stressed or mild heated leaves, in which presumably PS II was mainly affected, as listed bellow: 1. Similar activity of PS I for control and dehydrated leaves is shown by similar light saturation curves of the extent of cytochrome-f photooxidation by light-1. The variable fluorescence, in contrast, markedly decreases (17). 2. A mild heat treatment mainly affects PS II (18). We confirmed that the variable fluorescence indeed decreases by the treatment. Photosystem I was inhibited to a much lesser degree as judged by the relative efficiency of light-1 in cytochrome-f oxidation (not shown).
Following these initial observations, a clear cut case was provided by DCMU treatment, which does not affect PS I at the concentrations used. Complete inhibition by DCMU resulted in the same relative efficiency of far red light in cytochrome-f oxidation. The extent of the variable fluorescence was conserved, with fluorescence induction order of magnitude faster, as expected (not shown).
Fig. 1 shows examples of raw data for the oxygen evolution activity of a control and treated samples, in the absence and presence of saturating intensity of background light-1, to demonstrate the Emerson enhancement. Fig. 2

depicts plots of the Emerson enhancement ratio as a function of the extent of inhibition for dehydration, heat and DCMU treatments. In the case of dehydration the enhancement ratio E increased with the inhibition degree. In the cases of heat treatment and DCMU inhibition it decreased monotonically. The common feature, however, is its persistence in all cases, even with a very strong degree of inhibition (e.g. E = 1.16 for about 10% remaining activity).

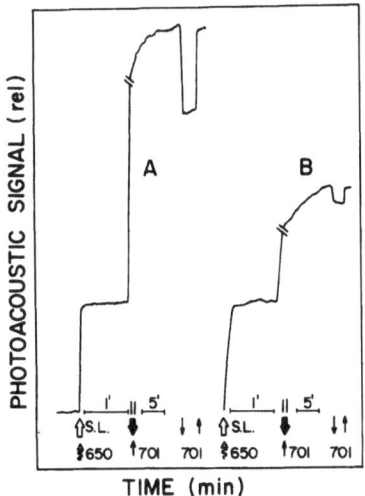

FIGURE 1. Raw data. In-phase photoacoustic signals from a control leaf (A) and heat treated leaf (B) - (2 min. at 45°C). Experimental details were: Modulated light - 650 nm, 19 Hz, 7.6 Wm^{-2} Background far-red light - 701 nm, 12.2 Wm^{-2} Background saturating light - white (<680 nm), 410 Wm^{-2} ⇑ S.L. (Background saturating light) on, ⬇ off; ⇕650 (modulated light) on; ↑ 701 (Far-red background light-1) on, ↓ off.

TIME (min)

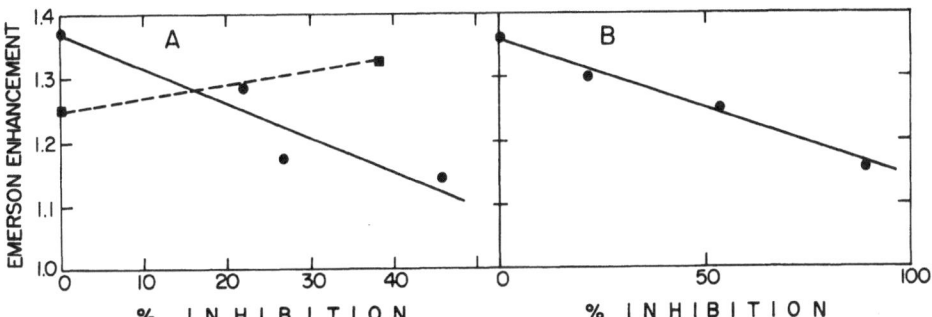

FIGURE 2. Emerson enhancement in state 1 as a function of the (%) degree of inhibition in the quantum yield for oxygen evolution.
A) —●— Leaf discs heated at 45° C in the dark for 0 (control), 1, 2 or 3 min. --- ■--- Leaf discs dehydrated in air for 0 (control) and 4 hours. Other experimental conditions as in Fig. 1.
B) DCMU-poisoned tobacco leaves. Leaf discs were infiltrated for 15 min with 0 (control) 10, 50 and 100 μM aqueous solutions of DCMU. Experimental conditions were similar to those in Fig. 1.

The persistence of the imbalance of active excitation in favor of PS II apparently contradicts the observation that PS II is strongly inhibited compared to PS I. The following argument shows that as PS II is gradually inhibited the Emerson enhancement is expected to disappear first, before the appearance of any noticeable effect on the oxygen evolution. Let's denote the modulated light distribution ratios for PS I and PS II by α and β, respectively, with $\beta > \alpha$. Let's denote by k the degree of inhibition in PS II. There are two ranges for k: For small degrees of inhibition such that $(1 - k)\ \beta > \alpha$, the enhancement ratio is given by the expression (13) $E = (1 - k)\ \beta/\alpha$, which decreases steadily as k increases. In this range of k the steady electron transport rate is limited by PS I, hence it is a constant, proportional to α - the same as the control rate. For larger degrees of inhibition such that $(1 - k)\ \beta < \alpha$ there is no enhancement ($E = 1$). The rate is now limited by PS II and decreases in proportion to $(1 - k)\ \beta$. The expectation from this theoretical argument is demonstrated in fig. 3A.

The above consideration applies only when the electron transfer between the photosystems is random (as symbolized in fig. 3A). If, however, PS I and PS II reaction centers are linked in pairs (as symbolyzed in fig. 3B), an inhibition in a specific PS II unit results in a total loss of electron transport of the linked chain altogether, including PS I. The percent inhibition would then be proportional to the fraction of inhibited chains, with the rest of the intact chains behaving normally. In this case the Emerson enhancement is not expected to change at all (fig. 3B).

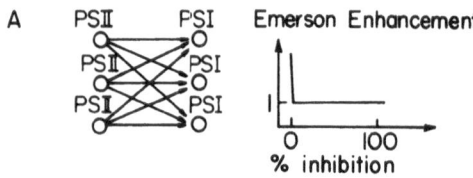

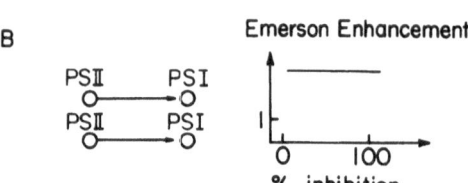

FIGURE 3. Models for the electron transport relation between the two photosystems (left) and the prediction from each model for the enhancement ratio as a function of the inhibition degree (right).

A) Random diffusion of electron equivalents from PS II to PS I.
B) Electron transfer between paired photosystems.

It was necessary to assure ourselves that the material as a whole is inhibited uniformly. For example, it could be thought that the remaining activity comes from macroscopic "pockets", which resisted inhibition due to local reasons. (i.e. large diffusion distance from the leaf disc edges through which DCMU penetrates, unequal dehydration, or combined thermal and photoinhibition due to the heterogeneous light intensity distribution and the resulting heterogenity in the effect). We have checked these possibilities: Small pieces cut from the leaf disc for the case of DCMU treatment showed uniform degree of inhibition for either central or peripheral regions; Heat treatment was done in complete darkness; The effect of dehydration is largely reversible (19), and during the long stay in the photoacoustic apparatus there is presumably uniform equilibration with the vapor pressure which will tend to homogenize the degree of inhibition.

The decrease in the enhancement ratio with the degree of inhibition is relatively insignificant compared to that expected from the model of Fig. 3A. Initially, therefore, one could choose the linked units model and excuse the drop in the enhancement by secondary and less relevant causes (e.g. incomplete adaptation to state 1 in the inhibited samples). It is possible however to consider a mixed model, in which part of the electron transport occurs in linked photosystems and part in isolated ones. We made calculations which predict indeed a gradual decrease of the enhancement, as in fig. 2B, depending on the ratio between linked and non-linked photosystems and possible different sensitivities to the inhibiting treatment. For reasons of space the details of the calculation cannot be brought here.* It follows that the results of fig. 3B are consistent with a mixed population of linked (80%) and non-linked (20%) photosystems, with linked PSII units somewhat more sensitive to DCMU then the non-linked ones. It is possible, and we did not take it into account, that the enhancement ratio can be influenced by changes in energy distribution due to regulatory effects. Still, this will modify the ratio of linked to non-linked units, but the neccessity to assume the existence of non-linked units will remain.

A previous analysis of the light saturation curves of the Emerson enhancement in state 1 pointed out that the photosystems are separated in terms of excitation transfer (13) at strict physiological conditions. One concludes that even if the photosystems are linked they nevertheless keep some distance away, thus preventing energy transfer between them. It is possible, however, that environmental perturbations may induce closer contact between the photosystems, which is perhaps the case e.g. at liquid nitrogen temperature, or with the addition of DCMU, conditions which were used to probe such energy transfer (11,12).

Admittedly, the above conclusion was initially considered with suspicion, in view of the heavy evidence brought to prove the contrary. However, a very recent report (20) demonstrates very significant whole chain electron transport activity in inverted granal vesicles, but only after addition of plastocyanine . Thus both PS I and PS II must coexist in these preparations, which were depleted from plastocyanine during their preparation. A close examination of the evidence in (6), which apparently supports the separation of the photosystems, reveals that in the granal partitions the population of PS I is rather almost comparable to that of PS II, by showing significant concentrations of CP-I and CP-Ia in the inverted particles (cf. there, table I), with a chlorophyll ratio for CP-II/(CP-I + CP-Ia) of only about 1.4 . If the ratio Chlorophyll/Reaction-center is larger in CP-II compared to CP-I the reaction-centers ratio PS II/PS I would approach closer to 1. Independently, there are reports (20,21) claiming that the amount of P_{700} in inverted thylakoids is close to that found for control chloroplasts - a result achieved by different assaying conditions compared to (6). Furthermore, an immunocytochemical study concludes that signficant amount of PS I occurs on granal partitions regions (22). Similar observation (23), confirming earlier report by biochemical analysis (24), demonstrates equal distribution of cyt b_6/f complex in stacked and unstacked regions, which is not consistent with the suggested role of plastoquinone (3) as a connecting agent between apressed and nonapressed regions.

The whole question should therefore be regarded at least as an open one. Indeed, from the point of view of one who accepts the concept of separated photosystems there is a need to provide an explanation to the Emerson enhancement effect totally different from that of the classical one, or else to introduce marked heterogenous behaviour towards inhibitors and stresses.

---

*Details of the calculation can be obtained directly from the authors, upon request

ACKNOWLEDGEMENT
This research was supported by the US-Israel Binational Agricultural Research and Development Fund (BARD). Grant no. I-388-81

REFERENCES
1. Anderson JM (1975) Biochim Biophys Acta 416: 191-235
2. Arntzen CJ (1978) In: Current Topics in Bioenergetics (Sanadi, R and Vernon, LP. eds.) Vol. 8, pp. 111-160, Acadmic Press, New York
3. Anderson JM (1981) FEBS Lett 124: 1-10.
4. Andersson B and Akerlund H-E (1978) Biochim Biophys Acta 503: 462-472
5. Andersson B Sundby C and Albertsson P-A (1980) Biochim Biophys Acta 599: 391-402
6. Andersson B and Anderson JM (1980) Biochim Biophys Acta 593: 427-440
7. Anderson JM and Melis A (1983) Proc Natl Acad Sci (US) 80: 745-749
8. Melis A (1985) Biochim Biophys Acta 808: 334-342.
9. Butler WL and Strasser RJ (1977) Proc Natl Acad Sci (US) 74: 3382-3385
10. Butler WL (1979) In: Chlorophyll Organization and Energy Transfer in Photosynthesis. Ciba Foundation Symp. 61 (new series) pp. 237-256, Excerpta Medica, Amsterdam
11. Butler WL and Kitajima M (1975) Biochim Biophys Acta 396: 72-85
12. Satoh K Strasser RJ and Butler WL (1976) Biochim Biophys Acta 440: 337-345
13. Canaani O and Malkin S (1984) Biochim Biophys Acta 766: 513-524.
14. Havaux M and Lannoye R (1985) J Agr Sci (Cambridge) 104: 501-504.
15. Bults G Horowitz BA Malkin S and Cahen D (1983). Biochim Biophys Acta 679: 452-465
16. Poulet P Cahen D and Malkin S (1983) Biochim Biophys Acta 724: 433-446
17. Havaux M Canaani O and Malkin S (1986) Submitted to Plant Physiology
18. Schreiber U and Armond PA (1978) Biochim Biophys Acta 502: 138-151
19. Havaux M Canaani O and Malkin S (1986) Submitted to Plant Physiology
20. Atta-Asafo-Adjey E and Dilley RA (1986) Archiv Biochem Biophys. In the Press
21. Tiemann R Renger G and Graeber P (1981) In: Photosynthesis III - Structure and Molecular Organization of the Photosynthetic Apparatus (Akoyunoglou, G, ed.) pp. 85-95, Balaban Int Sci Serv, Philadelphia
22. Vaughn KC Vierling E Duke SO and Alberte RS (1983) Plant Physiol 73: 203-207
23. Allred DR and Staehelin LA (l985) Plant Physiol 78: 199-202
24. Anderson JM (1982) FEBS Lett 138: 62-66

Photosynthesis Research 10: 297–302 (1986)
© *Martinus Nijhoff Publishers, Dordrecht*

OBSERVATION OF ENHANCEMENT AND STATE TRANSITIONS IN ISOLATED INTACT CHLOROPLASTS

P. HORTON AND P. LEE
Dept. of Biochemistry and Research Institute for Photosynthesis,
The University, Sheffield, S10 2TN, U.K.

ABSTRACT
Enhancement of photosynthesis by supplemental photosystem 1-enriched (707nm) light has been investigated in intact spinach chloroplasts by the simultaneous measurement of the rate of oxygen evolution, yield of chlorophyll fluorescence and quenching of 9-aminoacridine fluorescence. Chloroplasts reducing CO_2 showed a 75% increase in the rate of O_2 evolution after the addition of 707nm light, whereas if nitrite was used as substrate, an enhancement of only 20% was observed. Reduction of glycerate-3-phosphate was associated with a 40% enhancement by 707nm light. There appears to be a correlation between the degree of enhancement and the requirement for ATP in addition to reducing power. Prolonged illumination in 707nm light resulted in an elevation of enhancement whereas illumination with 650nm light caused a loss of enhancement, demonstrating the operation of state transitions in intact isolated chloroplasts.

INTRODUCTION
The observation of enhancement is compelling evidence for two independent photosystems co-operating in the light-induced assimilation of CO_2 (Myers, 1971). The two-photosystem 'Z' scheme provides a ready explanation of enhancement, since the two systems need to operate in series for linear electron flow to NADP (Myers, 1971). However, enhancement has not been consistently reported in studies on isolated chloroplasts. There are a number of reports showing enhancement of NADP reduction (Sun, 1972; Govindjee et al, 1964; Joliot et al, 1968), although the degree of enhancement is generally smaller than that shown by intact chloroplasts reducing CO_2 (Sinclair, 1972; Williams et al, 1976). In a comparative study, considerable enhancement was found during the observation of CO_2-dependent O_2 evolution but not for NADP reduction itself (McSwain and Arnon, 1968). There has been considerable debate about the reasons for these apparent discrepancies and they have generally been explained by differences in the measurement technique or by the use of incorrect ionic conditions. However, it has also been suggested that enhancement results from the necessity for cyclic electron flow, driven by PS1, to provide the 'extra' ATP to drive CO_2 reduction (McSwain and Arnon, 1968; McSwain and Arnon, 1972; Arnon, 1977). The obvious interaction between the stoichiometry of ATP and NADPH production and the operation of the reductive pentose phosphate pathway (for review, see Horton, 1985a) leads to the possibility that enhancement is also related to carbon assimilation.
The occurrence of enhancement is an indicator of the inability of the chloroplast to distribute absorbed radiation so that the required rates of excitation of PS2 and PS1 are brought about. However, both plants (Chow et al, 1981; Canaani et al, 1984) and green algae (Bonaventura and Myers, 1969) have been shown to have a mechanism that can, for example, increase the amount of energy transferred to PS1 when an excess arrives at PS2. This process, described as the "State 1 to State 2 transition", is brought about in higher plants and green algae by the phosphorylation of LHC-II (for reviews see Horton, 1983a). Phosphorylation is catalysed by a protein kinase and results in the decreased transfer of excitation absorbed by LHC-II to PS2 and an increase in the rate of excitation of PS1 relative to PS2. The protein kinase is activated when plastoquinone is reduced and since this will sense the relative rates of excitation of the photosystems, imbalance will be corrected and enhancement diminished. However, it is also clear

that the redox state of plastoquinone will respond to the NADPH level and hence the rate of CO_2 assimilation.   Therefore increased transfer to PS1 may result from an ATP deficit because NADPH will 'accumulate' and the protein kinase be activated (Horton, 1983a).   This increase in transfer may well make up the ATP deficit by enhancing cyclic electron transfer.   Recently, an additional finding has been that a low ΔpH, induced either by an uncoupler or by increased ATP consumption causes an increased level of LHC-II phosphorylation (Fernyhough et al, 1983; Fernyhough et al, 1984). This again suggests that the state transition may include a response to the demand for an appropriate ATP/NADPH ratio.   Indeed, if enhancement is in part related to ATP production, then a causative factor in its elimination during the state transition will also be the ATP status.

MATERIALS AND METHODS

Intact spinach chloroplasts were isolated as previously described (Walker, 1971) and assayed in a reaction medium containing sorbitol (0.33M), EDTA (2mM), $MgCl_2$ (5mM), $MnCl_2$ (1mM), Hepes (50mM), KH_2PO_4 (0.5mM), pyrophosphate (5.0mM) and catalase (220Uml^{-1}), adjusted to pH 7.6. For CO_2 reduction, the $NaHCO_3$ concentration was 10mM. $NaNO_2$ and glycerate-3-P were both 2.0mM. The simultaneous assay of chlorophyll fluorescence, 9-aminoacridine fluorescence and O_2 evolution used an apparatus described previously (Horton, 1983).   Two separate tungsten-halogen sources provided illumination at 707nm and 650nm (defined by Balzers B-40 interference filters) with intensities of 6 and 38 Wm^{-2} respectively.

RESULTS

Figure 1a shows the responses of intact chloroplasts supplied with HCO_3^- to successive illuminations with light of wavelength 650nm and 707+650nm. (With 707nm alone no measurable fluorescence rise or O_2 evolution was observed). Illumination with 650nm light, preferentially absorbed by PS2, caused O_2 evolution, fluorescence

A.   B.   C.

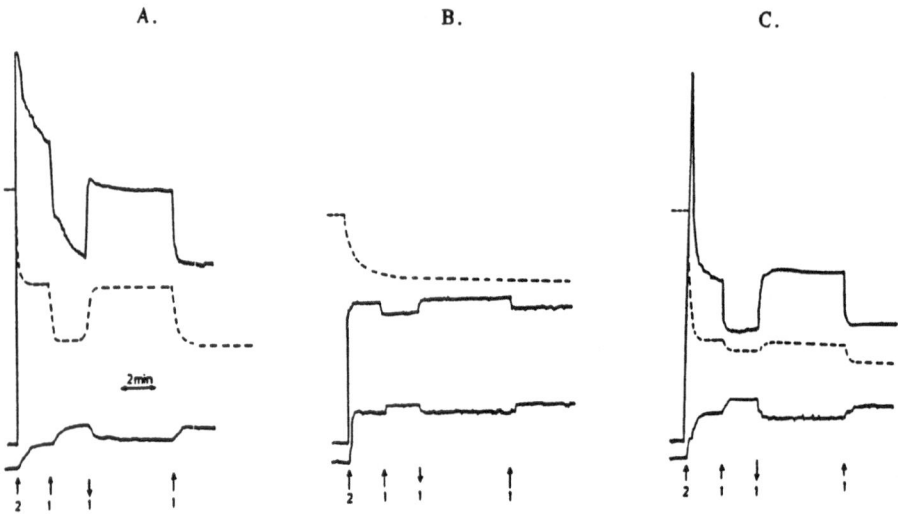

FIGURE 1. Measurement of the rate of O_2 evolution (lower curve), intensity of chlorophyll fluorescence (upper curve) and 9-aminoacridine fluorescence (dotted curve) in the presence of HCO_3^- (A), nitrite (B) and PGA (C). "2" and "1" refer to 650nm and 707nm light respectively. ↑ refers to light on and ↓ light off.

TABLE 1. Enhancement in chloroplasts reducing CO_2, PGA and nitrite. E_O was the rate of O_2 in 650+707nm together divided by the sum of the rates in 650nm and 707nm alone. E_f is the ratio of the $F_{650}-F_O/F_{650+707}-F_O$. Values are given ± SEM with 95% confidence limits obtained from between 7 and 12 replicates. Enhancement was assayed after illumination for 2-3 minutes in either 707+650nm or 650nm alone.

Substrate		Enhancement Value	
		E_O	E_f
CO_2		1.75 ± 0.08	1.51 ± 0.09
PGA		1.41 ± 0.08	1.44 ± 0.07
Nitrite		1.18 ± 0.04	1.11 ± 0.04

quenching and ΔpH formation. Addition of 707nm light, absorbed preferentially by PS1, induced a biphasic decline in fluorescence. The changes in chlorophyll fluorescence are mainly due to changes in the redox state of Q; the changes in ΔpH are small at these light levels and only just above the threshold above which changes in energy-dependent quenching occur (Horton, 1985a). Addition of 707nm light caused oxidation of Q, an increase in ΔpH and an increase in the rate of O_2 evolution. The changes in ΔpH were abrupt whilst those in the rate of O_2 evolution were slower than the fluorescence changes. This emphasises a complication in examining intact chloroplasts where increases in the rate of carbon assimilation are always associated with induction periods as levels of metabolites and enzyme activities adjust to optimise carbon flux (Walker, 1973). Measurement of the rates of O_2 evolution in the presence of 650nm and 650+707nm light indicated significant enhancement values (Table 1); The PS1-enriched 707nm light caused a stimulation of approx. 75% in the rate of O_2 evolution. Enhancement could also be seen in the changes in fluorescence level which indicated that approx. 50% more F_v is removed when only 650nm light is given compared to that seen upon supplementation with 707nm light.

Under exactly the same illumination conditions and using the same chloroplast preparation, the use of nitrite rather than CO_2 as the photosynthetic substrate produced markedly different results (Figure 1b). NO_2 reduction requires only reduced ferredoxin and not ATP and NADPH as does CO_2 fixation. Imposition or removal of 707nm light resulted in much smaller changes in the fluorescence level and the rate of O_2 evolution. Note also that the changes were abrupt and without the induction periods seen in Figure 1a. Calculated enhancement values are much smaller whether measured as the change of fluorescence or the rate of O_2 evolution (Table 1). It is of note, however, that enhancement *was* observed with an increase of approx. 20% in the rate of O_2 evolution seen in the presence of supplemental PS1-enriched light.

Chloroplasts supplied with glycerate-3-P rather than nitrite consume both ATP and NADPH. In this case higher enhancement values were restored and a behaviour pattern intermediate between that seen for CO_2 and NO_2 was observed. The fluorescence and rate of O_2 evolution now showed significant changes (Figure 1c) on imposition or removal of the 707nm light. In contrast to the behaviour shown when CO_2 was the substrate, changes in the rate and fluorescence were abrupt and devoid of the second slower phase associated with induction of CO_2 fixation (Walker, 1973).

Experiments to detect the transitions between State 1 (high enhancement) and State 2 (low enhancement) were undertaken. One problem is that the time constant for state transitions is of the same order as that for the induction of CO_2 fixation. A second problem is that even highly intact chloroplasts are prone to partial inactivation *in vitro* such that sustained rates of CO_2 fixation over long periods are not always obtainable. For these reasons, it was decided to examine chloroplasts in State 1 or 2 rather than trying to observe repetitive transitions between these states as is customary for *in vivo* studies. Thus, chloroplasts were illuminated for several minutes with either 707+650nm or 650nm alone and the effect of removing or adding the 707nm light was observed;

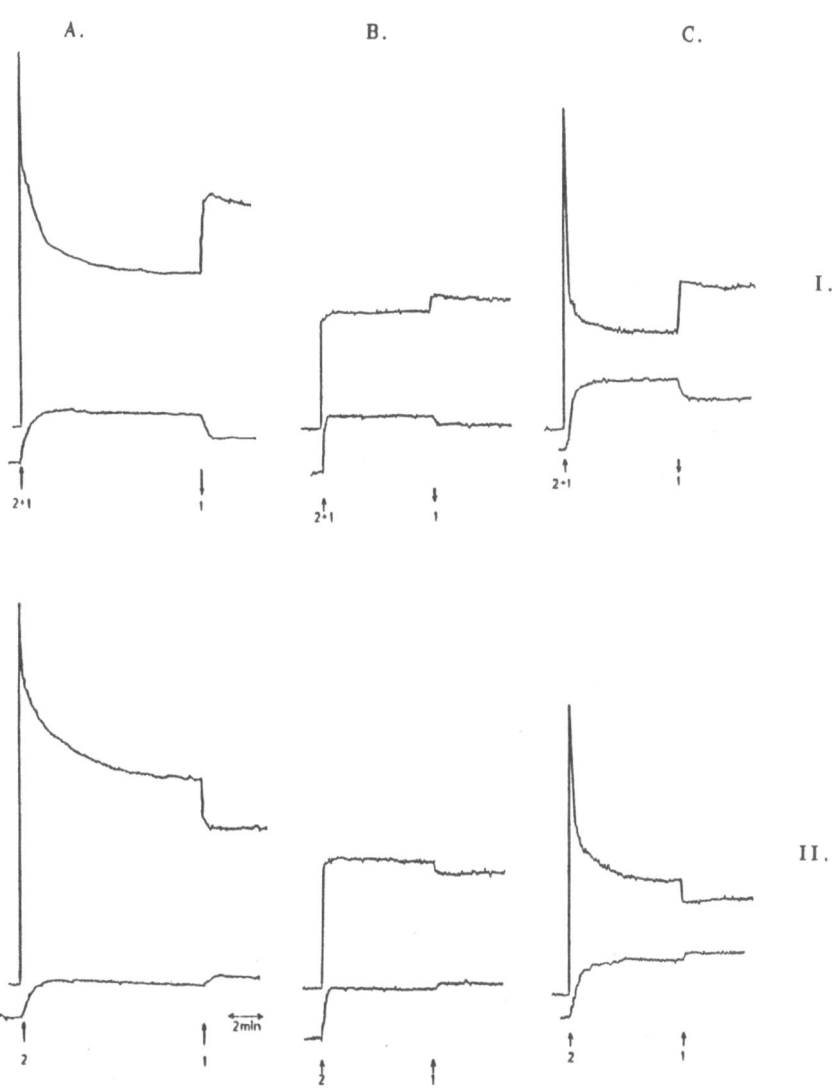

FIGURE 2. Measurement of chlorophyll fluorescence (lower curve) and rate of O_2 evolution (upper curve) in intact chloroplasts illuminated in the prescence HCO_3^- (A), NO_2^- (B) and PGA (C) to induce state 1 (I) with 707nm light and state 2 (II) with 650nm light.

after several minutes illumination with 707+650nm the chloroplasts are assumed to exist in State 1 and with 650nm alone, State 2.

With CO_2 as the acceptor, different enhancement values were obtained depending on whether the chloroplasts were given 650nm or 707+650nm light. In "State 1" an E_O

value of 1.5 was observed, decreasing to 1.3 in "State 2" (Figure 2A). When nitrite was present, the small enhancement effect was almost equal after illumination in 707+650nm light ($E_0 \approx 1.2$) or 650nm alone ($E_0 \approx 1.1$) whether assayed by fluorescence or O_2 evolution (Figure 2B). In contrast, in the presence of glycerate-3-P, large changes in enhancement were observed consistent with predicted transitions between States 1 and 2 (Figure 2C). After 6 minutes illumination with 650nm light, enhancement was very low, and imposition of 707nm light caused only a 10% increase in the rate of O_2 evolution, indicating State 2. In contrast, after illumination with 707+650nm light, the removal of the 707nm light resulted in a reduction of approximately 30% in the rate of O_2 evolution, indicative of State 1. The changes in fluorescence due to imposition or removal of the 707nm light are consistent with these transitions between State 1 and State 2.

After several minutes illumination in the presence of glycerate-3-P it was clear that differences in the distribution of excitation between PS2 and PS1 were found. Figure 3 shows the time course of the development of these differences. Enhancement ratios of approximately 1.3 were found if assays were performed after 1 minute's illumination in either 707+650nm or 650nm light. Continued illumination with PS2 light further lowered this ratio to approximately 1.1 within 5-6 minutes. Illumination with PS2 and PS1 light raised the enhancement ratio to over 1.6 in a slower process with a half-time of about 5 minutes. Thus dark-adapted chloroplasts start in a condition that is close to State 2 and the major part of the differences shown in Figure 2 results from a State 2 to 1 transition.

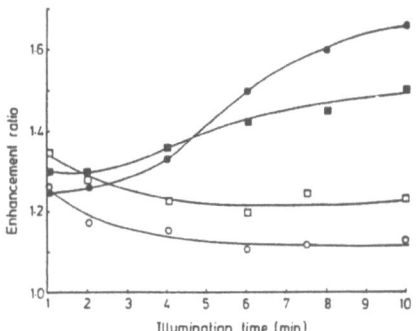

FIGURE 3. Enhancement ratios measured after varying periods of illumination with 650nm (O,□) or 707nm light (●,■) in the presence of PGA. Enhancement was measured for the rate of O_2 evolution (●,O) and fluorescence (■,□) as in Table 1.

DISCUSSION

The study of enhancement has, in earlier work using chloroplasts, lead to the suggestion that it relates to ATP synthesis and carbon assimilation in addition to linear electron flow to NADP (Sun, 1972; Govindjee *et al*, 1964; Joliot *et al*, 1968; Williams *et al*, 1976; McSwain and Arnon, 1968). Here, this suggestion has been clearly substantiated by a comparison between the extents of enhancement in intact chloroplasts reducing either CO_2, glycerate-3-P or nitrite. An explanation for the higher enhancement ratios found with CO_2 and glycerate-3-P compared to nitrite can be found in the requirement of ATP production for the reduction of CO_2 and glycerate-3-P. CO_2 fixation and glycerate-3-P reduction are dependent on a high ATP/ADP ratio and require additional ATP over that provided by linear electron flow to NADP (added glycerate-3-P will be converted to triose phosphate and then to ribulose-5-P, hence eliciting ATP consumption relative to NADPH consumption approaching the value of 1.5 required for CO_2 fixation). Conversely, nitrite reduction requires only the electrons from reduced ferredoxin and as such is independent of photophosphorylation. The enhancement values found here for CO_2 reduction (approx. 1.5-2.0) and for nitrite (approx. 1.1-1.3) are close to those found previously in separate studies of O_2 evolution in intact chloroplasts (Sinclair, 1972; Williams *et al*, 1976) and for NADP reduction by broken chloroplasts (Sun, 1972).

The loss of enhancement (State 1 to State 2) that occurs upon illumination with PS2-enriched light and the enhancement increase (State 2 to State 1) when PS1 is excessively excited has been shown here in intact chloroplasts for the first time. Previous work has used thylakoids supplemented with ATP and has relied solely on fluorescence analyses or LHC-II phosphorylation to implicate state changes (Horton and Black, 1981; Telfer et al, 1983). It is of note that chloroplasts prepared from spinach by usual procedures are predominantly in State 2; illumination with PS2 light only reduces enhancement by a small degree whereas PS2 and PS1 light cause a large increase in enhancement. The half-times of these changes (i.e. approximately 2 minutes for State 1-2 and 5 minutes for State 2-1) are equivalent to those reported for green algae and leaves (Chow et al, 1981; Canaani et al, 1984; Bonaventura and Myers, 1969). Since the protein kinase is reductively activated and the level of NADPH can often be high in intact chloroplasts in the dark (Takahama et al, 1981), it is not unexpected to find the chloroplasts partially in State 2 in darkness.

Recent work using maize chloroplasts has shown that the level of phosphorylation of LHC-II is increased when the ΔpH is lowered either by the addition of an ATP sink (Fernyhough et al, 1983) or by low uncoupler concentrations (Fernyhough et al, 1984). This effect was interpreted as supporting the hypothesis that state transitions are related, in part, to the demand for ATP via cyclic electron flow (Horton, 1985a,b). The observation that enhancement, too, is related to ATP demand and not just to linear electron flow is, of course, entirely consistent with this view; state transitions are changes in enhancement and if the latter is related to ATP demand then factors such as ΔpH and the ATP/ADP ratio should be important in the control of protein phosphorylation and excitation distribution. The importance of the relationship between thylakoids and their stromal and cellular environment should again be pointed out; the distribution of excitation and the turnover capacity of the two photosystems are inextricably linked to the metabolism of the plant cell (Horton, 1985a,b).

REFERENCES

1. Arnon DI (1977) In: Encyclopedia of Plant Physiol Photosynthesis 1 (Trebst, A. and Avron, M. eds.) vol.V pp 7-56, Springer Verlag, Berlin.
2. Bonaventura C and Myers J (1969) Biochim Biophys Acta 189: 366-383.
3. Canaani O, Barber J and Malkin S (1984) Proc Natl Acad Sc 81: 1614-1618.
4. Chow WS, Telfer A, Chapman DJ and Barber J (1981) Biochim Biophys Acta 638: 60-68.
5. Fernyhough P, Foyer CH and Horton P (1983) Biochim Biophys Acta 725: 155-161.
6. Fernyhough P, Foyer CH and Horton P (1984) FEBS Lett 176: 133-138.
7. Govindjee, Govindjee R and Hoch G (1964) Plant Physiol 39: 10-14.
8. Horton P and Black MT (1981) FEBS Lett 125: 193-196.
9. Horton P (1983a) FEBS Lett 152: 47-52.
10. Horton P (1983b) Proc Roy Soc Lond B 217: 405-415.
11. Horton P (1985a) In: Topics in Photosynthesis. Photosynthetic Mechanisms and the Environment (Barber, J. and Baker, N.R. eds.) pp135-187 Elsevier Biomedical Press.
12. Horton P (1985b) In: Regulation of Sources and Sinks in Crop Plants (Jeffcoat, B.,Hawkins, A.F. and Stead, A.D. eds.) pp19-33 Brit Pl Growth Reg Group, Bristol.
13. Joliot P, Joliot A and Kok B (1968) Biochim Biophys Acta 153: 635-652.
14. McSwain BD and Arnon DI (1968) Proc Natl Acad Sci USA 61: 989-996.
15. McSwain BD and Arnon DI (1972) Biochem Biophys Res Commun 49: 68-75.
16. Myers J (1971) Ann Rev Plant Physiol 22: 289-312.
17. Sinclair J (1972) Plant Physiol 50: 778-783.
18. Sun ASK (1972) Biochim Biophys Acta 256: 409-427.
19. Takahama U, Shimizu-Takahama M and Heber U (1981) Biochim Biophys Acta 637: 530-539.
20. Telfer A, Allen JF, Barber J and Bennett J (1983) Biochim Biophys Acta 722: 176-181.
21. Walker DA (1971) Methods in Enzymology 23: 211-220.
22. Walker DA (1973) New Phytologist 72: 209-235.
23. Williams WP, Salaman Z, Muallem A, Barber J and Mills J (1976) Biochim Biophys Acta 430: 300-311.

Photosynthesis Research 10: 303–308 (1986)
© Martinus Nijhoff Publishers, Dordrecht

ENERGY-DEPENDENT QUENCHING OF DARK-LEVEL CHLOROPHYLL FLUORES-
CENCE IN INTACT LEAVES.

W. BILGER AND U. SCHREIBER

Institut für Botanik und Pharmazeutische Biologie der Universität Würzburg, Mittlerer Dallenbergweg 64, D-8700 Würzburg, FRG

ABSTRACT
    A new type of modulation fluorometer was used in the study of energy-de-
pendent chlorophyll fluorescence quenching (q_E) in intact leaves. Under
conditions of strong energization of the thylakoid membrane (high light intensity, absence of CO_2) not only variable fluorescence, F_V, but also dark-
level fluorescence, F_0, was quenched, leading to definition of a quenching
coefficient, q_0. Information on q_0 was shown to be essential for correct
determination of photochemical (q_Q) and energy dependent quenching (q_E) by
the saturation pulse method. The relationship between q_E and q_0 was analysed over a range of light intensities at steady state conditions. q_E was
found to consist of two components, the second of which is linearly correlated with q_0. q_0 and the second component of q_E are interpreted to reflect
the state 1 – state 2 shift caused by LHC II phosphorylation.

INTRODUCTION
    Various quenching mechanisms determine chlorophyll fluorescence yield
in vivo (for recent reviews, see refs. 22, 21, 11, 25). Major quenching
components are Q-quenching (q_Q), due to photochemical energy conversion at
PS II reaction centers, and energy-dependent quenching (q_E), due to an increased rate of radiationless deexcitation upon "energization" of the
thylakoid membrane (22, 10, 8, 19). Differentiation between q_Q and q_E is
required for the interpretation of the complex fluorescence induction kinetics (Kautsky effect), which provides important information on the state
of the photosynthetic apparatus.
    Recently, there has been considerable progress in the attempt to differ-
entiate between q_Q and q_E (19, 8, 9, 23, 24, 14, 26, 27). By applying
short pulses of saturating light, it is possible to momentarily remove q_Q.
It has been assumed that any remaining quenching primarily reflects q_E,
(9, 23, 14, 26). For the calculation of q_Q and q_E, the basic assumption has
been made that only variable fluorescence (F_V) is affected by q_E. For
$q_Q = 1$, i.e., at the F_0-level, no energy-dependent quenching has been
assumed.
    In the present report it is shown that suppression of F_0 may occur under
conditions of strong energization. The relationship between energy-dependent quenching of F_V, (q_E), and of F_0, (q_0), is analysed. With this information on q_0 it is possible to derive more accurate values of q_Q and q_E.
Furthermore, the properties of q_0 suggest that it may be an indicator for
state 1 – state 2 shifts in vivo, due to LHC II phosphorylation (5, 16, 1,
15).

MATERIALS AND METHODS
    Experiments were carried out with detached leaves of Arbutus unedo and

of a number of other plants, grown at the Institute's Botanical Garden.
Leaf samples, the petiole of which was immersed in water, were exposed to
a stream of water-saturated air (30 l x h^{-1}) with or without CO_2.

Chlorophyll fluorescence was measured with a new type of modulation fluorometer (26, 27), which is now commercially available (PAM 101 Chlorophyll
Fluorometer; H. Walz, Effeltrich, Germany). This system monitors fluorescence yield with a weak modulated measuring beam (integrated intensity
1 mW x m^{-2}). Selective amplification of the modulated signal is not dis-
turbed by actinic illumination and application of saturation pulses. Light
intensities were measured with a LICOR 185 B Radiometer.

RESULTS AND DISCUSSION

Fig. 1 shows chlorophyll fluorescence induction curves of an <u>Arbutus</u> leaf
in the presence and absence of CO_2 monitored by a pulse-modulated measuring
beam (26, 27). In addition to continuous actinic illumination, there is
repetitive application (at 0.1 Hz) of short saturation pulses to differentiate between redox-dependent quenching (q_Q) and energy-dependent quenching
(q_E). Following a rapid build-up of q_E, only the sample in the presence of
CO_2 shows relaxation of q_E, as may be expected from the onset of Calvin
cycle activity (24, 26). When actinic illumination is switched off, there
is a substantial transient decrease of fluorescence below the original F_0-
level in the sample without CO_2 (see also inset of Fig. 1). In this sample,
steady state fluorescence also drops below the original F_0. The occurrence
of F_0-quenching has to be taken into account when the quenching coefficients, q_Q and q_E, are calculated. In Fig. 2 the definition of a new
quenching coefficient, q_0, and the consequences for q_Q and q_E determinations are depicted (for comparison see Fig. 3 in ref. 26). The assumption
is made that F_0-quenching reflects a decrease of absorbed energy directed
to PS II, formally equivalent to a decrease in excitation light intensity.
Hence, besides F_0, also variable fluorescence is correspondingly suppres-

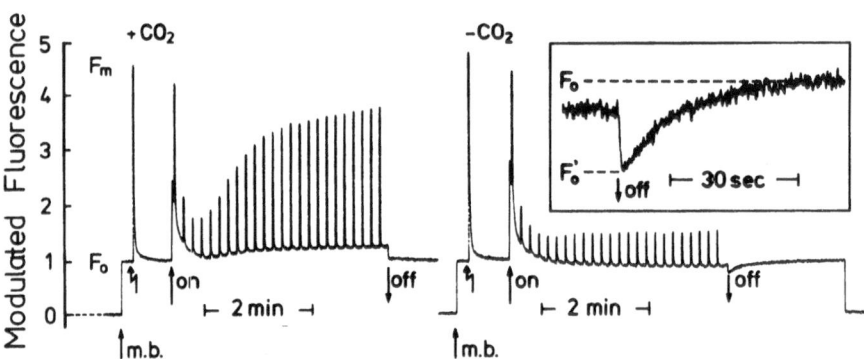

FIGURE 1.   Energy dependent quenching of F_v and F_0 in relation to CO_2
availability in intact leaves of <u>Arbutus</u> <u>unedo</u>.
Actinic intensity, 60 W/m². Saturation pulses, 1000 W/m² for 700 ms. The
arrows indicate switching on of the measuring beam (m.b.), application of
a saturation pulse ( ) for determination of F_m and switching on/off of
the actinic light ( ). In the (-CO_2) experiment, the light-off response
is also shown at 8-fold amplification in the inset.

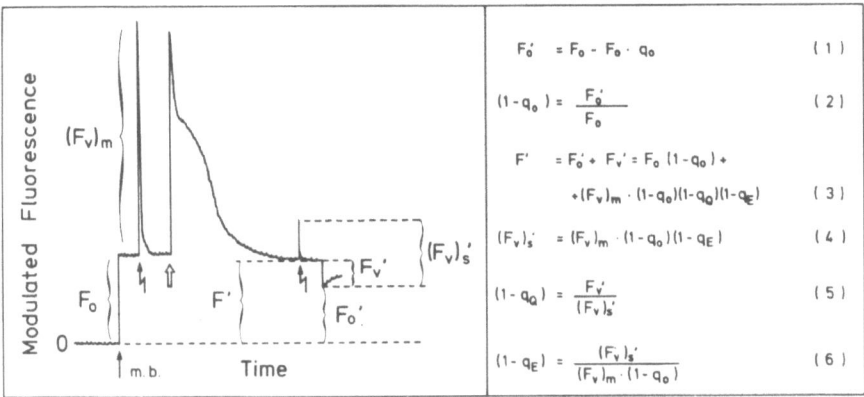

The equations shown in the figure:

$$F_o' = F_o - F_o \cdot q_o \qquad (1)$$

$$(1 - q_o) = \frac{F_o'}{F_o} \qquad (2)$$

$$F' = F_o' + F_v' = F_o \, (1 - q_o) + (F_v)_m \cdot (1 - q_o)(1 - q_o)(1 - q_E) \qquad (3)$$

$$(F_v)_s' = (F_v)_m \cdot (1 - q_o)(1 - q_E) \qquad (4)$$

$$(1 - q_Q) = \frac{F_v'}{(F_v)_s'} \qquad (5)$$

$$(1 - q_E) = \frac{(F_v)_s'}{(F_v)_m \cdot (1 - q_o)} \qquad (6)$$

FIGURE 2.   Definition of quenching coefficients under conditions of F_0-quenching.
Schematic reproduction of an experimental curve equivalent to that of Fig. 1 ($-CO_2$) the proportions of which were changed for illustration. See text for further explanation.

sed. As revealed by equations (5) and (6), the existence of F_0-quenching affects q_Q and q_E-determinations in two ways; First, the new F_0'-level has to be taken into account for evaluation F_v' and $(F_v)_s'$. Second, the maximal possible variable fluorescence (for $q_Q = 0$ and $q_E = 0$), which is not directly accessible, must be assumed to be $(F_v)_m' = (F_v)_m (1 - q_0)$.

Making use of equations (2), (5) and (6) (see Fig. 2), the different quenching coefficients may be calculated for the experiment of Fig. 1. Shortly before the actinic light was switched off, with CO_2 present, $q_0 = 0$, $q_Q = 0.91$ and $q_E = 0.20$; in absence of CO_2, $q_0 = 0.23$, $q_Q = 0.83$ and $q_E = 0.83$. In this example, ignorance of F_0-quenching would have led to an overestimation of q_E by about 15%; no meaningful value for q_Q could have been calculated with F_v going negative.

We have observed F_0-quenching in leaves of a large variety of different plants, under conditions which also produce strong energy dependent quenching of F_v. So far, there appears to be an upperlimit of $q_0 = 0.35$. In certain species, like Phaseolus vulgaris and Vicia faba, steady state fluorescence was well above F_0, and no F_0-quenching could be detected, even when strong energy quenching was induced.

The relationship between q_E, q_Q and q_0 in Arbutus unedo is shown in Fig. 3 A, B for steady state conditions at different light intensities in the absence of CO_2. It is apparent that substantial q_E accumulated before any q_0 was detected. The increase in q_E consisted of two distinct components, and the second component corresponded with the increase in q_0. The decrease in q_Q exhibited three successive transitions with increasing amplitude. In Fig. 3 B a plot of q_0 versus q_E is presented, which suggests a linear dependency between the second component of q_E and q_0. When this dependency was extrapolated to $q_0 = 0$, the separated first component of q_E amounted to 0.5.

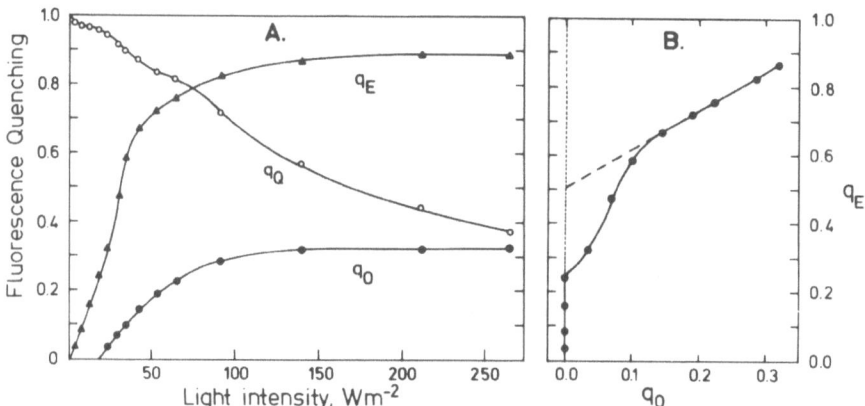

FIGURE 3. Light intensity dependency of the quenching coefficients in
Arbutus unedo, in absence of CO_2.
A. Plot of q_Q, q_E, and q_0 versus light intensity. B. Plot of q_E versus q_0.
At each light intensity, quenching was evaluated at steady-state illumi-
nation (about 5-min equilibration between successive light intensities):

We have considered the possibility that q_0 in our measurements might be
an artifact resulting from changes in effective light absorption or fluorescence reabsorption, accompanying light-induced changes in optical properties of the leaves. Indeed, there is close correlation between F_0-
quenching and the increase of apparent absorbance of green light ("light
scattering") (Bilger, Heber and Schreiber, manuscript in preparation). However, we do not consider q_0 to be simply a manifestation of altered leaf
absorption for the following reasons: First, fluorescence was measured from
the leaf surface and the 650 nm measuring light can be assumed to be
strongly absorbed in the surface layer of leaf chloroplasts. If leaf optical changes would mimic fluorescence quenching, this should be substantially enhanced if fluorescence is measured through the leaf. In reality, q_0
was found to be identical when fluorescence was measured from the surface
or through the leaf (not shown in the figures). Second, large increases in
"light scattering" were induced in the dark by changes in leaf water content without observation of any effect on F_0, measured in the same way as
in the experiments described above (not shown in the figures). Hence, the
observed changes of q_0 should be considered real.
    The tripartite model of Butler (12, 13, 28), as well as recent insights
into the topology of the thylakoid membrane (2 - 4) and into state 1 -
state 2 shifts by LHC II phosphorylation (5, 16, 15, 17, 20) may provide
satisfactory interpretations for the presented results. According to Butler
(12, 13), F_0 is determined by the extent of PS II excitation when all PS II
traps are open. A decrease of F_0 (at conditions of constant light absorbance) does suggest a decrease of energy transfer from the LHC II to PS II,
which was found to take place upon LHC II phosphorylation (15, 20). There
is lateral movement of this pigment complex away from high fluorescence
PS II in the partition region towards low fluorescence PS I in the margin

region of the thylakoids. Initiation of protein phosphorylation requires reduction of the plastoquinone pool (6, 18, 29). Indeed, the data of Fig. 3 confirm that development of q_Q is retarded until q_Q begins to decline. In this context, it should be considered, that a small amount of Q_A^- (corresponding to $qQ = 0.97$) is in equilibrium with reduced secondary acceptor Q_B^- which will accumulate even in presence of an oxidized PQ-pool (7, 30, 31). Further decrease in qq reflects the accumulation of centers in the state $Q_A^- \ Q_B^=$ when reoxidation is prevented due to reduction of the PQ-pool. The postulated state shift results in stimulated PS I activity and reduced PS II activity, releasing electron pressure on PQ and possibly causing the inflection in the light-intensity dependency of q_Q (Fig. 3 A) around 70 W/m^2. A correlative study of F_0-quenching and low temperature fluorescence spectroscopy will be required to determine whether or not this interpretation of q_Q by LHC II phosphorylation can be substantiated.

In conclusion, changes in q_Q and q_E, which have become readily accessible by the new modulation fluorometer, may provide a means of assessing energy dependent state shifts in intact leaves. Such state shifts play an important role in the regulation of photosynthesis.

ACKNOWLEDGEMENTS
We wish to thank Ulrich Heber for stimulating discussions. Support by the Deutsche Forschungsgemeinschaft is gratefully acknowledged.

REFERENCES

1. Allen JF and Bennett J (1983) FEBS Lett 123: 67-70
2. Andersson B and Anderson JM (1980) Biochim Biophys Acta 593: 427-440
3. Anderson JM (1981) FEBS Lett 124: 1-10
4. Barber J and Chow WS (1979) FEBS Lett 105: 5-10
5. Bennett J, Steinback KE and Arntzen CJ (1980) Proc Natl Acad Sci USA 77: 5253-5257
6. Bennett J (1983) Biochem J 212: 1-13
7. Bouges-Bocquet B (1984) Biochim Biophys Acta 314: 250-256
8. Bradbury M and Baker NR (1981) Biochim Biophys Acta 635: 542-551
9. Bradbury M and Baker NR (1984) Biochim Biophys Acta 765: 275-281
10. Briantais JM, Vernotte C, Picaud M and Krause GH (1980) Biochim Biophys Acta 591: 198-202
11. Briantais JM, Vernotte C, Krause GH and Weis E (1986) In Light Emission by Plants and Bacteria (Govindjee, Amesz J and Fork DC, eds.) Academic Press, in press, New York
12. Butler WL and Kitajima M (1975) Proc 3rd Congr Photosynth, Vol. I, pp. 13-24
13. Butler WL (1977) In Encyclopedia of Plant Physiology (Trebst A and Avron M eds.) Vol. V, pp. 149-167, Springer Verlag, Heidelberg
14. Dietz KJ, Schreiber U and Heber U (1985) Planta 166: 219-226
15. Hodges M and Barber J (1983) Plant Physiol 72: 1119-1122
16. Horton P and Black MT (1980) FEBS Lett 119: 141-144
17. Horton P and Black MT (1981) Biochim Biophys Acta 635: 53-62
18. Horton P, Allen JF, Black MT and Bennett J (1981) FEBS Lett 125: 193-196
19. Krause GH, Briantais JM and Vernotte C (1983) Biochim Biophys Acta 723: 169-175
20. Krause GH, Behrend U (1983) Biochim Biophys Acta 723: 176-181
21. Krause GH and Weis E (1984) Photosynth Res 5: 139-157
22. Lavorel J and Etienne AL (1977) In Primary Processes of Photosynthesis (Barber J, ed.) pp. 202-268 Elsevier, Amsterdam

23. Quick P and Horton P (1984) Proc R Soc Lond 220: 361-370
24. Quick P and Horton P (1984) Proc R Soc Lond 220: 371-382
25. Renger G and Schreiber U (1986) In Light Emission by Plants and Bacteria (Govindjee, Amesz J and Fork DC, eds.) Academic Press, in press, New York
26. Schreiber U, Bilger W and Schliwa U (1985) Photosynth Res, in press
27. Schreiber U (1986) Special Issue of Photosynth Res, in press
28. Strasser RJ and Butler WL (1977) Biochim Biophys Acta 462: 295-306
29. Telfer A, Bottin H, Barber J and Mathis P (1984) Biochim Biophys Acta 764: 324-330
30. Thielen APG and van Gorkom HJ (1981) FEBS Lett 129: 57-61
31. Velthuys BR and Amesz J (1974) Biochim Biophys Acta 333: 85-94

Note added after manuscript submission:

Very recently Malkin, Telfer and Barber (Biochim. Biophys. Acta 848, 48-. 57, 1986) reported on fluorescence measurements with a conventional modulation technique, suggesting changes in absorption cross-section of the two photosystems accompanying changes in F_0 and F_V during Light 1 - Light 2 induced state changes. While our findings do agree with the conclusions of these authors, it should be noted that there is a principal difference in the definition of F_0 in the approach of Malkin and co-workers as compared to our study. These authors determine "F_0" using a measuring light of 7 W/m² which is 7000 times stronger than what was applied here, and which is sufficiently intensive to induce appreciable variable fluorescence. Hence, the observed changes in "F_0" could well represent, at least in part, changes in F_V.Even if,as suggested by Malkin (Isr. J. Chem. 21, 306-315, 1981), the F_V which can not be quenched by Light 1 originated from "unconnected PS II centers", it remains uncertain to what extent the relative size of this population of centers may vary during state changes. Such variation would cause changes in F_V mimicking changes in "F_0". Furthermore, besides of non-photochemical quenching induced by a state 1 - state 2 transition, also changes due to membrane energization (q_E) should be considered. Therefore, we believe that for definite demonstration of true changes in absorption cross-section the actual F_0 has to be measured.

Photosynthesis Research 10: 309–318 (1986)
© *Martinus Nijhoff Publishers, Dordrecht*

MECHANISMS OF CHLOROPHYLL FLUORESCENCE REVISITED: PROMPT OR DELAYED EMISSION FROM PHOTOSYSTEM II WITH CLOSED REACTION CENTERS ?

G.H. SCHATZ AND A.R. HOLZWARTH

Max-Planck-Institut für Strahlenchemie, Stiftstr. 34 - 36, D-4330 Mülheim a.d. Ruhr, West Germany

Keywords: Charge recombination, Charge separation, Exciton decay, Fluorescence kinetics, Kinetic model, Photosystem II.

Abstract. This paper proposes a model which correlates the exciton decay kinetics observed in picosecond fluorescence studies with the primary processes of charge separation in the reaction center of photosystem II. We conclude that the experimental results from green algae and chloroplasts from higher plants are inconsistent with the concept that delayed luminescence after charge recombination should account for the long-lived (approx. 2 ns) fluorescence decay component of closed photosystem II centers. Instead, we show that the experimental data are in agreement with a model in which the long-lived fluorescence is also prompt fluorescence. The model suggests furthermore that the rate constant of primary charge separation is regulated by the oxidation state of the quinone acceptor Q_A.

1. Introduction

A hypothesis promoted by Klimov et al. (see (1,2) for summarizing reviews) postulates a charge recombination (P680$^+$ Pheo$^-$ Q_A^- ⟶ P680 Pheo Q_A^-) between the oxidized primary donor, P680$^+$, and the reduced primary acceptor, Pheo$^-$, in reaction centers (RC) of photosystem II (PS II) with a reduced first quinone acceptor, Q_A^-. It was suggested that this charge recombination process is associated with light emission from reexcited chlorophylls (Chl). This emission has been termed delayed (in contrast to prompt) fluorescence or recombination luminescence and was considered as origin of the 3 to 5 fold increase in Chl fluorescence yield which is associated with the reduction of Q_A (3).

The Klimov hypothesis has also been adopted to explain the slow (approx. 2 ns) component of time-resolved fluorescence studies (4,5). However, several recent experimental findings cannot be reconciled with this concept. In this paper, after summarizing the conflicting aspects, we present a kinetic model which permits a quantitative test of the compatibility of the Klimov hypothesis with experimental results of fluorescence decay kinetics.

2. Results of picosecond fluorescence studies.

Picosecond Chl fluorescence experiments with green algae and higher plant chloroplasts reported during the past years have commonly been analyzed in terms of a sum of 3 exponentials: a short-lived component ($\tau(1/e) \approx 50$-150 ps), a middle component ($\tau(1/e) \approx 450$-750 ps) and a long-lived one ($\tau(1/e) = 1400$-2300 ps) (see reviews (5-7)). From many studies screening the influences of various parameters (e.g. redox potential, light intensity, inhibitor concentration) on these decay components, one result was unambiguous: the increase in the yield of the long-lived component upon reduction of Q_A. The yield of the short-lived component was found to decrease upon reduction of Q_A in green algae (8). With spinach chloroplasts such behaviour of the rapid decay was not always (9,10) so evident.

Recently, the spectral resolution of fluorescence decay components allowed us to separate PS I- from PS II-emissions at room temperature in Chlorella (11) and Scenedesmus (12). Furthermore, the kinetic analysis for the resolution of more than 3 exponentials has been improved. The results can be summarized as follows:

i.   Emission from PS I is red-shifted with respect to that from PS II. It decays with $\tau(1/e) \approx 80$ ps. Both, amplitude and lifetime are not affected by the reduction of Q_A.

ii. All other fluorescence decay components, attributed to PS II, show identical spectral shapes, with maxima at 685 nm, strongly suggesting an origin from the same chromophore(s).

iii. Contributions to the corresponding time resolved excitation spectra by chlorophyll b (11) show that the light harvesting complex (LHC) has no separate emission at variance with earlier interpretations (4,8,13).

iv. With open centers (F_o) PS II emission is characterized by τ_1 (1/e)=200-300 ps and τ_2(1/e)=500-650 ps, with closed centers (F_{max}) by τ_3(1/e)=1150-1350 ps and τ_4(1/e)= 2200-2400 ps.

v. τ_1 and τ_4 can be attributed to open and closed PS II$_\alpha$-centers, respectively, and τ_2 and τ_3 to the corresponding states of PS II$_\beta$-centers (7,12), as proposed in (14).

vi. A quantitative comparison (12) shows that the sums of the __amplitudes__ (not yields) of the decay components in the F_o state equal those in the F_{max} state, i.e. $(g_1+g_4)=(g_2+g_3)$.

Thus, in any one of the extreme states (completely open or closed) the decay of each type of center (α or β) can be characterized by a single exponential! In intermediate states all four of these PS II contributions should be present together (not yet resolved experimentally). Upon closing a reaction center, the increase in amplitude of the long-lived component is counterbalanced by the concomitant decrease in amplitude of the short-lived decay component in each type of center. It is proposed that upon reduction of Q_A the decay components with τ_3 and τ_4 replace those with τ_1 and τ_2. Such a pairwise complementary relationship between the amplitudes of the short- and the long-lived components in open and closed centers, respectively, was already found in (8). However, owing to the limitations by the conventional three-exponential analysis at a single emission wavelength, the β-center and the PS I contributions had not yet been resolved at that time. It was then recognized by Butler and coworkers (14) that the middle component of the data in (8) might well have reflected this second type of PS II centers. Their heterogeneous bipartite model (14,15) predicted a discrimination between the contributions by α- and β-centers.

We want to emphasize that the above mentioned pairwise complementary relationship between the amplitudes of the decay components represents the decisive experimental result which conflicts with the hypothesis of recombination luminescence by the following reasoning: The fast phases observed with open reaction centers are interpreted (8,11) (15,16) to reflect the overall energy migration / charge separation process. The mechanism of energy migration in the antenna is unlikely to be affected when the RC becomes closed (15). Therefore, the substitution of the fast by

the slow phase upon closing PS II (11) is assumed to reflect a diminished
efficiency of primary charge separation in the closed RC. The alternative
interpretation of the slow phase as a charge recombination process (1,2,4)
necessarily requires a preceding efficient charge-separation. The latter
should be indicated by the presence of both, the fast plus the slow phase
under F_{max}-conditions. This is in contrast to the above mentioned
experiments (11,12), which show the absence of fast PS II-phases under such
conditions (within an accuracy of 5%). This qualitative argument will now
be corroborated by a kinetic model which takes into account both processes,
charge separation and charge recombination.

3. Kinetic model: energy partition between chlorophyll-excitons and the radical pair [P680$^+$ Pheo$^-$]

For each of the PS II units (α and β) we propose a model which assumes: (a)
Chlorophylls of the core and of the LHC form a tightly coupled antenna
domain. (b) Trapping of an exciton by P680 from the antenna chlorophylls is
reversible (shallow trap) and reaches equilibrium within a time short compared to the overall exciton decay time. Thus, the probability for P680
being excited is statistically small (approx. 1/200 in PS II$_\alpha$ units), because inversely proportional to the antenna size. Under such conditions,
the apparent rate of primary charge separation is trap limited, i.e. approximately proportional to the probability of P680 being excited (17).
This model is represented by the following scheme :

$$A^* \underset{k_{-1}}{\overset{k_1}{\rightleftarrows}} B \overset{k_2}{\longrightarrow} C$$

$$k_a = k_d + k_{rad} \downarrow$$

$$A$$

(scheme A)

A denotes a strongly coupled domain of chlorophylls (antenna chlorophylls
plus P680), B is the singlet radical pair [P680$^+$ Pheo$^-$], and C represents
the product(s) of primary charge stabilization and/or loss processes
(including triplet formation and decay to the ground state). Rate constants
k represent: radiative decay (k_{rad}), radiationless decay (k_d), primary
charge separation (k_1), charge recombination to the excited state (k_{-1}) and
processes of charge stabilization, triplet formation and recombination to
the ground state (k_2). The latter processes are assumed to be irreversible
on the nanosecond time scale. In this model different states (open or
closed) of the RC can be represented by different sets of rate constants.
With open RCs a high yield of charge stabilization requires $k_2 \gg k_1 \gg k_{-1}$.
With closed RCs, electron transfer to Q_A^- (B to C) is blocked and the value
of k_2 reduced as compared to open RCs. The model should meet the following
criteria in the extreme states, F_o and F_{max}, of PS II$_\alpha$ units: (a) (almost)
monoexponential decay kinetics; (b) lifetimes of about 200 ps and about 2
ns, respectively; and (c) about identical amplitudes of the corresponding
decay components.

The differential equations for this model are given in the appendix. The
time dependence of A^*, the fluorescence emitter, results in:

$$[A^*](t) = [A^*]_o \{g_a \exp(- t/\tau_a) + g_b \exp(- t/\tau_b)\} \tag{1}$$

where the amplitude factors (g_i) and lifetimes (τ_i) are functions of the
rate constants k (see appendix). Eq. (1) generally describes a biexponential kinetics for a given set of rate constants, both for completely

open and completely closed RCs, obviously is in contrast to experimental findings (see above, items iv. and v.). Nevertheless, this model possibly describes the real situation if one of the two exponential terms becomes so fast or so small in amplitude that it escapes experimental detection. This possibility is examined by the following numerical solutions.

4. Numerical results

Upper and lower limits for values of the rate constants are chosen as follows: For $k_A = k_d + k_{rad}$ we suggest the range 0.2 ns^{-1} $\leq k_A \leq 0.8$ ns^{-1}, corresponding to the lifetimes of Chl* in organic solvents (5-6 ns (18)) and in isolated Chl-protein complexes (19), respectively. In our model, the apparent rate constant for charge separation is limited by the trapping time of excitons correponding to the experimentally observed fast (≈ 200 ps) PS II fluorescence decay in open RCs. Therefore, we use the value $k_1 = 5$ ns^{-1}. In a photosystem with an antenna of ≈ 200 Chl/RC such a value for exciton trapping would correspond to a time of ≥ 1 ps for the step P680 Pheo$\rightarrow$P680$^+$Pheo$^-$. This is in the same order of magnitude as the values of bacterial RCs. Plausible ranges for values of k_2 are considered to be 5 ns^{-1} $\leq k_2 \leq 20$ ns^{-1} for open and 0 ns^{-1} $\leq k_2 \leq 0.5$ ns^{-1} for closed RCs, respectively.

Table I shows the data calculated for four qualitatively different cases of the model:

In CASES A and B (F_0 and F_{max}), the value of $k_{-1} = 0.35$ ns^{-1} is chosen corresponding to the suggested time constant of approx. 3 ns for charge recombination in closed RCs (1,20). For open centers (CASE A) an almost monoexponential fast fluorescence decay can be predicted assuming high charge stabilization rates. In this case the mentioned criteria are well met (+ signs in Table I).

CASE B assumes that closing of the RC will affect only the rate constant of charge stabilization. Formation of the radical pair [P680$^+$ Pheo Q$_A^-$] is possible with high yields provided that [P680 PheoQ$^-$] be still photochemically active as suggested by Klimov (see e.g. (1)). It is evident, that CASE B fails to fit the experimental data (- in Table I), since the contributions by the fast phases remain dominant. Hence, the numerical results corroborate the qualitative argument against the recombination concept. Therefore, we propose that reduction of Q_A must also affect the rate of charge separation, k_1.

As shown in CASE C, a tenfold decrease of k_1 results in a much larger contribution by a 2.2 ns component and in an equivalent decrease of the fast phase. An additional increase of the charge recombination rate by a factor of 10 to 20 predicts fluorescence kinetics close to that observed. In this case the yield of radical pair formation is considerably lowered. This is not unreasonable, since the electric field due to the negative charge on Q_A may affect both rate constants, k_1 and k_{-1}. It has been reported that the Chl fluorescence yield can be reversibly increased from F_0 to F_{max} when an artificial membrane potential is applied by external electric field pulses (21). The fluorescence yield calculated for CASE C is about tenfold increased over that for open RCs (CASE A), as expected for PS II$_\alpha$ units. For PS II$_\beta$ centers, this factor should be smaller as judged from the lifetime ratio $\tau_3^\beta/\tau_2 \approx 2$. Altogether, the F_{max}/F_0-ratios result in an average of 4-5 known for photosynthetic membranes. The total fluorescence yield, Φ_F, in photosynthetic membranes is approximately 3% for open (22) and about 12% for closed centers. Similar values are predicted by our model

313

Table I

Calculation of exciton decay kinetics, fluorescence yield and radical pair formation as predicted by the model from scheme A for different sets of rate constants k. Amplitude factors (g) and lifetimes (τ) have been calculated according to eq.(1), the total fluorescence yield (Φ_F) according to eq. (A.14) using k_{rad}^{-1} = 15 ns (36). RP_{max}, the maximal relative concentration of the radical pair [P680$^+$Pheo$^-$] and t_{max}, the time when RP_{max} is reached, have been calculated according to eqs. (A.15) and (A.16). The amplitude factors g are always given in relation to [A$_0^*$]. The + and – signs designate agreement or conflict of calculated amplitude (g) and lifetime (τ) data with experimental values.

case	RC state	k_A ns^{-1}	k_1 ns^{-1}	k_{-1} ns^{-1}	k_2 ns^{-1}	g_a %	τ_a ps	g_b %	τ_b ps	Φ_F %	RP_{max} %	t_{max} ps	criteria for g	criteria for τ	remarks
A	open	0.2	5	0.35	5	47	151	53	253	1.37	35.3	194		+	
					20	0.7	49	99.3	197	1.3	15.4	91	+	+	
		0.4	5	0.35	5	51	149	49	247	1.31	34.6	190		+	
					20	0.8	49	99.2	189	1.26	15.3	89	+	+	
B	closed	0.2	5	0.35	0	93.9	180	6.1	79106	33.4	89	1101	–	–	charge separation not affected, high yield of RP,
					0.5	92.7	180	7.3	2086	2.12	71.3	482	–	–	
		0.4	5	0.35	0	94.3	175	5.7	40897	16.7	85	957	–	–	(Klimov model)
					0.5	93.2	174	6.8	2027	2.0	69	466	–	–	
C	closed	0.4	0.5	0.35	0.5	53	773	47	2193	9.6	21.9	1245	–		charge separation affected, low yield of RP,
			0.5	3.5	0.5	12	223	88	2426	14.4	8.8	586		+	
			0.5	7	0.5	6.5	125	93.5	2460	15.4	5.3	393	+	+	
D	open	0.4	50	0.35	5	99.2	20	0.8	201	0.14	75.5	51	–		k_2-limit
	closed	0.4	50	0.35	0.5	99.3	20	0.7	2003	0.22	94.1	92	–		(Breton model)
E	open	0.8	1.0	0.35	20	0.1	49	99.9	561	3.74	3.9	131	+	+	PS II β-centers
	closed	0.8	0.1	3.5	0.5	3.3	243	96.7	1264	8.2	1.6	497	+	+	

(see Table I, CASES A and B).

CASE D is adapted to a hypothesis by Breton (23). It assumes that the fluorescence kinetics at F_0 reflect the charge stabilization rather than the charge separation process. In this case, the decay of A^* is limited by k_2 and all preceding steps have to be much faster. Assuming that the calculated fast decay phase ($\approx$ 20 ps) is too rapid to be detected experimentally, this model would indeed fit the observed lifetimes. However, several severe problems remain: the predicted fluorescence yield would be far too low and almost invariable. Furthermore, the apparent time for charge separation ($\approx$ 20 ps) is surprisingly short: it is shorter than the expected first passage time which is required for an exciton to pass through a 200 Chl/RC antenna towards the trap, as extrapolated from recent studies on the PS I antenna (16). In addition, the time corresponding to the elementary step of charge separation ought to be shorter than approximately 0.1 ps.

We have tried to find also rate parameters describing the fluorescence decay kinetics attributed to β-centers (CASE E). They are quite well mimicked by introducing two changes with respect to the α-centers: an increase in k_A for spillover of excitation energy from β-centers towards PS I (24), and a decrease in k_1 for a lower photochemical efficiency.

5. Discussion

The combination of CASES A and C, which describe best the observed fluorescence kinetics, can be characterized as follows: excitons are equilibrated rapidly (too fast to be resolved by current single photon timing instruments) between antenna- and RC-Chl. In open RCs equilibrated excitons are mainly used for photochemistry and decay with k_1, the limiting apparent rate constant of charge separation. The primary radical pair is stabilized at a rate probably faster than that for apparent charge separation. Thus, most of the excitation energy is very rapidly ($k_2 \gtrsim k_1$) converted into free energy of stabilized products. Consequently, the relative radical pair concentration will not exceed 35 % at any time (see RP_{max} in Table I). In closed RCs the yield of charge separation is most likely reduced by reduction of k_1 and increase of k_{-1}. Due to the lower probability of charge separation, the disappearance of the antenna excitons is no longer controlled by the k_1 process but rather by intrinsic deactivation processes for antenna Chl^* (k_A). We note that already a tenfold decrease of k_1 suffices to increase the fluorescence yield by a factor of almost 5 (e.g. from 2.0 % to 9.6 %, see Table I). Our calculations show furthermore, that in closed RCs relative concentrations of [P680$^+$ Pheo$^-$] in the range of 5-10% (e.g. 5.3% or 8.8%) may be obtained. Its maximal value is reached at 400-600 ps after excitation (e.g. 393 ps or 586 ps). The decay of [P680$^+$ Pheo$^-$] is controlled by $\tau_b = 1/\gamma_b$ (see eq A.4). Experimentally observed processes as e.g. spin polarized triplet formation (25,26) and photoinduced accumulation of Pheo$^-$, both proceeding via [P680$^+$Pheo$^-$Q$_A^-$], can be accommodated by this model. Pheo$^-$ formation is reported with yields between 0.002 and 0.01 (27,28) corresponding to a competitive P680$^+$ reduction for about 1/10 to 1/50 of the maximal radical pair concentration, RP_{max}, in CASE B. Based on the radical pair lifetime of approx. 2 ns this suggests a P680$^+$ reduction within 20 ns $\leq \tau(1/e) \leq$ 100 ns under conditions of low redox potential (E $\leq$ -430 mV).

In our model, as expressed in scheme A, the relative [P680$^+$ Pheo$^-$] concentration depends on the antenna size. This is a consequence of the dependence of the excitation probability for P680 on the number of Chl taking part in exciton equilibration. The apparent rate constant of charge separation, k_1, shows the same dependence. Reduction of the antenna size would lead to a reciprocal increase in k_1 and, as is evident from the data in Table I, to higher temporary [P680$^+$ Pheo$^-$] concentrations. In the extreme case of isolated RCs, the probability of radical pair formation approaches values close to 1, even when the RCs are closed. Therefore, high yields of radical pair formation and significant recombination may be expected only in isolated RCs and in particles with a very small antenna. For photosystems in vivo, however, these processes appear to be insignificant. This indicates that there is no correlation of the slow fluorescence decay component from algae and chloroplasts from higher plants with a process requiring radical pair formation, be it reexcitation of Chl or of Pheo (29). It is suggested that a large antenna may serve two principal purposes: optimized light capturing properties by high absorption cross section, and protection of the RC against loss processes by enabling efficient photochemistry at low levels of radical pair formation.

We also emphasize the fact that the rise of [P680$^+$ Pheo$^-$] (B) is not associated with a rise of Chl (A) in any of the model cases (only positive values of amplitudes g_i). Recently, a 200 ps risetime of a 1.2 ns fluorescence decay has been reported and taken as evidence for recombination luminescence in Chlorella (30). This result was obtained after subtraction of the fluorescence kinetics with open RCs from those with (partially) closed RCs, while the individual kinetics did not show any rise term. We doubt that such subtraction of exciton kinetics arising from two different RC-states can provide information about processes characteristic for one of these states (F_{max}). In our model the socalled variable fluorescence yield, $F_{var} = F_{max} - F_0$, does not originate from a special fluorescing species, which would add to the emission from open RCs as they become closed. Rather, the increased fluorescence yield is due to the increase of the average lifetime of the same emitters! Therefore, F_{var} is an operational term from CW measurements and does not represent an identifiable part of picosecond time resolved fluorescence.

The modified bipartite model (14,15) and our model have in common that both explain long-lived fluorescence components without using the Klimov mechanism, since both include excitation transfer from P680 back to the strongly coupled antenna. The differences are the following: the bipartite model refers to exciton partition between antenna- and RC-Chl; it does not include the reversibility of primary photochemistry. It results in a biexponential decay kinetic for each type of center in any one of the states, F_0 and F_{max} (15), and it fails to simulate the decrease in amplitude of the fast PS II phase upon closing the RC. If extended to account also for charge recombination, more complex, i.e. threeexponential, kinetics would be expected. Our model focuses on the primary photochemical steps of energy partition between P680 and the radical pair. Antenna properties are still important; they are included not as explicit parameters but as an implicit probability function for the excitation of P680. With parameters accounting for photophysical as well as photochemical properties of PS II RCs, this model describes apparently monoexponential exciton decay kinetics and thus meets the criteria given above.
One of the possibilities to test the different cases of the model is to determine the activation energies of the rate limiting step (k_{-1}). It is expected to be positive for any contribution by recombination luminescence,

while being close to zero for prompt fluorescence. Chl fluorescence yields have been observed as a function of temperature at F_0 (2,31,32), F_{max} (32) and F_0', after light induced quenching of F_{max} upon photoaccumulation of Pheo$^-$ (2,31). Results were contradictory: In (31) and (2) an activation energy of 0.04-0.08 eV was found, while in (32) and (33) it was concluded that there is no evidence for such an activation energy. More information should be obtained from measurements of nanosecond and picosecond absorption changes, which could directly monitor the radical pair. The only such report (20) showed a 4 ns decay with a difference spectrum attributed to [P680$^+$ Pheo$^-$] in closed RCs. It was observed in a PS II particle (TSF 2a) of small antenna size. But the decay phase was associated with a low yield of radical pairs. It was assumed to reflect the same process as a slow component of fluorescence decay measurements presented in the same paper, but obtained with a less purified PS II preparation (TSF 2, with a typically threefold larger Chl/RC ratio than TSF 2a (34)). In view of these facts a straightforward correlation of the two kinetics should deserve more careful experiments.

Appendix

The differential equations for the kinetic system in scheme A are:

$$d[A^*]/dt = -(k_A+k_1)[A^*] + k_{-1}[B] \tag{A.1}$$

$$d[B]/dt = k_1[A^*]-(k_{-1}+k_2)[B] \tag{A.2}$$

With the initial boundary values for t=0

$[A^*]_0 = [A_0^*]$ and $[B]_0 = 0$ they have the following solutions (35)

$$[A^*] = [A_0^*]\{g_a \exp(-\gamma_a t) + g_b \exp(-\gamma_b t)\} \tag{A.3}$$

$$[B] = [A_0^*]\{k_1/(\gamma_a-\gamma_b)\}\{-\exp(-\gamma_a t) + \exp(-\gamma_b t)\} \tag{A.4}$$

where
$$g_a = (X-\gamma_b)/(\gamma_a-\gamma_b) \tag{A.5}$$

$$g_b = (\gamma_a-X)/(\gamma_a-\gamma_b) \tag{A.6}$$

$$\gamma_a = 1/2 \{(X+Y) + [(X-Y)^2 + 4 k_1 k_{-1}]^{1/2}\} \tag{A.7}$$

$$\gamma_b = 1/2 \{(X+Y) - [(X-Y)^2 + 4 k_1 k_{-1}]^{1/2}\} \tag{A.8}$$

$$\tau_a = 1/\gamma_a \tag{A.9}$$

$$\tau_b = 1/\gamma_b \tag{A.10}$$

$$X = k_A + k_1 \tag{A.11}$$

$$Y = k_{-1} + k_2 \tag{A.12}$$

The fluorescence yield is given as:

$$\Phi_F = 1/[A_0^*] \int_0^\infty [A^*] k_{rad} \, dt \tag{A.13}$$

With eq. (A.3) and assuming k_{rad} to be constant one obtains

$$\Phi_F = k_{rad}[g_a\tau_a + g_b\tau_b] \tag{A.14}$$

The maximal value of the relative concentration of the radical pair $[P680^+$ Pheo$^-]$, RP_{max}, is obtained after setting $d[B]/dt = 0$ as:

$$[B]_{max} = \frac{k_1}{\gamma_a - \gamma_b} \left[(\gamma_b/\gamma_a)^{\gamma_b/(\gamma_a - \gamma_b)} - (\gamma_b/\gamma_a)^{\gamma_a/(\gamma_a - \gamma_b)} \right] \tag{A.15}$$

at the time

$$t_{max} = \ln(\gamma_a/\gamma_b)/(\gamma_a - \gamma_b) \tag{A.16}$$

Acknowledgement

We are grateful to Dr. H.-P. Schuchmann for translating reference (33), published in Russian. We thank Prof. K. Sauer for critically reading the manuscript and Prof. K. Schaffner for his interest and support.

References

1. Klimov,VV and Krasnovsky,AA (1982) Biophys. 27 :186-198
2. Klimov,VV (1984) in Adv. in Photosynthesis Research; Proc. 6th Int. Congr. Photosyn. 1 :131-138
3. Duysens,LNM and Sweers,H (1963) Microalgae and Photosynth Bact :353-372
4. Haehnel,W, Nairn,JA, Reisberg,P, and Sauer,K (1982) Biochim. Biophys. Acta 680 :161-173
5. Karukstis,KK and Sauer,K (1983) J. Cell. Biochem. 23 :131-158
6. Holzwarth,AR (1985) in Encyclopedia Plant Physiology: Photosynthetic Membranes Eds: LA Staehelin and CJ Arntzen, in press, Springer, Berlin
7. Holzwarth,AR (1986) in Topics in Photosynthesis Ed: J Barber, in press, Elsevier, Amsterdam
8. Haehnel,W, Holzwarth,AR, and Wendler,J (1983) Photochem. Photobiol. 37 :435-443
9. Karukstis,KK and Sauer,K (1983) Biochim. Biophys. Acta 725 :246-253
10. Karukstis,KK and Sauer,K (1983) Biochim. Biophys. Acta 722 :364-371
11. Holzwarth,AR, Wendler,J, and Haehnel,W (1985) Biochim. Biophys. Acta 807 :155-167
12. Kaminski,J (1985) Acta Phys. Pol. A 67 :679-700
13. Nairn,JA, Haehnel,W, Reisberg,P, and Sauer,K (1982) Biochim. Biophys. Acta 682 :420-429
14. Butler,WL, Magde,D, and Berens,SJ (1983) Proc. Natl. Acad. Sci. USA 80 :7510-7514
15. Berens,SJ, Scheele,J, Butler,WL, and Magde,D (1985) Photochem. Photobiol. 42 :59-68
16. Gulotty,RJ, Mets,L, Alberte,RS, and Fleming,GR (1985) Photochem. Photobiol. 41 :487-496
17. Pearlstein,RM (1982) in Photosynthesis 1: Energy Conversion by Plants and Bacteria :293-331
18. Kaplanova,M and Parma,L (1984) Gen. Physiol. Biophys. 3 :127-134
19. Lotshaw,WT, Alberte,RS, and Fleming,GR (1982) Biochim. Biophys. Acta 682 :75-85
20. Shuvalov,VA, Klimov,VV, Dolan,E, Parson,WW, and Ke,B (1980) FEBS Lett. 118 :279-282
21. Meiburg,RF, van Gorkom,HJ, and van Dorssen,RJ (1983) Biochim. Biophys. Acta 724 :352-358
22. Latimer,P, Bannister,TT, and Rabinowitch,E (1956) Science 124 :585-586
23. Breton,J (1983) FEBS Lett. 159 :1-5
24. Butler,WL (1978) Ann. Rev. Plant Physiol. 29 :345-378
25. Rutherford,AW (1983) in The Oxygen Evolving System of Photosynth. :63-69
26. Evans,MCW, Atkinson,YE, and Ford,RC (1985) Biochim. Biophys. Acta 806 :247-254
27. Klimov,VV, Klevanik,AV, Shuvalov,VA, and Krasnovsky,AA (1977) FEBS Lett. 82 :183-186
28. Diner,BA and Delosme,R (1983) Biochim. Biophys. Acta 722 :452-459
29. Breton,J (1982) FEBS Lett. 147 :16-20
30. Mauzerall,DC (1985) Biochim. Biophys. Acta 809 :11-16
31. Klimov,VV, Allakhverdiev,SI, and Pashchenko,VZ (1978) Dokl. Akad. Nauk. SSSR. 242 :1204-1207
32. Mathis,P (1984) in Adv.in Photosynthesis Research; Proc. 6. Int. Congr. Photosynth. 1 :155-158
33. Voznyak,VM, Sevdimaliev,RM, and Kim,VA (1984) Stud. Biophys. 99 :59-66
34. Klimov,VV, Ke,B, and Dolan,E (1980) FEBS Lett. 118 :123-126
35. Demas,JN (1983) in Excited State Lifetime Measurement. :1-273
36. Borisov,AY and Il´ina,MD (1971) Biochem. USSR 36 :693-695

Photosynthesis Research 10: 319–325 (1986)
© Martinus Nijhoff Publishers, Dordrecht

EVIDENCE THAT THE VARIABLE CHLOROPHYLL FLUORESCENCE IN CHLAMYDOMONAS
REINHARDTII IS NOT RECOMBINATION LUMINESCENCE

I. MOYA[a], M. HODGES[a], J-M. BRIANTAIS[a] AND G. HERVO[b]

[a]Laboratoire de Photosynthèse, CNRS, 91190 Gif sur Yvette et Laboratoire
pour l'Utilisation du Rayonnement Electromagnétique (LURE) Université
Paris XI, Bât. 209C, 91405 Orsay, France

[b]Service de Biophysique, CEN de Saclay, BP2, 91191 Gif sur Yvette, France

Key words: Chlorophyll fluorescence, photon counting, lifetime,
            charge recombination, photosystem II, mutant.

Abstract: Room temperature single photon timing measurements on intact
Chlamydomonas reinhardtii cells at low excitation energies have been analysed using a four exponential kinetic model. Closing the PSII reaction centres produced two major variable lifetime and two minor constant lifetime
components. The yield of each component mirrored the changes in lifetime.
Such observations indicate the presence of well-connected PSII centres favoring excitation energy transfer. A Chlamydomonas mutant lacking PSII reaction centre proteins exhibited decay components equivalent to those seen at
F_M in the wild-type. A titration of in vivo fluorescence, in both the mutant
and wild-type algae, using DNB, produced decay components similar to those
seen on opening PSII reaction centres. Such observations indicate that the
luminescence hypothesis for the origin of the long-lived lifetime component
is not the case.

INTRODUCTION
    Energy transfer among PSII units is predicted from several types of ob-
servations including: the shape of the relationship between the steady-state rate of O_2 emission and the number of open RCII (1); the sigmoidal shape
of the chlorophyll fluorescence induction rise in the presence of DCMU (2),
(however, see also (3,4) for restrictions on the validity of this criterion); the constancy of the ratio $<\tau>/\Phi$ during the fluorescence induction rise in algae (where $<\tau>$ =average lifetime, Φ = fluorescence yield) (5-7). This
observation means that the major part of the fluorescence emission at F_0 is
homogeneous to the variable part (F_V).
    Using a single photon counting technique and ps excitation, the fluores-
cence kinetics can be resolved into a sum of at least three exponential components (8). Applying these techniques, an almost constant $<\tau>/\Phi$ ratio was
still observed when RCII became closed but it was judged fortuitous because
the fluorescence increase seemed to be caused mainly by the increase in the
yield of a slow component whose lifetime (1-2.2 ns) exhibited only minor
changes (9). This finding was consistent with the hypothesis that F_V was a
fast luminescence arising from charge recombination between $P_{680}^{\ddagger}$ and I^- when

Abbreviations: DCMU = 3-(3,4-dichlorophenyl)-1,1-dimethyl urea; DNB = m,Di-
nitrobenzene; PSII = photosystem II; RCII = PSII reaction centre; I^- = reduced pheophytin; Q_A = primary stable electron acceptor of PSII; Chl = chlorophyll; LHCII = light harvesting Chla/b protein complex of PSII; F_0 = initial
fluorescence level; F_M = maximum fluorescence level; F_V = variable fluorescence (F_M-F_0); ps = picosecond; ns = nanosecond.

Q_A was reduced (10). However, Haehnel et al. later reported that in Chlorella cells, the yield of the fast (130 ps) and the slow (1.4-2 ns) components exhibited an opposite dependence on the state of RCII; consequently, they judged the recombination luminescence mechanism difficult to reconcile with their results (11).

The finding of similar kinetic components in chloroplasts of a wild-type and a PSII-lacking mutant of corn, the fluorescence of which was not affected by light intensity, suggested that the slow component could be kinetically controlled by the decay processes of the LHCII and did not necessarily have to originate from charge recombination in the RCII (12).

We present in this report new results on the fluorescence decay decomposition obtained with an improved single photon counting fluorimeter having an instrumental response function of 60 ps (FWHM), which is several times shorter than in previous work. Special attention has been taken to produce a homogeneous sample illumination and to analyse well defined fluorescence levels, especially a "true" F_0. By comparing the change in the lifetimes produced by either photochemistry or DNB (an external quencher of chlorophyll fluorescence) in wild-type and PSII-lacking Chlamydomonas cells, the origin of the variable fluorescence has been questioned.

MATERIAL AND METHODS

Chlamydomonas reinhardtii was grown as previously described (13) and diluted in the growth medium to 20 μg Chl/ml for experimentation. The fluorescence measurements were carried out at 90° to the excitation beam in a 2mm x 2mm cuvette located at the entrance slit of a monochromator (3nm bandwidth). To measure the F_0, the algae were circulated by a peristaltic pump at 20 ml/s. The F_M was obtained using stationary samples in the presence of 20 μM DCMU. Intermediate fluorescence levels were produced by circulating either dark-adapted or preilluminated algae (in which case all RCII are initially closed) at different flow rates to produce different levels of Q_A^-. The titration of fluorescence with DNB was carried out under F_M conditions.

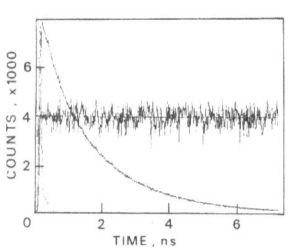

The single photon timing apparatus consisted of a mode-locked and cavity dumped dye laser system (dye DCM) synchronously pumped by an Ar^+ laser. This system is wavelength tunable (610-700 nm) and provides 10-15 ps pulses at up to 4 MHz. Emitted photons were detected by a microchannel plate photomultiplier (Hamamatsu R1564U) which has a S20 spectral response. The output pulses were passed to a constant fraction discriminator through a 24 db gain wide band amplifier (1 GHz). The instrumental response function was 60-70 ps FWHM

Figure 1. The excitation profile and the fluorescence decay of C. reinhadtii measured at 685 nm under F_M conditions. The best fit of a four exponential decay model is superimposed with the experimental decay. The weighted difference between the curves is shown in the center of the plot (scale + 10 to - 10). The component lifetimes (and relative amplitudes) were 2.000 (68 %), 1.204 (26 %), 0.320 (4 %) and 0.052 (2 %) ns, and the average lifetime was 1.693 ns. The Ki^2 was approximately 1.0.

(see Fig. 1) when scattered light was examined. The instrumental function and the fluorescence decay were consecutively recorded in a 512 channel memory group of a multichannel analyser and transferred to a HP 9836 computer. Deconvolution into a sum of exponentials was carried out by a least squares program using the Marquardt search algorithm for the non-linear parameters. Quality of fits were judged by the reduced Ki^2 criterion and the plot of the weighted residuals (see 11) which are shown in Fig. 1 along with the instrumental response function to scattered light and the chlorophyll fluorescence decay of C. reinhardtii measured at F_M. The apparatus and deconvolution program were checked using the dye oxazine which gave a single exponential decay with a lifetime of 785 ps and a Ki^2 of 1.0.

RESULTS
Decay kinetics and component characterization of the wild-type
The fluorescence decays of C. reinhardtii monitored at 685 nm with 625 nm excitation have been recorded at several different fluorescent states between F_0 and F_M. The ratio between the average fluorescence lifetime (calculated from the deconvoluted parameters) and the total Φ_{ex} remains constant (Fig. 2) in agreement with the previously reported data obtained by either phase fluorimetry (7) or single photon counting (9).

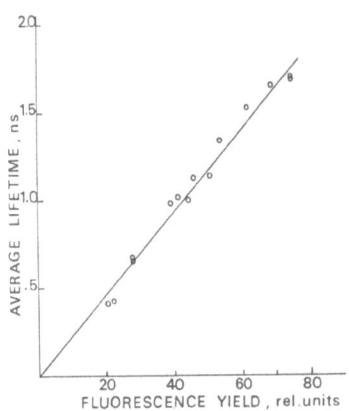

This result means that the data presented in Fig. 3 and 4 are equivalent to component lifetime versus fluorescence yield. The fluorescence decays were deconvoluted into a sum of exponentials. In general at least three free running components are required to fit the data and produce Ki^2 values between 1.0 and 1.2. However, from time-resolved emission spectra, the expected constant lifetime relationship with emission wavelength was not observed especially at F_M, when three components were used, inferring that such a model was perhaps inadequate. The decays were therefore deconvoluted using a free-running four component model, in which case the expected relationship between lifetime and emission wavelength was observed (data not shown).

Figure 2. The average lifetime of C. reinhardtii as a function of total chlorophyll fluorescence yield. The different fluorescence levels were achieved by varying the proportion of open reaction centers.

Figure 3a shows the relationship between the component lifetime and the average lifetime for wild-type C. reinhardtii between F_0 and F_M. There are two striking observations: first the lack of a long-lived component at F_0, at variance with previously published results for algae (8,9,11) and second, the proportional increase in the lifetimes of two components with fluorescence yield on increasing the number of closed RCII (as indicated by the average lifetime, see Fig. 2). The lifetimes of the two remaining components stayed almost insensitive to PSII trap closure with lifetimes of approximately 50 and 250 ps. Their total weight contribution to F_0 was 22 % (not shown). As for the corresponding lifetime changes, only the yields of the two slow components are significantly altered by the same order of magnitude as their lifetimes (data not shown).

The 50 ps component has an emission maximum at 695-705 nm (not shown) and arises from PSI pigments (see 14,15). The 250 ps decay and the two variable components peak at 685 nm and are present in PSII particles prepared according to the protocol of Berthold et al. (16) (not shown). Such properties infer that these components are related to PSII.

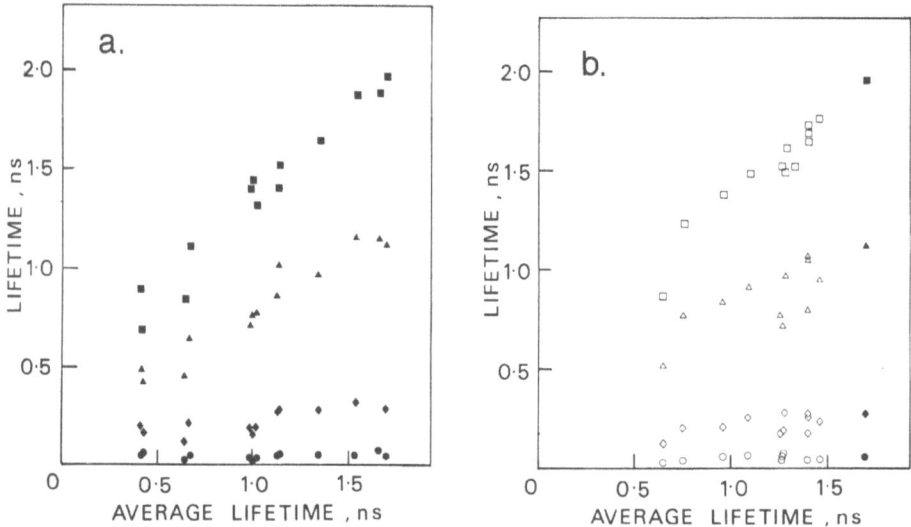

Figure 3. Variations in component lifetimes in the wild-type C. reinhardtii brought about by either a) photochemistry or b) increasing additions of DNB (the closed symbols are for minus DNB).

Decay kinetics of a mutant lacking PSII

F139 is a mutant of C. reinhardtii having no PSII activity and in which the 45 kDa PSII reaction centre protein (17) and the 42 and 33 kDa proteins associated with PSII are absent (13). The PSI and LHCII content seem unaltered, and it exhibits a fluorescence yield which is independent of light intensity (i.e. no variable fluorescence). The decomposition of the fluorescence decay into four exponential components yields an average lifetime in the same order of magnitude as that of the wild-type at F_M (see Table 1). In addition, the deconvoluted lifetime parameters are practically the same as the wild-type (compare Fig. 3b with Fig. 4 and see Table 1). It is obvious in the mutant that the slow component can not be attributed to the recombination luminescence mechanism. However, the fluorescence components found in the F139 mutant may have origins different from those seen in the wild-type. In order to check this point, we have investigated, in both the mutant and wild-type cells, the effect of an external quencher known to have a specific effect on the F_V.

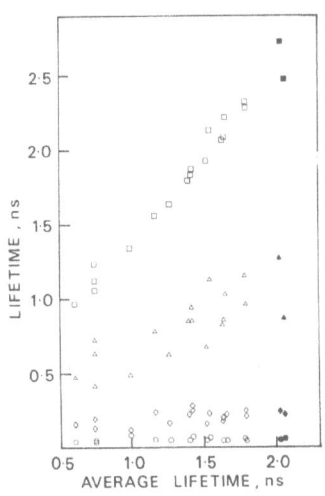

Figure 4. Variations in component
lifetimes in the F139 mutant brought
about by increasing concentrations
of DNB (open symbols) and in the
absence of DNB (closed symbols).

TABLE I

Lifetime (ns) and amplitude values
A in % of the fluorescence decays from
the wild-type and PSII-lacking mutant
of C. reinhardtii measured at F_M.

Wild type $\langle\tau\rangle = 1.69$		Mutant $\langle\tau\rangle = 2.03$	
τ	A	τ	A
0.052	2	0.042	4
0.32	4	0.23	7
1.20	26	1.28	25
2.20	68	2.64	64

Effect of DNB on the decay kinetics

DNB is a well known quencher of Chl fluorescence in vitro as well as in
vivo. It has been shown that, at concentrations $< 10^{-4}$M, DNB has a specific
quenching action on F_V, whereas F_0 is unaltered (18). Previous lifetime measurements have shown a parallel decrease in both average lifetime and total fluorescence yield for this concentration range (19). The same trend is
observed in this work (not shown). However, as a result of the increased
time resolution, we observe a specific decrease in the lifetimes of the two
slow, variable components (Fig. 3b). The same trend is observed for the
yields of these components (not shown). Lifetimes and yields of the two
fast, constant components are unaffected by the treatment. A comparison of
Fig. 3a with 3b shows that in the wild-type, the effect produced by the addition of low concentrations of DNB on lifetime is identical to the effect
obtained by partial quenching by open RCII.

Figure 4 shows that the addition of low concentrations of DNB to the
PSII mutant generates a specific quenching of the two slow lifetime components, as in the wild-type, inferring that the origin of the fluorescence
emission is the same in both cases.

DISCUSSION

Variable fluorescence in green algae and higher plants has been interpreted as recombination luminescence between P_{680}^+ and I^- when Q_A becomes
reduced (10). This followed the appearance of a 4 ns fluorescence decay component in PSII particles when Q_A was chemically reduced and which mirrored
the disappearance of reduced pheophytin as detected by absorption techniques
(10). Because an earlier decomposition of the fluorescence decay into three
exponential components yielded a slow lifetime component (1.1-2.2 ns) whose

lifetime exhibited a small change when RCII became closed but whose yield increased by 20-60 fold, it was proposed that the long-lived component arose from charge recombination. As a consequence of the increased time resolution of our ps. fluorimeter and the precautions taken to analyze a "true" F_0 level, a slow lifetime component is never observed at F_0. This result is independent of any deconvolution problems.

By introducing four exponential decay components, we find that variable chlorophyll fluorescence corresponds to the increase in two slow lifetime components which exhibit a 4-6 fold increase in both lifetime and quantum yield. These observations infer that PSII is organized in a well-connected system favoring good energy transfer between RCII's as previously proposed (1-7). The data of Fig. 3a show that the long-lived decay can not arise directly from charge recombination as this would lead to a constant lifetime component. Furthermore, if recombination is the mechanism at the basis of the reexcitation of the antenna pigments when PSII becomes closed to photochemistry, our results imply that the recombination time must be very short (< 100 ps) and not 2-4 ns as previously suggested (9-11), as we would otherwise not expect to observe the parallel increase in both lifetime and yield.

The fluorescence quencher, DNB, was therefore used to see if the longlived components exhibited by the PSII-lacking mutant (see Table I) had similar origins as the decay components of the wild-type. In both types of algae, DNB has no action on the fast constant components but mimics the quenching by open RCII on the variable components of the wild-type cells. The similar effect of DNB on both the wild-type and F139 mutant cells (compare Fig. 3b with Fig. 4) infers that the origin of the two slow components is the same in the two cases. As a consequence of the absence of RCII in the mutant, the slow component can not be attributed to the recombination luminescence mechanism in the alga.

A recent report in this field (20) claimed that a lag (i.e. a component having a negative amplitude) is seen in the variable fluorescence of <u>Chlorella</u> cells. This observation was found, along with the other results, to be consistent with the slow component arising from recombination luminescence. With our ps. fluorimeter which exhibits a ten-fold better resolution, a negative amplitude component has never been found. Our deconvolution program which has been tested with synthetic data in the presence of noise, can find negative as well as positive amplitudes. Therefore, the deconvolution can not be the reason for the difference. Such negative components can be explained, however, by the fact that, in 20, the variable fluorescence decay was produced by substracting the F_0 decay from that at F_M. As 70-80% of the F_0 fluorescence is homogeneous with the variable fluorescence, such a difference decay, as carried out in (20), will lead to at least four exponential components of which two will have negative amplitudes arising from the substraction.

Acknowledgements
   M. Hodges wishes to thank the Royal Society (London) for financial support. This work was also supported by grant SAV 7738 CNRS-CEA.

REFERENCES
1. Joliot P and Joliot A (1964) C R Acad Sci Paris Ser D 258: 4622-4625
2. Delosme R (1967) Biochim Biophys Acta 143: 108-128
3. Paillotin G (1967) J Theor Biol 58: 237-252
4. Sorokin EM (1985) Photobiochem Photobiophys 9: 3-19
5. Tumerman LA and Sorokin EM (1967) Mol Biol 1: 628-638
6. Briantais JM, Merkelo H and Govindjee (1972) Photosynthetica 6: 133-141

7. Moya I,Govindjee,Vernotte C and Briantais JM (1977) FEBS Lett 75: 13-18
8. Karukstis KK and Sauer K (1983) J Cell Biochem 23: 131-158
9. Haehnel W,Nairn JA, Reisberg P and Sauer K (1982) Biochim Biophys Acta 680: 161-173
10. Klimov VV, Allakhverdiev SI and Paschenco VZ (1978) Dokl Akad Nauk 242: 1204-1207
11. Haehnel W, Holzwart AR and Wendler J (1983) Photochem Photobiol 37: 435-443
12. Green BR, Karukstis KK and Sauer K (1984) Biochim Biophys Acta 767: 574-581
13. Maroc J and Garnier J (1981) Biochim Biophys Acta 637: 473-480
14. Yamazaki I, Mimuro M, Tamai N, Yamazaki T and Fujita V (1965) FEBS Lett 179: 65-68
15. Holzwarth AR, Wendler J and Haehnel W (1985) Biochim Biophys Acta 807: 155-167
16. Berthold DA, Babcock GT and Yocum CF (1981) FEBS Lett. 134: 231-234
17. Nakatani HY, Ke B, Dolan E and Arntzen CJ (1984) Biochim Biophys Acta 765: 347-352
18. Etienne AL, Lemasson C and Lavorel J (1974) Biochim Biophys Acta 333: 288-300
19. Moya I (1979) Thesis, Univ Paris XI Orsay France
20. Mauzerall DC (1985) Biochim Biophys Acta 809: 11-16

Photosynthesis Research 10: 327–333 (1986)
© Martinus Nijhoff Publishers, Dordrecht

RADIATIONLESS TRANSITIONS AS A PROTECTION MECHANISM AGAINST PHOTOINHIBITION IN HIGHER PLANTS AND A RED ALGA[1,2].

DAVID C. FORK, SALIL BOSE, and STEPHEN K. HERBERT
Department of Plant Biology, Carnegie Institution of
Washington, Stanford, California 94305 (U.S.A.)

ABSTRACT
Exposure of the red alga Porphyra perforata or leaves of Phytolacca americana and Echinodorus sp. to white light equivalent to full sunlight for short periods induced large decreases of variable fluorescence measured at 695 nm at 77K. This change was not produced by photoinhibition but rather appeared to result from an increase in the rate constant of radiationless transition in the reaction centers of photosystem II. It is proposed that this increase is related to the formation of the high energy state which serves as a photoprotective mechanism in plants.

1. INTRODUCTION

Although photosynthesis requires light for its proper functioning certain components and partial reactions of the process can be damaged (photoinhibited) by excess light (for a review see 16). Some plants, particularly those that live in high light environments, seem to have developed mechanisms for photoprotection. Thus, leaves of certain plants can avoid excess irradiation by turning away from the light. In other plants chloroplasts are able to reorient themselves so as to absorb less light. It is becoming clear that plants apparently also employ biophysical and biochemical mechanisms to protect themselves against photodamage. A case in point is the intertidal red alga Porphyra perforata that has apparently developed several strategies to avoid photoinhibition (2, 15, 17–20).

In this study we have identified another process that apparently functions to limit photodamage in Porphyra as well as in leaves of higher plants by increasing the rate by which energy is dissipated via radiationless transitions in the reaction centers of photosystem II (PSII). This process seems to be associated with the formation of the high-energy state in photosynthesis (10).

2. PROCEDURE

2.1. Materials and Methods

Gametophytic plants of the red alga Porphyra perforata were collected in the field and maintained in the laboratory under vigorous aeration in f/2 Guillard's enriched seawater (14) changed twice weekly in lighting of 25 µE m^{-2}s^{-1} provided by cool-white fluorescent lamps operated on a 12:12 hr cycle. The temperature was held at 12°C. Samples for experiments were cut from these growing plants.

Leaves of the aquatic higher plant Echinodorus sp. (Amazon sword plant) were maintained in aquaria under fluorescent lighting of 14 µE m^{-2}s^{-1} at 27°C. Phytolacca americana (pokeweed) leaves were obtained from plants cultivated in the field.

Time courses of chlorophyll (Chl) fluorescence were measured using a fiber optic system to excite and collect the fluorescent light (3). O$_2$ evolution was measured with a Clark-type oxygen electrode (Rank Brothers).

3. RESULTS

3.1. Red Algae

Fig. 1 shows the kinetics of the fluorescence rise in Porphyra at 20°C in the presence of 3-(3,4-dichlorophenyl)-1,1-dimethylurea (DCMU) for dark adapted samples and for samples exposed to white light for 30 sec. This light treatment produced about an 80% loss of variable fluorescence yield (F_m-F_o) as compared to the dark controls.

The loss of variable fluorescence at physiological temperatures after light treatment was also seen at 77K. Fig. 2 compares the kinetics of the fluorescence rise at 695 nm at 77K of control and of Porphyra exposed to white light for 5 min as described for Fig. 1. Variable fluorescence decreased 60% after this treatment. F_o in the light-treated sample was essentially unchanged.

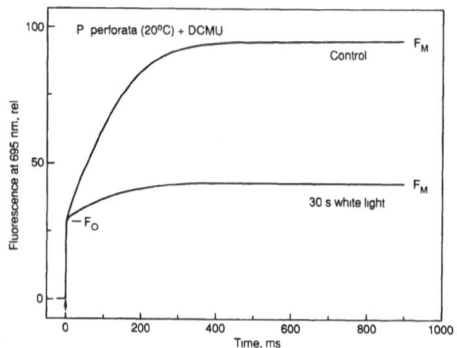

FIGURE 1. Time courses at 20°C of Chl a fluorescence transients in Porphyra perforata measured in the presence of 20 µM DCMU. The light-treated sample was exposed for 30s to white light (2000 µE m^{-2}s^{-1}), incubated in the dark in seawater containing DCMU for 2 min, and then illuminated with green actinic light (defined by Corning colored glass filters 4-96 and 3-69) that had an intensity of 3.3 µE m^{-2}s^{-1}.

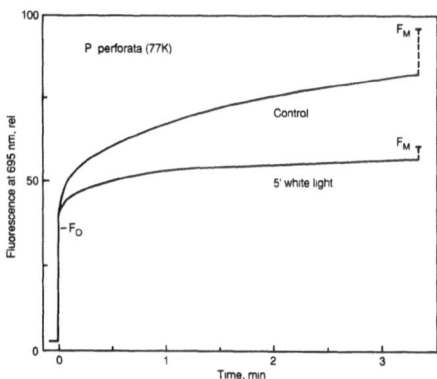

FIGURE 2. Time courses at 77K of Chl a fluorescence transients at 695 nm in Porphyra measured in the absence of DCMU. The light treated sample was exposed for 5 min to white light as described in Fig. 1, kept in the dark for 2 min and frozen in the dark. The green actinic light described in Fig. 2 was used at an intensity of 0.7 µE m^{-2}s^{-1}.

Butler and co-workers (1, 6) have shown that $F_I = fF_{II}$ when the PSII traps are reduced at 77K and that a plot of F_I versus F_{II} produces a straight line:

$$F_I/F_{II} = kT(II{\rightarrow}I) \; \phi_{FI}/k_{FII},$$

where $kT(II{\rightarrow}I)$ represents the rate constant for energy transfer (spillover) from PSII to PSI, ϕ_{FI} is the fluorescence yield of PSI as defined by the rate constants for competing ways of deactivating excited chlorophyll (Chl) via fluorescence or radiationless decay, and k_{FII} the rate constant for fluorescence from PS II.

Extrapolation of the straight line back to the Y-axis (where $F_{II} = 0$) yields the fraction of the light that is distributed initially to PSI (or α) where $\alpha + \beta = 1$. The fraction of light distributed initially to photosystem II is denoted by β. Fig. 3 shows the X-Y plot relating F695 and F735 in dark adapted or light-treated Porphyra illuminated at 77K. As noted already by Ley and Butler (13), the traces obtained using red algae are non-linear. Ley and Butler (1977) suggested that the initial non-linearity could be explained on the basis of part of the fluorescence of variable yield at 735 nm was controlled by closure of photosystem I (PSI) reaction centers. We found, as did Ley and Butler, that the oxidation of P700 at 77K with light absorbed largely by PSI (440 nm or far-red light) that did not alter the state of the PSII reaction centers eliminated the initial curvature of the traces and produced a straight line plot (data not shown).

It can be seen from examination of Fig. 3 that 15 sec of white light treatment produced a marked increase in the slope of the curve compared to that of the dark control (the slopes were drawn so as to avoid the initial non-linear part). Back extrapolation of these lines intersected the Y-axis at nearly the same point. At 695 nm F_v decreased by about 40% while the parameters at 735 nm were not much affected.

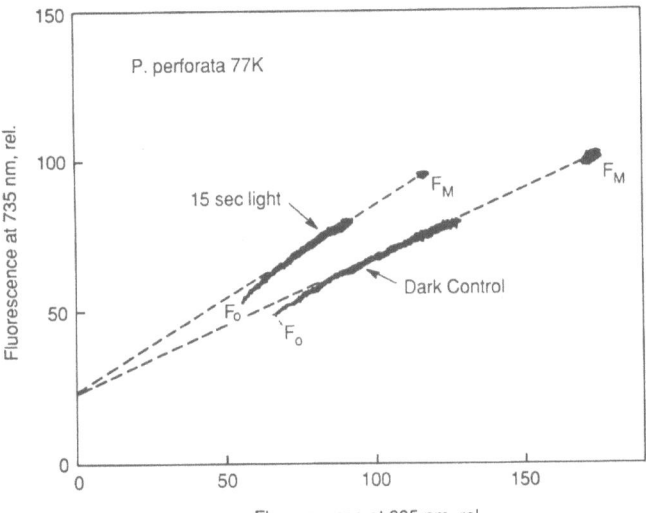

FIGURE 3. Fluorescence at 735 nm as a function of fluorescence at 695 nm measured at 77K in dark adapted and light-treated Porphyra perforata illuminated at 77K. The light-treated sample was illuminated for 15 sec in white light (2000 µE m^{-2}s^{-1}), returned to the dark for 2 min and then frozen in the dark. Green actinic light as described in Fig. 1 (0.7 µE m^{-2}s^{-1}) was used.

Table 1 summarizes measurements made on fluorescence and O_2 evolution (light limited) in <u>Porphyra</u>. Treatment with white light for up to 10 min did not reduce photosynthetic efficiency yet induced marked changes in the fluorescence parameters. Although F_v expressed as (F_v/F_m) decreased significantly after light treatment there was little effect on F_o.

3.2 <u>Higher plants</u>. Fig. 4 compares the relationship of F695 versus F735 at 77K in dark-adapted control leaves of <u>Phytolacca</u> and in leaves given white light equivalent to approximately full sunlight for 20 min. The insert of Fig. 4 shows the kinetics of the fluorescence increase at 695 nm measured at 77K before and after a 5 min exposure to white light. As with <u>Porphyra</u>, pretreatment of <u>Phytolacca</u> with white light induced about 50% lowering of F_v. Unlike the results obtained with <u>Porphyra</u>, the relationship between F735 and F695 in <u>Phytolacca</u> both before and after light treatment was linear. White light pretreatment had no effect on the slope or on α (or β).

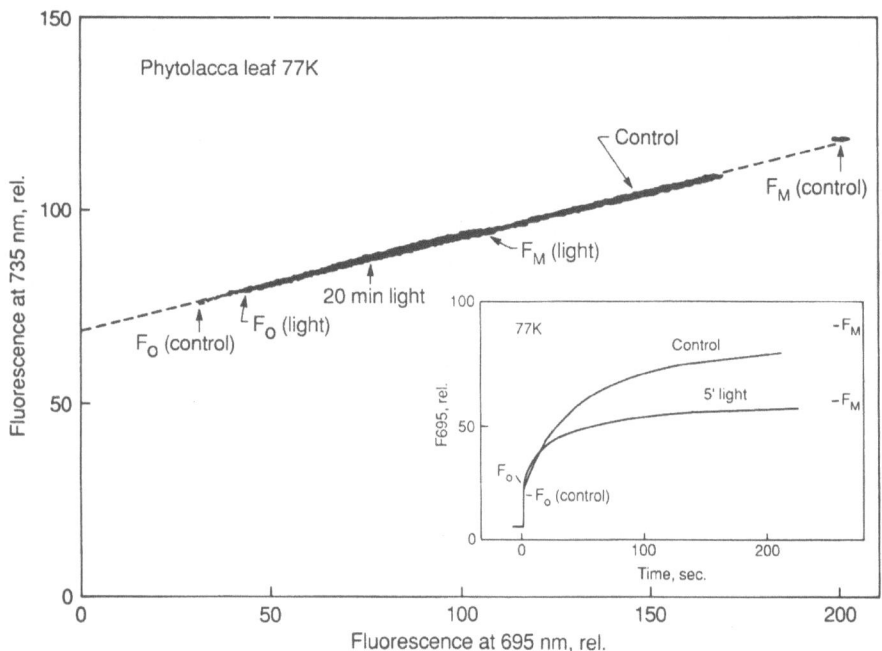

FIGURE 4. Fluorescence at 735 nm as a function of fluorescence at 695 nm measured at 77K in dark adapted and light treated leaves of pokeweed (<u>Phytolacca americana</u>) illuminated at 77K. The light treated sample was given 20 min white light (2000 µE $m^{-2}s^{-1}$), 2 min dark and then frozen in the dark. Green actinic light as described in Fig. 1 (0.7 µE $m^{-2}s^{-1}$). The insert gives an example of the kinetics at 77K of the fluorescence increase from F_o to F_m in dark adapted and light treated leaves.

Table 2 compares the effect of light treatment from 1 to 40 min on O_2 evolution and fluorescence parameters in leaves of <u>Phytolacca</u> and <u>Echinodorus</u>. It can be noted that the slopes and Y intercepts were nearly identical after all of the light treatment periods. The values of F_o increased after short periods of light treatment and then decreased again to the control level.

TABLE 1. The effect of exposure to white light on the rate of light-limited O_2 evolution and on parameters of fluorescence at 695 nm measured at 77K in Porphyra perforata. Discs were cut from thalli and kept in darkness for 5 min before freezing or illuminated for various periods with white light (2000 $\mu E\ m^{-2}\ s^{-1}$), returned to darkness for 1 min, and frozen in the dark to 77K or used to measure activity of O_2 evolution. The fluorescence parameters at 695 nm were taken from the X-Y plots measured as described in the text. Fluorescence was excited with green light as described in the Materials and Methods (intensity: 3.3 $\mu E\ m^{-2}\ s^{-1}$). O_2 evolution was measured with white light (equivalent to 40 $\mu E\ m^{-2}\ s^{-1}$ in a band from 400-700 nm). Slope (m), Y intercept (c).

Light Exposure (min)	F_o	F_m	F_v/F_m	m	c	O_2 Evolution (% Control)
0	70	162	0.568	0.41	22	100
1	81	137	0.408	0.62	20	98.6
5	72	119	0.395	0.73	28	99.3
10	78	137	0.430	0.62	22	97.5
30	77	120	0.358	0.69	25	93.9
45	70	103	0.320	0.76	20	86.4

TABLE 2. The effect of exposure to white light on O_2 evolution and on parameters of fluorescence at 695 nm measured at 77K in Phytolacca americana (Pokeweed) and in Echinodorus sp. (Amazon sword plant). Leaf discs from both plants were kept in darkness for 5 min before freezing or illuminated for various periods with white light (2000 $\mu E\ m^{-1}s^{-1}$), returned to darkness for 1 min and frozen in the dark to 77K or used to measure activity of O_2 evolution. The fluorescence parameters at 695 nm were taken from the X-Y plots measured as described in the text. Slope (m), Y intercept (c).

	Light treatment (min.)	F_o	F_m	F_v/F_m	m	c	O_2 evolution (relative)
Phytolacca	0	28	159	0.82	0.23	56	--
	1	45	108	0.58	0.25	52	--
	5	39	112	0.65	0.22	54	--
	10	36	112	0.67	0.23	55	--
	20	36	104	0.65	0.23	55	--
	40	35	89	0.60	0.24	53	--
Echinodorus sp.	0	47	137	0.65	0.32	45	100
	1	78	117	0.33	0.35	44	98
	3	--	--	--	--	--	96
	5	68	94	0.27	0.35	46	96
	10	60	80	0.25	0.32	42	92
	20	47	64	0.26	0.32	45	81
	40	44	55	0.20	0.34	34	71

4. DISCUSSION

Marked decreases of variable fluorescence are seen both in light-treated red algae and higher plant leaves. Porphyra is adapted to an environment in which it experiences repeated cycles of drying (often in combination with salinity increase) during periods of full sunlight. It would not be expected, therefore, that an exposure of a few minutes to a light intensity equivalent to that of full sunlight would photoinhibit the light-limited rate of O_2 evolution and indeed no such inhibition was seen (Table 1). However, light exposures as short as 15 s (Fig. 4) induced marked decreases in F_v at 695 nm.

Part of this decline in F_v695 in Porphyra may be accounted for on the basis of a State 1 – State 2 transition since the increased slope of the line relating F695 and F735 after light treatment suggests, according to the model of Butler (6), an increased transfer from PSII to PSI. However light treatment had no effect on α, the fraction of light delivered initially to PSI. In an experiment (not shown) where we measured the PSI fluorescence at 742 nm (4) both in dark control and light-treated samples using blue actinic light (442 nm) absorbed almost exclusively by Chl a of PSI we saw an increase of about 14% suggesting that part of the increased slope after light treatment seen in Fig. 3 may be attributed to an increase in ϕ_{PSI} (15).

Unlike Porphyra, leaves of higher plants showed no slope changes in the F_I/F_{II} plots (Fig. 4). Thus in leaves which showed a significant decrease of F_v there was no change in the rate constant for energy transfer from PSII to PSI (kT(II—>I)), nor was there any change in the fraction (α) of the light distributed initially to PSI and therefore the fraction (β) to PSII. In leaves of Echinodorus it was seen that a short treatment in light decreased F_v but did not photoinhibit the light limited rate of O_2 evolution (Table 2). Although not demonstrated directly here, the same probably holds true for pokeweed, a sun plant which is unlikely to be photoinhibited by a 5 min exposure to a light intensity equivalent to that of full sunlight.

An explanation for the decreased F_v seen at 695 nm in leaves of higher plants and, perhaps, in Porphyra also can be sought on the basis of an increase in the rate constant for radiationless transition whereby quanta are dissipated via a mechanism involving production of heat. Such dissipation may serve as a means of photoprotection in plants whose photosynthetic rates are light saturated but whose photosynthetic apparatus still continues to receive excess quanta. Since the decrease in fluorescence yield is observed only in the variable part and not in F_o, it appears that an increase in the rate constant for radiationless transition has taken place in the reaction centers rather than in the bulk chlorophyll.

It seems reasonable to explain the decreased F_v seen in light-treated plants which are not photoinhibited on the basis of the fluorescence decrease ('quenching') induced upon the formation of a high energy state (HES) as a result of proton uptake. According to present interpretations (11, 12) the formation of a ΔpH induces a fluorescence decrease as a result of increased thermal de-excitation. It is known that the light induced fluorescence decrease is prevented by the uncoupler carbonylcyanide m-chlorophenylhydrazone (CCCP) (7-9). We found that the decrease in F_v caused by white light treatment is inhibited in the presence of CCCP in both Porphyra and Echinodorus (data not shown). This supports the idea that the decreased F_v in light treated plants is mostly due to the formation of the high-energy state.

A light-induced increase in radiationless transition may represent a method to eliminate excess excitation energy from PSII and protect this photosystem from photodynamic damage. Red algae differ from higher plants in that they show both a slope change and a decrease in the length of the line segment in the X-Y plot (Fig. 3). Thus, these algae seem to regulate excitation energy in PSII by increasing the rate constant for energy transfer from PSII to PSI (state changes) as well as by increasing the amount of radiationless transition. Both mechanisms may be more needed in the red algae than in higher plants because the overlap of absorption of pigments associated with PSI and PSII is large in higher plants but very little in red algae. In red algae state changes may be more useful as a means of balancing excitation energy than in higher plants. Thorne et al. (21) have suggested that light-induced changes in scattering represent a mechanism for the regulation of excitation energy between the two photosystems of leaves.

Since its first use to study photoinhibition in bean (5), the measurement of F_v at 77K has become a commonly used measure of this phenomenon (16). It is clear from the results presented here that F_v can be decreased by other processes in addition to the decrease induced by photoinhibition (5). It is necessary, therefore, to be able to estimate the contribution of the high energy state, state transitions and perhaps other processes to the decrease of F_v before correct conclusions can be drawn concerning photoinhibition from measurements of F_v at low temperature.

1. CIW DPB Publ. No. #809.
2. This paper is dedicated to the memory of Warren Butler.

REFERENCES

1. Butler WL (1978) Ann Rev Plant Physiol 29:345-378
2. Fork DC and Öquist G (1981) Z Pflanzenphysiol 104:385-393
3. Fork DC, Ford GA and Catanzaro B (1978) Carnegie Inst Year Book 78:196-199
4. Fork DC, Öquist G and Hoch GE (1982) Plant Sci Lett 24:249-254
5. Fork DC, Öquist G and Powles SB (1981) Carnegie Inst Wash Year Book 80:52-57
6. Kitajima M and Butler WL (1975) Biochim Biophys Acta 408:297-305
7. Krause GH (1973) Biochim Biophys Acta 292:715-728
8. Krause GH (1974) Biochim Biophys Acta 333:310-313
9. Krause GH (1978) Planta 138:73-78
10. Krause GH and Weis E (1984) Photosyn Res 5:139-157
11. Krause GH, Briantais J-M and Vernotte C (1983) Biochim Biophys Acta 723:169-175
12. Krause GH, Vernotte C and Briantais J-M (1982) Biochim Biophys Acta 679:116-124
13. Ley AC and Butler WL (1977) In Photosynthetic Organelles, Miyachi S, Katoh S, Fujita Y and Shibata K (eds) Japanese Society of Plant Physiologists, Kyoto
14. McLachlan J (1973) In: Stein JR (ed) Handbook of Phycological Methods pp 25-51 Cambridge University Press, London
15. Öquist G and Fork DC (1982) Physiol Plant 56:56-62
16. Powles SB (1984) Ann Rev Plant Physiol 35:15-44
17. Satoh K and Fork DC (1983) Biochim Biophys Acta 722:190-196
18. Satoh K and Fork DC (1983) Photosyn Res 4:61-70
19. Satoh K and Fork DC (1983) Plant Physiol 71:673-676
20. Satoh K, Smith CM and Fork DC (1983) Plant Physiol 73:643-647
21. Thorne SW, Duniec JT and Lee JA (1983) Photobiochem Photobiophys 5:71-78

III.   Reaction Centers; Primary Photochemistry; and Early Acceptors and Donors

Figure 6. Photo shows Warren L. Butler, Jack Myers, Bessel Kok, Barbara Zilinskas, Among other .... discussions on primary reactions of alcohol.

Figure 7. Warren Butler, Wolfgang Junge, and Lila Butler .... photo taken at Warren's home in California.

Photosynthesis Research 10: 337–346 (1986)
© *Martinus Nijhoff Publishers, Dordrecht*

ELECTRON DONORS AND ACCEPTORS IN PHOTOSYNTHETIC REACTION CENTERS

J. AMESZ and L.N.M. DUYSENS
Department of Biophysics, Huygens Laboratory of the State University,
P.O. Box 9504, 2300 RA Leiden, The Netherlands

1. SUMMARY

A review is given of primary and associated electron transport reactions in various divisions of photosynthetic bacteria and in the two photosystems of plant photosynthesis. Two types of electron acceptor chains are distinguished: type 'Q', found in purple bacteria, Chloroflexus and system II of oxygenic photosynthesis and type 'F', found in green sulfur bacteria, Heliobacterium and photosystem I. Secondary donor reactions are discussed in relation to plant photosystem II.

2. INTRODUCTION

There is a large body of evidence (see ref. 62 for a review) that in all photosynthetic organisms the primary photochemical reaction consists of the transfer of an electron from an excited chlorophyll (Chl) $\underline{a}$ or bacteriochlorophyll (BChl) $\underline{a}$ or $\underline{b}$ molecule (P), to an acceptor, usually called I. The electron is subsequently transferred from I^- to other acceptors. In each electron transfer step a more stable (longer lived) charge separation is established. Additional stabilization occurs by transfer of an electron from a donor molecule (D), e.g. a cytochrome, to oxidized P, P^+.

These reactions thus can be written as:

$$D\ P^*I\ X \xrightarrow{1} D\ P^+I^-X \xrightarrow{2} D\ P^+I\ X^- \xrightarrow{3} D^+P\ I\ X^-$$

Reactions 1, 2 and 3 occur, depending upon the species, in times of about 10 ps or less, of 50 - 600 ps, and of 0.1 - 200 µs, respectively. The multistep electron transfer is required to transfer the electron extremely rapidly across the membrane, a relatively large distance of 5 nm. The rapidity is necessary for a good quantum yield: the electron transfer rate should be much higher than the decay rate of P^* which is of the order of $10^9\ s^{-1}$. Rapid electron transfer is possible if the transferring molecules are so close to each other that the electron wavefunctions overlap. This requires three or more molecules (such as Chl) to bridge the gap across the membrane. In addition, the large transmembrane distance slows down the back transfer of the electron from X^- to D^+, so that these compounds have sufficient time to be restored to X and D by the subsequent dark reactions of photosynthesis.

This short review will mainly deal with the characteristics of the electron acceptor chain in reaction center complexes of various photosynthetic systems and organisms, with emphasis on recent developments. Donor reactions will be only discussed in relation to photosystem II of oxygenic photosynthesis. For more extensive reviews the reader is referred to refs. 2,4,5,17,36,62,67 and 83.

Dedicated to the memory of Warren L. Butler

3. THE TYPE Q ACCEPTOR CHAIN

As will be discussed in this and the next section, the primary electron acceptor chains found in various divisions of photosynthetic bacteria and in the two photosystems of plant photosynthesis may be divided into two types, which we call Q and F, that can be distinguished on basis of the chemical nature and midpoint potentials of the electron acceptors involved. Type Q is found in purple bacteria, <u>Chloroflexus</u> and system II of oxygenic photosynthesis. It may be summarized by the following set of reactions

$$P^*I \ Q_A Q_B \rightarrow P^+I^- Q_A Q_B \rightarrow P^+I \ Q_A^- \ Q_B \rightarrow P^+I \ Q_A Q_B^- \ ,$$

where I is pheophytin (Pheo) <u>a</u> or bacteriopheophytin (BPheo) <u>a</u> or <u>b</u> and Q_A and Q_B are bound quinones (ubiquinone, menaquinone or plastoquinone). Q_A and Q_B together serve to reduce Q_B to the quinol by two one-electron transfers from I^-, accompanied by the uptake of protons. Reduced Q_B is then restored to the oxidized state, presumably by exchange with a quinone molecule from a large pool in the membrane (19,29,88).

3.1. Purple bacteria and Chloroflexus

The primary and secondary electron transport reactions have been studied much more extensively in purple bacteria than in other organisms. The main reason was the availability of isolated reaction center complexes, obtained from cytoplasmic membranes of various species of purple bacteria by means of detergents. These reaction center complexes have been characterized in terms of their chemical constituents (for reviews see refs. 32 and 58). They all contain four BChl <u>a</u> or <u>b</u> and two BPheo <u>a</u> or <u>b</u> molecules, one molecule of carotenoid (except for some mutants like <u>Rhodobacter</u> (<u>Rhodopseudomonas</u>) <u>sphaeroides</u> R-26) and, depending on preparation and species, one non-heme iron or manganese (68) and one or two quinone molecules. The use of isolated reaction centers enabled the application of flash spectroscopy and various electron spin resonance techniques with better precision than with intact cells or isolated membranes. These studies need not be reviewed here; the reader is again referred to refs. 17 and 62.

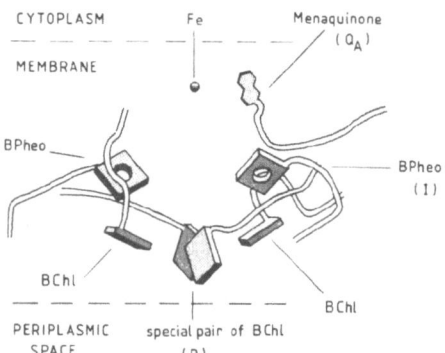

FIGURE 1. Arrangement of reaction center components as determined by X-ray diffraction analysis of a crystalline reaction center complex of <u>Rps. viridis</u>. Redrawn from ref. 20. The figure was kindly provided to us by Dr. H.J. van Gorkom.

A new and important step in photosynthesis research is the crystallization of the reaction center complex of Rhodopseudomonas viridis (51, 52), and the subsequent X-ray analysis of its structure (20,21). Together with the primary structure of the constituent polypeptides (see 1,53), the results of the X-ray analysis provide a three-dimensional picture of the organization of the reaction center that rapidly becomes more detailed.

The often debated (e.g. 24,61) question of whether the primary electron donor P is a (B)Chl dimer has been resolved at least for Rps. viridis by the X-ray data (20). The density distribution, at 3 A resolution, showed the location and orientation of the four BChl, two BPheo, and one menaquinone (Figure 1) that are contained in the reaction center. Two BChl molecules are close together, at a center-to-center distance of 7 A, and apparently represent the P dimer. The arrangement of BChl and BPheo shows roughly a two-fold rotation symmetry about an axis perpendicular to the membrane (9) and passing through the middle of P. Presumably, electron transfer occurs from excited P, P*, via the BChl (73) and BPheo (I) at the right hand side of Figure 1 to the menaquinone (Q_A) (66). The BChl at the left hand side can be removed or modified without affecting the quantum efficiency for electron transfer (50,74). The ubiquinone that serves as secondary electron acceptor Q_B is missing in the crystals. The protein moiety of the reaction center complex shows ten helical stretches of amino acid residues (21) that belong to the so-called L and M subunits (58).

Recent evidence has shown that Chloroflexus aurantiacus, which was classified as a green bacterium (63) on the basis of its morphological

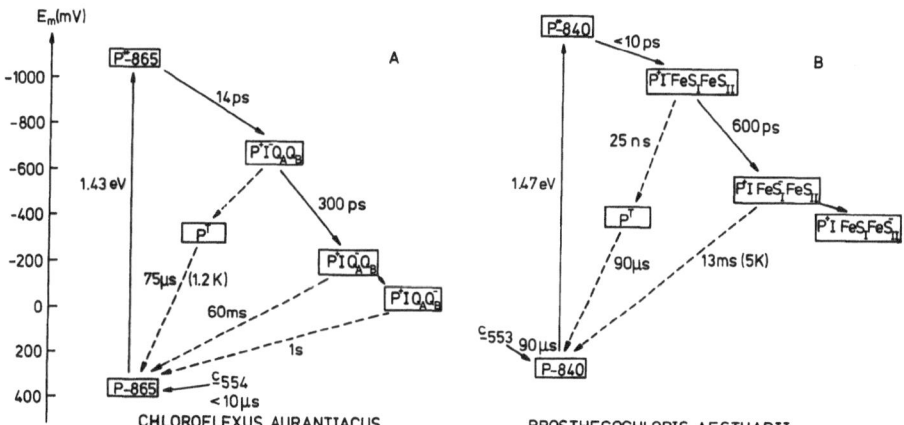

FIGURE 2. Energy scheme of electron transfer kinetics in the reaction center of (A) the green filamentous bacterium C. aurantiacus and (B) the green sulfur bacterium P. aestuarii. The excited primary electron donor P is indicated by P*, excitation by vertical arrows. Downward pointing solid arrows indicate main electron transport pathways; broken arrows are "back reactions" that become predominant when main electron transport is blocked. P^T, triplet state of P; FeS, iron sulfur center; c, cytochrome c; the other symbols are defined in the text. Unless otherwise indicated, the time constants refer to room temperature; those for the primary charge separation are taken from unpublished experiments of V.A. Shuvalov and coworkers. Scheme (A) with somewhat different time constants and energy levels applies also to purple bacteria.

characteristics and pigment composition (it has chlorosomes and contains BChl c as its major pigment), is photochemically quite similar to purple bacteria, and contains a 'type Q' electron transport chain in contrast to the green sulfur bacteria that will be discussed in the next section (Figure 2A). Isolated reaction centers have been obtained (64) that are optically and structurally similar to those of purple bacteria (25,74,85, 87). However, the pigment composition is different: the reaction center contains three instead of two BPheo a molecules, one of which replaces BChl a (4). Measurements of rapid absorption changes in these reaction centers by means of flash spectroscopy have shown the reduction of BPheo a (39,40). In about 300 ps the electron is transferred from BPheo a to the secondary acceptor, which is menaquinone (86).

3.2. Photosystem II

The first evidence for an electron acceptor Q (Q_A) for photosystem II (PS II) came from studies on quenching of Chl a fluorescence (27). Cramer and Butler (18) titrated the yield of fluorescence and obtained two mid-point potentials, at -30 and -300 mV. (For a review of more recent titration experiments, see ref. 89).

Experiments to determine the chemical identity of Q at first provided somewhat confusing evidence. Butler and coworkers (13,14,16,31,41) showed that an absorption change near 550 nm, discovered by Knaff and Arnon (43,44) and called C-550, shows all the characteristics of Q (see also ref. 84), but measurements in the ultraviolet region indicated that Q would be a plastoquinone molecule that is reduced to the semiquinone anion upon illumination (82,90). Plastoquinones do not absorb at 550 nm, and subsequent investigations have shown that the apparent paradox can be ex-

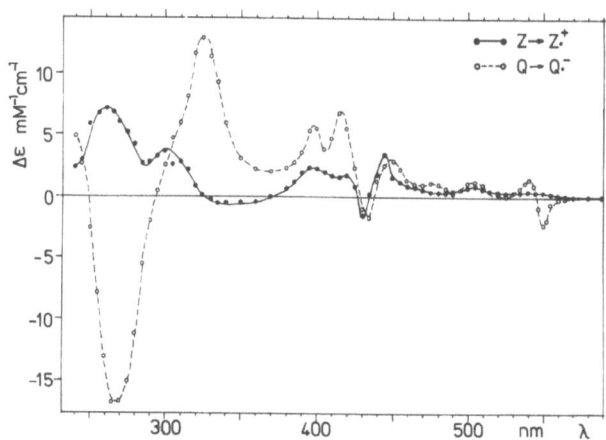

FIGURE 3. Broken line: absorption difference spectrum of the reduction of Q_A in a purified photosystem II preparation. Below about 350 nm the spectrum is due to changes in the absorption spectrum of plastoquinone upon reduction. The absorbance changes at 400 - 440 and 540 - 550 nm (C-550) are due to band shifts of Pheo a. Solid line: absorption difference spectrum due to oxidation of the secondary donor Z. In the ultraviolet the spectrum is presumably caused by the oxidation of a plastoquinol molecule, whereas the band shift near 430 nm is probably due to Chl a, perhaps P 680. Note that C-550 is absent in this spectrum (22).

plained by a shift of the absorption spectrum of Pheo a caused by the negative charge on Q^- (70,83). Figure 3 shows a recent difference spectrum of the absorbance changes due to Q_A reduction in a purified PS II preparation (22).

Until recently evidence for the function of Pheo a as primary electron acceptor in PS II was based on experiments in which $\overline{Pheo^-a}$ was allowed to photoaccumulate in strong light under reducing conditions (42; see also ref. 83). The lack of isolated reaction centers and of apparatus of sufficient signal-to-noise ratio precluded the use of picosecond difference spectrocopy. However, Nuijs and coworkers (54,57) recently succeeded in obtaining flash-induced absorption difference spectra of a PS II preparation with a time resolution of less than 50 ps. The spectra not only showed the generation and decay of excited antenna Chl, but also of absorbance changes due to P^+680, the oxidized primary electron donor, and of $Pheo^-a$. It was concluded that electron transfer for $Pheo^-a$ to Q_A occurs with a time constant (1/e) of 270 ps in these preparations. Together with the older evidence, these results establish that I in PS II is indeed Pheo a.

Another point of similarity between PS II and the photosystem of purple bacteria is the homology between the primary structures of the L and M subunits of the reaction center of purple bacteria and of the so-called D_1 and D_2 (33 kDa) proteins of PS II (see refs. 1,46,79). Although a possible functional similarity is still a matter of controversy, this indicates a common evolutionary origin for these photosystems.

4. THE TYPE F ELECTRON TRANSPORT CHAIN
4.1. Photosystem I

P 700, the primary electron donor of photosystem I (PS I) of algae and higher plants is probably a dimer (see discussion in ref. 67) of Chl a or a related pigment (26). Its midpoint potential is about the same as that of the primary electron donor of purple bacteria (~ 0.48 V (67)). However, redox titrations of the photochemical activity of PS I by Ke (37) and Lozier and Butler (47) gave midpoint potentials of about -500 mV for the acceptor side, indicating that the PS I acceptors are considerably more electronegative than those of purple bacteria or of PS II. ESR and optical studies have shown that PS I contains three bound iron sulfur centers (30,38,48,69,78) called F_X, F_A and F_B, with midpoint potentials of about -700, -550 and -590 mV, respectively (67). Since most of the data concerning these iron-sulfur centers have been obtained from photoaccumulation experiments, in which the reduced forms were trapped at low temperature, it is not clear if F_X, F_A and F_B operate in a linear electron transfer chain and in which order. For a discussion we refer to ref. 67.

Absorption difference spectroscopy in the ns and sub-ns region indicates that the primary electron acceptor is a Chl a molecule absorbing around 693 nm, which is bleached upon reduction. In highly enriched PS I particles prepared by detergent treatment the bleaching has been observed to reverse again in about 200 ps after a flash, presumably by electron transfer to a subsequent acceptor (72), but recent experiments indicate that this reaction is several times faster under more native conditions (54).

Photoaccumulation experiments under reducing conditions have suggested the possible involvement of two early electron acceptors (6,34). The involvement of the first of these, which shows a bleaching at 670 nm ascribed to reduction of Chl a (49,78) was not confirmed by flash spectroscopy (54), whereas the identity of the other one is still unclear. No absorbance changes were detected in the visible region upon its reduction

(49). Q-band EPR spectra are compatible with the suggestion that it may be a quinone (67).

4.2. Green sulfur bacteria and Heliobacterium

Considerable advances were made in recent years concerning the electron transport pathway in green sulfur bacteria. Because of the large antenna contained in the chlorosomes of these bacteria (typically 1000 - 2000 BChl c or d molecules per reaction center) all these experiments were done with isolated membranes from which the chlorosomes had been removed or with pigment protein complexes derived from these membranes.

Upon illumination such preparations showed the oxidation of the primary donor P, called P 840 in these organisms (60,76) and, under reducing conditions, the photoaccumulation of reduced iron sulfur centers, as in PS I (77). Information about the identity of the primary electron acceptor has only recently become available (55,81). Absorption difference spectra of membranes of Prosthecochloris aestuarii measured shortly after a 35 ps laser flash showed, in addition to the formation of P^+ and of excited states in the antenna, a bleaching near 670 nm. This bleaching disappeared again in about 600 ns, and could be ascribed to reduction of I followed by re-oxidation by electron transfer to the next electron acceptor. At first I was tentatively identified as BPheo c, but a recent quantitative analysis of the pigment content of membranes of P. aestuarii (8) has shown that the acceptor can only be a BChl c-like pigment. A scheme of electron transfer in P. aestuarii is given in Figure 2B.

The recently discovered photosynthetic bacterium Heliobacterium chlorum (35) contains BChl g as major pigment (11). Neither its pigment composition, nor its structure, provides a clue to its taxonomic position as it does not seem to posess chlorosomes or invaginations of the cytoplasmic membrane (35). However, its electron transport pathway appears to be similar to that of green sulfur bacteria. The primary electron donor of Heliobacterium is probably a dimer of BChl g, called P 798 (33,65). Flash spectroscopy of isolated membranes indicates that, like in those of P. aestuarii, reduction of the primary acceptor causes a bleaching at 670 nm, suggesting that I is likewise a BChl c-like compound (56). There is evidence that the second acceptor has a midpoint potential of about -500 mV (56) and that the electron transport chain may contain one or more iron-sulfur centers (12,65).

5. THE PRIMARY DONOR COMPLEX OF PHOTOSYSTEM II

Analogous to the primary donors in bacteria and in PS I, the primary donor of PS II is a special Chl a. Its long-wave absorption band at 680 nm is bleached upon the formation of the P 680 cation radical; the cation has a weak broad absorption band around 820 nm.

Most experiments relating to P and its donor complex were performed with chloroplast preparations in which oxygen evolution did not occur (e.g. after Tris-treatment (see ref. 91), at low temperature, or with unintentionally damaged preparations). In most of these preparations electron transfer from the oxygen evolving complex (OEC) to P was interrupted. The evidence for the existence and function of P is largely based on circumstantial evidence obtained in various laboratories (for reviews see refs. 7,83). Here we will mainly discuss experiments with active OEC's.

Oxidation and reduction of P have been studied not only by changes in absorbance but also by changes in fluorescence yield. Okayama and Butler (59, see also ref. 15) observed that, at liquid nitrogen temperature, the fluorescence rise upon light-induced Q_A^- (C-550) formation was strongly decreased, when the chloroplast suspension was frozen in the presence of ferricyanide. They proposed that the dark reduction of some photooxidized component at the oxidizing site of PS II, possibly P^+, was prevented or retarded, and that this oxidized component quenched the Chl _a_ fluorescence. The hypothesis that P^+ would be a quencher of fluorescence was supported by measurements of Duysens and coworkers (28), who studied the relation between luminescence and fluorescence in the green alga Chlorella pyrenoidosa. A low fluorescence yield after a flash was found to be correlated with the occurrence of a 15 μs luminescence component, caused by the back reaction $P^+Q_A^- \rightarrow P^*Q_A$.

Studies of the re-reduction of P^+ by the secondary electron donor Z have shown that the rate of this reaction depends on the redox state (S-state, ref. 45) of the OEC. Measurements of absorption (80) and of fluorescence (75) changes showed that in the singly oxidized state (S_1) the reaction has a time constant of about 30 ns; in higher S-states, the re-reduction of P^+ is slowed down to several hundred ns (75). More precise, recent measurements of the kinetics at 820 nm (10,71) indicated that in most reaction centers P^+ was reduced in the states S_0 and S_1 with $t_{1/2} \approx 23$ ns. In the higher oxidation states S_2 and S_3, two phases with equal amplitudes with $t_{1/2} \approx 50$ and 260 ns were seen (10). A slow phase (of 35 μs) occurred in about 15 % of the reaction centers in the states S_2 and S_3. It was proposed that the S-state dependence of the kinetics was caused by the Coulombic force of a positive charge in the OEC in the states S_2 and S_3. Two donors in series, which we call Z_2 and Z_1, were postulated to operate between the OEC and P (see also ref. 83 for a discussion). The positive charge in S_2 or S_3 slows down the rate of electron transfer to P^+. In the microsecond time region after a flash, the positive charge occurs in most centers on Z (Z_1 or Z_2), but in the states S_1 and S_2 the 'equilibrium' charge distribution is such that, at any moment, in some reaction centers the charge occurs on P; P^+ decay reflects the reduction of Z^+, which can be observed as a decay in the 10 - 40 μs region, both in the 820 nm absorption (71) and in the yield of Chl _a_ fluorescence and luminescence (28).

From ESR data, mainly obtained under conditions in which electron transfer from the OEC to P^+ was blocked, Babcock and coworkers (3) concluded that a PS II donor (possibly Z_1) is a bound plastoquinol, which is oxidized by P^+ to its cation radical. The same conclusion was obtained by Dekker and coworkers (22) from the optical difference spectrum (Figure 3). Further studies of Z by means of absorption difference spectroscopy (23), ESR and luminescence may well enable a more complete elucidation of the donor reactions.

REFERENCES

1. Amesz J (1985) Progr. Bot. 47: 87-104
2. Amesz J and Knaff DB (1986) In: Environmental Microbiology of Anaerobes (Zehnder, A.J.B., ed.), in the press, John Wiley, New York
3. Babcock GT, Buttner WJ, Ghanotakis DF, O'Malley PJ, Yerkes CT and Yocum CF (1984) In: Advances in Photosynthesis Research (Sybesma, C, ed.), Vol. I, pp. 243-252, Martinus Nijhoff/Dr W Junk Publishers, The Hague
4. Blankenship RE (1984) Photochem Photobiol 40: 801-806
5. Blankenship RE (1985) Photosynth Res 6: 317-333
6. Bonnerjea J and Evans MCW (1982) FEBS Lett 148: 313-316
7. Bouges-Bocquet B (1980) Biochim Biophys Acta 594: 85-103

8. Braumann T, Vasmel H, Grimme LH and Amesz J (1986) Biochim Biophys Acta 848: 83-91
9. Breton, J. (1985) Biochim Biophys Acta 810: 235-245
10. Brettel K, Schlodder E and Witt HT (1984) Biochim Biophys Acta 766: 403-418
11. Brockmann H and Lipinski A (1983) Arch Microbiol 136: 17-19
12. Brok H, Vasmel H, Horikx JTG and Hoff AJ (1986) FEBS Lett 194: 322-326
13. Butler WL (1972) Proc Natl Acad Sci USA 69: 3420-3422
14. Butler WL (1983) Acc Chem Res 6: 177-184
15. Butler WL (1972) Biophys J 12: 851-857
16. Butler WL, Visser JWM and Simons HL (1973) Biochim Biophys Acta 292: 140-151
17. Cogdell RJ (1983) Annu Rev Plant Physiol 34: 21-45
18. Cramer WA and Butler WL (1969) Biochim Biophys Acta 172: 503-510
19. Crofts AR and Wraight CA (1983) Biochim Biophys Acta 726: 149-185
20. Deisenhofer J, Epp O, Miki K, Huber R and Michel H (1984) J Mol Biol 180: 385-398
21. Deisenhofer J, Epp O, Miki K, Huber R and Michel H (1985) Nature 318: 618-624
22. Dekker JP, Brok M and van Gorkom HJ (1984) In: Advances in Photosynthesis Research (Sybesma, C, ed.), Vol. I, pp 171-174, Martinus Nijhoff/Dr W Junk Publishers, The Hague
23. Dekker JP, van Gorkom HJ, Brok M and Ouwehand L (1984) Biochim Biophys Acta 764: 301-309
24. den Blanken HJ and Hoff AJ (1982) Biochim Biophys Acta 681: 365-374
25. den Blanken HJ, Vasmel H, Jongenelis APJM, Hoff AJ and Amesz J (1983) FEBS Lett 161: 185-189
26. Dornemann D and Senger H (1981) FEBS Lett 126: 323-327
27. Duysens LNM and Sweers HE (1963) In: Studies on Microalgae and Photosynthetic Bacteria. Special Issue of Plant Cell Physiol, pp. 353-372, University of Tokyo Press, Tokyo
28. Duysens LNM, den Haan GA and van Best JA (1974) In: Proc. 3rd Int Congr Photosynthesis (Avron, M, ed.) Vol. I, pp 1-13, Elsevier, Amsterdam
29. Evans MCW (1985) Physiol Végét 23: 563-569
30. Evans MCW, Telfer A and Lord AV (1972) Biochim Biophys Acta 267: 530-537
31. Erixon K and Butler WL (1971) Biochim Biophys Acta 234: 381-389
32. Feher G and Okamura MY (1978) In: The Photosynthetic Bacteria (Clayton RK and Sistrom WR, eds.) pp. 349-386, Plenum Press, New York
33. Fuller RC, Sprague SG, Gest H and Blankenship RE (1985) FEBS Lett 182: 345-349
34. Gast P, Swarthoff T, Ebskamp FCR and Hoff AJ (1983) Biochim Biophys Acta 722: 163-175
35. Gest H, and Favinger JL (1983) Arch Microbiol 136: 11-16
36. Hoff AJ (1984) Quart Rev Biophys 17: 153-282
37. Ke B (1972) Arch Biochem Biophys 152: 70-77
38. Ke B (1973) Biochim Biophys Acta 301: 1-33
39. Kirmaier C, Holten D, Feick R and Blankenship RE (1983) FEBS Lett 158: 73-78
40. Kirmaier C, Holten D, Mancino LJ and Blankenship RE (1984) Biochim Biophys Acta 765: 138-146
41. Kitajima M and Butler WL (1973) Biochim Biophys Acta 325: 558-564
42. Klimov VV and Krasnovskii AA (1982) Biophysics (USSR) 27: 186-198
43. Knaff DB and Arnon DI (1969) Proc Natl Acad Sci USA 63: 956-962
44. Knaff DB and Arnon DI (1969) Proc Natl Acad Sci USA 63: 963-969

45. Kok B, Forbush B and McGloin MP (1970) Photochem Photobiol 11: 457-475
46. Kyle DJ (1985) Photochem Photobiol 47:107-116
47. Lozier RH and Butler WL (1974) Biochim Biophys Acta 333: 460-464
48. Malkin R and Bearden AJ (1978) Biochim Biophys Acta 505: 147-181
49. Mansfield RW and Evans MCW (1985) FEBS Lett 190: 237-241
50. Maroti P, Kirmaier C, Wraight C, Holten, D and Pearlstein, RM (1985) Biochim Biophys Acta 810: 132-139
51. Michel H (1982) J Mol Biol 158: 567-572
52. Michel H (1983) Trends Biochem Sci 8: 56-59
53. Michel H, Weyer KA, Gruenberg H and Lottspeich F (1985) EMBO J 4: 1667-1672
54. Nuijs AM (1986) Doctoral thesis, University of Leiden, The Netherlands
55. Nuijs AM, van Bochove AC, Joppe HLP, Duysens LNM and Amesz J (1985) Biochim Biophys Acta 807: 24-34
56. Nuijs AM, van Dorssen RJ, Duysens LNM and Amesz, J (1985) Proc Natl Acad Sci USA 82: 6856-6868
57. Nuijs AM, van Gorkom HJ, Plijter JJ and Duysens LNM (1986) Biochim Biophys Acta, in the press
58. Okamura MY, Feher G and Nelson N (1982) In: Photosynthesis, Vol. 1, Energy Conversion by Plants and Bacteria (Govindjee, ed.) pp.195-272, Academic Press, New York
59. Okyama S and Butler WL (1972) Biochim Biophys Acta 267: 523-529
60. Olson JM (1980) Biochim Biophys Acta 594: 33-51
61. O'Malley PJ and Babcock GT (1984) Proc Natl Acad Sci USA 8: 1098-1101
62. Parson WW and Ke B (1982) In: Photosynthesis, Vol. 1, Energy Conversion by Plants and Bacteria (Govindjee, ed.) pp. 331-385, Academic Press, New York
63. Pierson BK and Castenholz RW (1974) Arch Mikrobiol 100: 5-24
64. Pierson BK and Thornber JP (1983) Proc Natl Acad Sci USA 80: 80-84
65. Prince RC, Gest H and Blankenship RE (1985) Biochim Biophys Acta 810: 377-384
66. Robert B, Lutz M and Tiede DM (1985) FEBS Lett 183: 326-330
67. Rutherford AW and Heathcote P (1985) Photosynth Res 6: 295-316
68. Rutherford AW, Agalides I and Reiss-Husson F (1985) FEBS Lett 182: 151-157
69. Sauer K, Mathis P, Acker S and van Best JA (1979) Biochim Biophys Acta 545: 466-472
70. Schatz GH and van Gorkom HJ (1985) Biochim Biophys Acta 810: 283-294
71. Schlodder E, Brettel K and Witt HT (1985) Biochim Biophys Acta 808: 123-131
72. Shuvalov VA, Klevanik AV, Sharkov AV, Kryukov PG and Ke B (1979) FEBS Lett 107: 313-316
73. Shuvalov VA and Duysens LNM (1986) Proc Natl Acad Sci USA, in the press
74. Shuvalov VA, Shkuropatov AYa, Kulakova CM, Ismailov MA and Shkuropatova VA (1986) Biochim Biophys Acta, in the press
75. Sonneveld A, Rademaker H and Duysens LNM (1979) Biochim Biophys Acta 548: 536-551
76. Swarthoff T, van der Veek-Horsley KM and Amesz J (1981) Biochim Biophys Acta 635: 1-12
77. Swarthoff T, Gast P, Hoff AJ and Amesz J (1981) FEBS Lett 130: 93-98
78. Swarthoff T, Gast P, Amesz J and Buisman HP (1982) FEBS Lett 146: 201-207
79. Trebst AV and Draber W (1986) Photosynth Res, this issue
80. van Best JA, and Mathis P (1978) Biochim Biophys Acta 503: 178-188

81. van Bochove AC, Swarthoff T, Kingma H, Hof RM, van Grondelle R, Duysens LNM and Amesz J (1984) Biochim Biophys Acta 764: 343-346
82. van Gorkom HJ (1974) Biochim Biophys Acta 347: 439-442
83. van Gorkom HJ (1985) Photosynth Res 6: 97-112
84. van Gorkom HJ, Tamminga JJ, Haveman J and van der Linden IK (1974) Biochim Biophys Acta 347: 417-438
85. Vasmel H (1986) Doctoral thesis, University of Leiden, The Netherlands
86. Vasmel H and Amesz J (1983) Biochim Biophys Acta 724: 118-122
87. Vasmel H, Meiburg RF, Kramer HJM, de Vos LJ and Amesz J (1983) Biochim Biophys Acta 724: 333-339
88. Velthuys BR (1982) In: Function of Quinones in Energy Conserving Systems (Trumpower, BL, ed.) pp. 401-408, Academic Press, New York
89. Vermaas WFJ and Govindjee (1981) Photochem Photobiol 34: 775-793
90. Witt K (1973) FEBS Lett 38: 116-118
91. Yamashita T and Butler WL (1968) Plant Physiol 43: 1978-1986

Photosynthesis Research 10: 347–354 (1986)
© Martinus Nijhoff Publishers, Dordrecht

INFLUENCE OF MAGNETIC FIELDS ON THE P-870 TRIPLET STATE IN RPS.
SPHAEROIDES REACTION CENTERS

M.H. VIDAL,[*] P. SETIF and P. MATHIS
Service de Biophysique, Centre d'Etudes Nucléaires de Saclay, F-91191
Gif-sur-Yvette Cedex, France

Key words : Photosynthesis ; Purple bacteria ; Reaction center ; Magnetic
field ; Quinone reduction ; Triplet state

Abstract

Magnetic fields influence two properties of the P-870 triplet state
observed in Rps. sphaeroides reaction centers : the yield of formation and
the kinetics of decay. These effects have been studied in reaction centers
which were prepared in three different states : state Q_A^-, state Q_A^{2-} and
state $(- Q_A)$ (Q_A depleted). The triplet yields decrease with increasing
magnetic fields, with B½'s of about 140, 41 and 57 Gauss, respectively.
The half-time of ^{3}P-870 decay is not influenced by the field in state Q_A^- ;
it increases at increasing fields, in state Q_A^- and state $(- Q_A)$, with
the same B½ as the triplet yield. These results are discussed in the
framework of current theories of the radical-pair dynamics and of the
mechanism of triplet decay.

*Département de Physico-Chimie, Service de Chimie Moléculaire, Centre
d'Etudes Nucléaires de Saclay. UA331 (CNRS)

Abbreviations : I, primary electron acceptor ; LDAO, lauryldimethylamine
oxide ; P-870, primary electron donor ; Q_A, first quinone acceptor ; SDS,
sodium dodecylsulfate ; YAG, Yttrium Aluminum Garnet.

Introduction

Many recent studies on the reaction center of photosynthetic bacteria
have been very fruitful, providing good models for higher plant and algal
photosystems. A precise understanding of structural and functional
properties of reaction centers still raises many important questions whose
answers require the combined use of many approaches. Among these questions
are the dynamic properties of the primary radical pair $(P-870^+ - I^-)$,
where P-870 is the primary electron donor and I is the bacteriopheophytin
molecule involved in electron transfer. Following light excitation, the
radical pair is initially formed in a singlet state, and it changes by
rapid electron transfer from I^- to the primary quinone Q_A. When Q_A is
prereduced or removed from the reaction center, a substantial fraction of
the radical pairs (depending particularly on temperature) ends in a
localized triplet state, ^{3}P-870. The amount of triplet formed is also
influenced by a magnetic field (2,8), a property which, together with the
unequal population of the triplet sublevels (9,19) and with the results of
nanosecond studies (5,10,11,20), led to rather detailed physical
descriptions of the radical pair dynamics and of the magnetic interactions
between the reaction center partners (3,6). Several experimental aspects
of the behavior of the radical pair and of the localized triplet, however,

are still not understood, such as the influence of many experimental conditions on the triplet yield and lifetime : presence or absence of Q_A, single or double reduction of Q_A, temperature, etc.

In this paper, we report our experiments on the effect of a magnetic field on the yield and lifetime of ^{3}P-870 in reaction centers from the carotenoid-less R-26 mutant of Rps. sphaeroides, in which Q_A is singly or doubly reduced, or removed. Our results complement a recent work by Chidsey et al. (4) ; they largely support their conclusions but also raise a few additional questions.

Materials and Methods

Cells of Rhodopseudomonas sphaeroides strain R-26 were grown under anaerobic conditions, and reaction centers were prepared after solubilization of the membranes with LDAO (1). Excess LDAO was removed by dialysis against 0.05 % Triton X-100 in 50 mM Tris-HCl, pH 8.0. Reaction centers depleted of the primary ubiquinone Q_A were prepared by incubation with 4 % LDAO and 10 mM o-phenanthroline, following Okamura et al. (12). For spectroscopic measurements the reaction centers were dissolved in 50 mM Tris-HCl, pH 8.0, in a 10 x 10 mm square cuvette, at a concentration of around 0.004 mM. All experiments were performed at room temperature (21°C).

Flash-induced absorption changes at 870 nm were measured essentially as in Schenck et al (16), with a time resolution of 1 MHz. The measuring light was filtered by two interference filters with 3 nm bandwidth placed between the lamp and the cuvette, and in front of the photodetector. Two types of actinic flashes were used : a 20 ns (full width at half maximum) flash (broadband around 595 nm, obtained by pumping a dye laser with a frequency-doubled YAG laser pulse) at a repetition rate of 0.2 Hz, or a 30 ps pulse from a YAG picosecond laser (532 nm) at a repetition rate of 1 Hz. The flash energy was attenuated in order to excite at most 10 % of the reaction centers in the cuvette. The cuvette was inserted in a laboratory made Helmholtz coil, providing magnetic fields up to 480 Gauss. The intensity of the magnetic field versus the dC-current through the coil was calibrated with a Hall probe. Experiments at a given field were always made alternately with experiments at zero field, to look for possible changes in the sample. A decrease in the signal size at zero field (less than 15 %), without any modification in the kinetics, was observed in a few cases in the presence of dithionite and was presumably due to sample ageing. Between two control experiments at zero field, the signal size was corrected by the appropriate correction factor.

Results

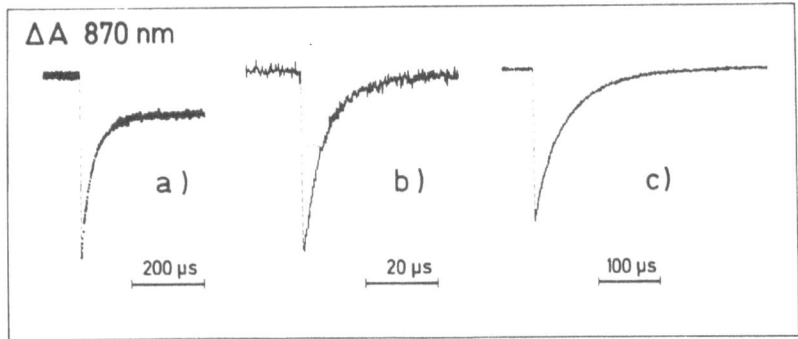

Fig. 1 Kinetics of flash-induced absorption changes at 870 nm, at zero
magnetic field, with : a/ quinone-depleted reaction centers (reaction
center concentration : 0.003 mM, ΔA max = 2.6 x 10^{-3} ; addition of 0.3 mM
ascorbate; average of 10 flashes), b/ reaction centers in the state Q_A
(reaction center concentration : 0.0025 mM, ΔA max = 1.0 x 10^{-3} ;
addition of 0.6 mg.ml^{-1} sodium dithionite ; average of 40 flashes) and c/
reaction centers in the state Q_A^{2-} (reaction center concentration :
0.0025 mM, ΔA max = 1.8 x 10^{-3} ; addition of 2 mg.ml^{-1} sodium dithionite,
without air ; average of 40 flashes). Experiment a : 20-ns laser ;
experiments b,c : 30-ps laser.

Quinone-depleted Reaction Centers

    When reaction centers are depleted of the primary quinone Q_A, the
primary photochemistry produces the state (P-870$^+$ - I$^-$) which leads to
^{3}P-870 by recombination (5,11,14,15). This behavior is illustrated in Fig.
1a : a bleaching at 870 nm recovers biphasically : a fast phase with t½ =
18 us attributable to the decay of ^{3}P, and a very slow phase which
presumably takes place in reaction centers where Q_A is still present.
Application of a magnetic field induces two effects : a decrease in the
amount of ^{3}P formed to 52 % of the maximum, and a slowing down of the
decay rate, as shown in Fig. 2 (t½ = 22 us at high field). Both effects of
the field practically saturate at 140 Gauss, and the half-effect field is
B½ = 57 Gauss. The slowly decaying signal is not influenced by magnetic
fields.
    Addition of sodium dithionite (3 mM) leads to a disappearance of the
slow phase (not shown). The fast phase has about the same size as without
dithionite as it is expected considering the low yield of triplet state
formation at room temperature (6) and the magnetic field effect is the
same (Fig. 2). The absorption change recovery, however, is somewhat
slower : t½ = 20 us for B = 0, t½ = 28 us at high field.

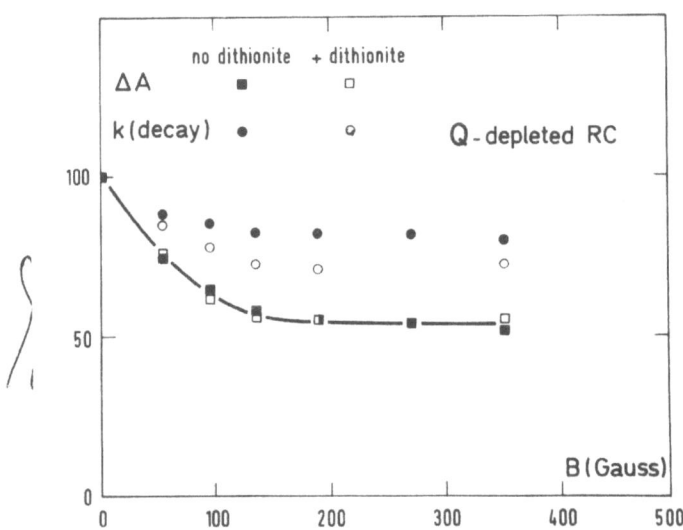

Fig. 2 Effect of magnetic field on the absorption change at 870 nm induced by 20-ns laser flashes in quinone-depleted reaction centers (concentration: 0.0032 mM). Addition of 0.3 mM ascorbate (full symbols) or 0.5 mg.ml^{-1} dithionite (open symbols). Average of 10 flashes at each field value.
Squares : relative magnitude of the absorption change decaying between t = 0 and 0.2 ms (ΔA = 100 corresponds to 3.2 x 10^{-3} for both treatments). Circles: relative rate constant of absorption change recovery (k = 100 corresponds to 39.10^3 s^{-1} with ascorbate and 35.10^3 s^{-1} with dithionite).

The slowing down of ^{3}P decay by sodium dithionite is unexpected, since quinone-depleted reaction centers have no constituent susceptible of reacting with dithionite. One possible effect of sodium dithionite is to decrease the oxygen concentration in the cuvette : O_2 might be a quencher of ^{3}P in quinone-depleted reaction centers. Shuvalov and Parson (17) observed a quenching of ^{3}P by O_2 in Rps. viridis reaction centers, but not in Rps. sphaeroides. However, their reaction centers were treated in a different manner, i.e. extraction of Q_A by SDS, and ^{3}P has a different behavior in their material (halftime of 60 us instead of 18 us).

Reaction Centers in the state Q_{A-}^-

Reaction centers supplemented with a low concentration of sodium dithionite give a flash-induced bleaching at 870 nm (Fig. 1) which recovers exponentially with t½ = 5.5 us in agreement with previous reports

(13,16,18). In some cases a small slow phase of variable amplitude was also present (not shown). Presumably it is formed in reaction centers where Q_A was not reduced. That slow phase has not been considered in the analysis of our results.

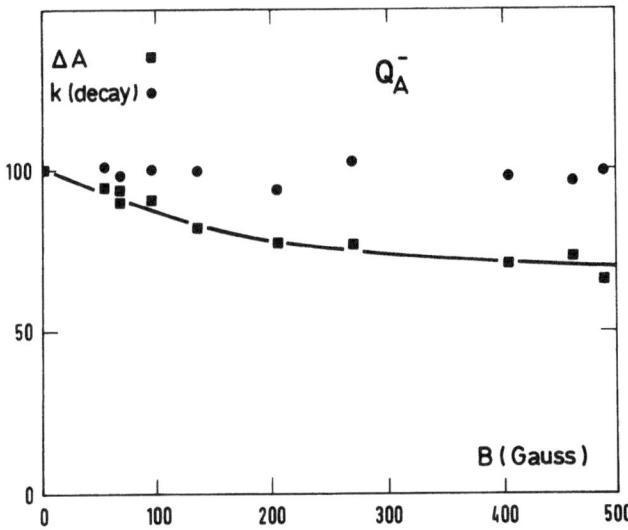

Fig. 3 Effect of magnetic field on the absorption change at 870 nm induced by 30-ps laser flashes in reaction centers (0.002 mM) with singly reduced Q_A (addition of 0.6 mg.ml^{-1} sodium dithionite). Average of 40 flashes at each field value. Squares : magnitude of absorption change at zero time, as appreciated with a 1-MHz bandwidth (A = 100 correspond to 0.8 x 10^{-3}). Circles : relative rate constant of absorption change recovery (k = 100 corresponds to 13 x 10^4 s^{-1}).

Application of a magnetic field decreases the triplet yield (Fig. 3) to 70 % of its maximum at 480 Gauss. If we consider that the effect is saturated at 480 Gauss, we obtain a B½ of 140 Gauss. It seems more probable, however, that saturation is not achieved and that B½ is larger. By contrast with the case of quinone-depleted reaction centers, magnetic fields have no influence on the triplet decay kinetics (Fig. 3).

Reaction Centers in the state Q_A^{2-}

When reaction centers are left to incubate with a higher sodium dithionite concentration (12 mM), the kinetics of absorption recovery at 870 nm become slower (Fig. 1). Under these conditions, Q_A was probably doubly reduced (17), as it was checked in the following way : after incubation with sodium dithionite, the sample was dialyzed against buffer for a few hours and then shown to behave identically as in the state Q_A^- (same signal size and kinetics of decay), thus indicating that the quinone molecule remains in the reaction center. In that state the triplet yield is decreased by a magnetic field (Fig. 4). A well-defined saturated effect is observed, with a yield at 60 % of the zero-field value and a B½ of 41 Gauss. The triplet decay is slowed down in parallel, with t½ = 36 us for B = O and 49 us at high field, and a B½ of 41 Gauss. It is noticeable that, at low laser excitation, the bleaching at 870 nm was about twice bigger in the state Q_A^{2-} than in the state Q_A^-, indicating a twice higher yield of triplet formation.

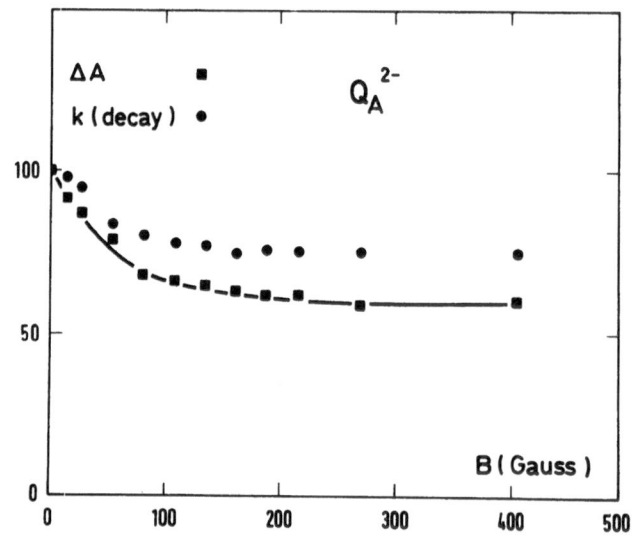

Fig. 4 Effect of magnetic field on the absorption changes at 870 nm induced by 30-ps laser flashes in reaction centers (0.002 mM) with doubly reduced Q_A (addition of 2 mg.ml^{-1} and incubation). Average of 20 flashes at each field value. Squares : magnitude of absorption change at zero time (ΔA = 100 correspond to 1.6×10^{-3}). Circles : relative rate constant of absorption change recovery (k = 100 corresponds to 20×10^3 s^{-1}).

Table : Magnetic field effects on the P-870 triplet state in different states of the reaction centers

Measurement / State of reaction centers	$(-Q_A)$	Q_A^-	Q_A^{2-}
t½ at zero field	18 us	5.5 us	36 us
t½ at high field	22 us	5.5 us	49 us
Ratio of the signal size at high field to the signal size at zero field	52 %	70 %	60 %
Half-effect field B½ (Gauss)	57 G	$\geq$140 G	41 G

Discussion

The present work deals with two aspects of the influence of magnetic fields on the P-870 triplet state : yield of formation and rate of decay. The yield of formation has been studied by several groups, unfortunately, under rather different experimental conditions (biological material, state of Q_A, saturation level of the actinic light). Our results were obtained with a single source of reaction centers, at a low laser exciting light where the size of A was linear with laser energy, with three different states of Q_A. Our results are in good agreement with published data concerning quinone-depleted reaction centers (5,11,14) : the B½ value is rather low (57 Gauss ; reported values range from 33 to 45 Gauss) and the triplet yield is strongly reduced (to about 50 % of its maximum value) at high field. As shown in this work, the field effect is very similar in the state that we interpret as Q_A^{2-}, for which magnetic field effects have not been reported in the literature. It is worth noting that in both cases (Q^{2-} or no quinone) there is no paramagnetic species with which the primary radical pair (P-870$^+$, I$^-$) could interact in the reaction center.

The situation is quite different in the state Q_A^-, which is paramagnetic, and for which we find a much higher B½ value of at least 140 Gauss. The maximum effect on the yield is smaller (30 % decrease at our highest field), but this might reflect an absence of saturation. The data in the literature, unfortunately, are rather incoherent. The oldest data by Blankenship et al (2) are fully consistent with ours, but more recent results indicated a low B½ in reaction centers (around 50 Gauss in ref. 10, and 50- 100 Gauss in ref. 8). In ref. 8 and 10, however, no information was given on the ^{3}P lifetime, and it is quite possible that some reaction centers were in the states Q_A^{2-} or $(-Q_A)$.

As shown in Figs. 2 and 4, the kinetics of ^{3}P decay show the same field dependence as the triplet yield, when the quinone is absent or doubly reduced. This identical behavior indicates that a common mechanism is at the origin of both properties. A field-induced splitting of the three sublevels in the triplet state of the radical-pair nicely accounts for the field dependence of the triplet yield (3,6). Chidsey et al (4) recently suggested that the same phenomenon is responsible for the slower triplet decay in the presence of a magnetic field, making the hypothesis that ^{3}P-870 decays mostly, at room temperature, by thermally activated repopulation of the triplet radical pair. Our results are consistent with these proposals, although we find a weaker field effect on the decay of ^{3}P-870, after quinone removal or in the state Q_A^{2-}, than the factor of two reported by Chidsey et al (4). However, it appears that the magnetic field does not influence the rate of ^{3}P-870 decay in the state Q_A^- (4, and this paper) whereas the triplet yield is decreased, but to a lesser extent and at higher magnetic fields than in the two other states. Our experiments show that this different behavior is neither due to the use of different preparations of reaction centers, nor to differences in conditions of light excitation. The presence of a paramagnetic quinone seems to play an important role in the yield of formation and decay kinetics of the P-870 triplet state (4, 17 and this paper). This can be due either to magnetic interactions between ^{3}P-870 and Q_A (21) or to magnetic interactions between an intermediate state, presumably the radical pair state (P-870$^+$ - I$^-$), and Q_A^--Fe. Such magnetic interactions have been extensively studied in reactions centers where the iron is decoupled from the quinone by detergent treatment (22-24). In any case, the scheme for ^{3}P-870 decay by repopulation of the biradical state, proposed by Chidsey et al (4), cannot account in its present form for the decay of ^{3}P-870 in the state Q_A^-.

References

1. Berger G, Tiede MD and Breton J. (1984) Biochem. Biophys. Res. Comm. 121:47-54
2. Blankenship RE, Schaafsma TJ and Parson, WW (1977) Biochim. Biophys. Acta 461:297-305
3. Boxer SG, Chidsey CED and Roelofs, MG (1983) Ann. Rev. Phys. Chem. 34:389-417
4. Chidsey CED, Takiff L, Goldstein, RA and Boxer SG (1985) Proc. Natl. Acad. Sci. USA 82:6850-6854
5. Chidsey CED, Kirmaier C, Holten D and Boxer, SG (1984) Biochim. Biophys. Acta 766:424-437
6. Hoff AJ (1981) Quart. Rev. Biophys. 14:599-665
7. Hoff AJ and Hore P (1984) Chem. Phys. Lett. 108:104-110
8. Hoff AJ, Rademaker H, van Grondelle R and Duysens LNM (1977) Biochim. Biophys. Acta 460:547-554
9. Leigh JS and Dutton PL (1974) Biochim. Biophys. Acta 357:67-77
10. Michel-Beyerle ME, Scheer H, Seidlitz H, Tempus D and Haberkorn, R (1979) FEBS Lett. 100:9-12
11. Ogrodnik A, Krüger HW, Orthuber H, Haberkorn R and Michel-Beyerle ME (1982) Biophys. J. 39:91-99
12. Okamura MY, Isaacson RA and Feher G (1975) Proc. Natl. Acad. Sci. USA 72:3491-3495
13. Parson WW, Clayton RK and Cogdell RJ (1975) Biochim. Biophys. Acta 387:265-278
14. Roelofs MG, Chidsey CED and Boxer SG (1982) Chem. Phys. Lett. 87:582-588
15. Schenck CC, Blankenship RE and Parson WW (1982) Biochim. Biophys. Acta 680:44-59
16. Schenck CC, Mathis P and Lutz M (1984) Photochem. Photobiol. 39:407-417
17. Shuvalov VA and Parson WW (1981) Biochim. Biophys. Acta 638:50-59
18. Shuvalov VA and Parson WW (1981) Proc. Natl. Acad. Sci. USA 78:957-961
19. Thurnauer MC, Katz JJ and Norris JR (1975) Proc. Natl. Acad. Sci. USA 72:3270-3274
20. Wasielewski MR, Bock CH, Bowman MK and Norris JR (1983) Nature 303:520-522
21. De Groot A, Lous EJ and Hoff AJ (1985) Biochim. Biophys. Acta 808:13-20
22. Gast P and Hoff AJ (1979) Biochim. Biophys. Acta 548:520-535
23. Gast P, Mushlin RA and Hoff AJ (1982) J. Phys. Chem. 86:2886-2891
24. Gast P, De Groot A and Hoff AJ (1983) Biochim. Biophys. Acta 723:52-58

Photosynthesis Research 10: 355–361 (1986)
© *Martinus Nijhoff Publishers, Dordrecht*

PHOTOREDUCTION OF PHEOPHYTIN IN PHOTOSYSTEM II
OF THE WHOLE CELLS OF GREEN ALGAE AND CYANOBACTERIA

V.V. KLIMOV, S.I. ALLAKHVERDIEV and V.G. LADYGIN

Institute of Soil Science and Photosynthesis, USSR Academy of
Sciences, Pushchino, 142292 USSR

Key words. Pheophytin; Photosystem II; Reaction center, Water
splitting; $NADP^+$ reduction (Algae cells)

Abstract. Photoreduction of Pheophytin 'a' (Pheo) accompanied by
a decrease in the chlorophyll fluorescence yield is observed in photosystem II (PS II) of the whole cells of green algae Chlamydomonas reinhardii (a wild type and a mutant lacking both photosystem I
and chlorophyll 'b'), Chlorella pyrenoidosa, Scenedesmus obliquus
and cyanobacteria Phormidium laminosum, Anabaena variabilis and
Cynechococcus elongatus under anaerobic conditions created by
means of the glucose–glucoseoxidase–catalase. The photoreaction is
activated by the addition of 1 μM CCCP, inhibited by 10 μM DCMU
and reactivated upon subsequent addition of either ascorbate or dithionite. Oxidized NADP, benzyl viologen and methyl viologen accelerate dark oxidation of the reduced Pheo indicating that they are
able to accept an electron from $Pheo^-$ in PS II.
  The data on both photoreduction of Pheo in the intact cells in
the absence of exogenous reductants, when electron donation to reaction centers of PS II occurs only from water, and the inhibition
of this photoreaction by DCMU, show that the Pheo photoreduction
is sensitized by the reaction centers of PS II and probably occurs
as a result of the electron donation from the water –splitting system
being in the state S_3, to $[P^{+\cdot}_{680}\ Pheo^-]$ Q^- producing the long-lived state S_0 $[P_{680}\ Pheo]\ Q^-$ and O_2.

Introduction. Pheophytin 'a' (Pheo) participates in primary photosynthetic reactions acting as an intermediary electron acceptor in reaction centers (RC) of photosystem II (PS II) between the primary
electron donor, chlorophyll P_{680}, and the primary ('stable') electron
acceptor Q (a complex of plastoquinone, with Fe) [1]. If Q is prereduced in chloroplasts or PS II particles in the dark (with dithionite) reversible photoaccumulation of the state $[P_{680}Pheo]$ Q^- is
observed as a result of electron donation from a secondary electron
donor to $[P^{+\cdot}_{680}\ Pheo^-]$ Q^- [1]. This photoreaction is accompanied by a decrease of chlorophyll fluorescence yield which is, in fact,
the disappearance of the nanosecond recombination luminescence
arising from charge recombination in the pair $[P^{+\cdot}_{680}\ Pheo^-]$ [1].
The relation of Pheo photoreduction to phototransfer of electron in RC of PS II is indicated by many observations [1]
though one cannot completely exclude the fact that some Pheo
outside the RC (formed during isolation of PS II or chloroplasts)
may react in the light with the strong reductant dithionite.
Recently photoreduction of Pheo has been shown in chloroplasts
and PS II particles in the absence of dithionite under anaerobic

conditions [2]. Here we demonstrate reversible photoreduction of
Pheo in the RC of PS II in the whole cells of green algae and
cyanobacteria under anaerobic conditions.

MATERIALS AND METHODS

The green algae Chlorella pyrenoidosa and Scenedesmus obliquus
and cyanobacteria Phormidium laminosum, Anabaena variabilis and
Cynechococcus elongatus were grown in liquid media during 3-6
days; Chlamydomonas reinhardii, a wild type and a mutant ACC-1
lacking photosystem I and Chl 'b', were grown on an agar-agar medium [3]. The rate of oxygen evolution under illumination of the mutant in the presence of 0.5 mM phenyl p-benzoquinone was about
280 μmol O_2 per mg Chl per h. During the experiments the algae
and cyanobacteria were suspended in the medium containing 50 mM
HEPES (pH 6.7), 35 mM NaCl/5 mM $MgCl_2$ and 5 mM KCl. Anaerobic conditions were created by the addition of glucose (10 mM),
glucose oxidase (approx. 50 U/ml) and catalase (approx. 1000 U/ml)
to a tightly closed 1-cm cuvette [2,5].

Photoinduced absorbance changes (ΔA) and changes of chlorophyll fluorescence yield (ΔF) with $\lambda > 670$ nm were measured with
the phosphoroscopic set-up as earlier [2].

RESULTS AND DISCUSSION

Chlamydomonas mutant ACC-1 lacking PS 1 and Chl 'b'

Under aerobic conditions the yield of fluorescence (F) excited in
the ACC-1 mutant by a low-intensity measuring light is increased
from the level F_0 to the level F_{max} equal to $F_0 + \Delta F$ when actinic
light 2 (which excites mainly PS II) is added (Fig. 1,A). The fluorescence increase reflects photoreduction of Q [1,2,4]. Further illumination leads to a slow decrease of F (Fig.1). After switching
off the actinic light F returns to the level F_0 showing that the weak
measuring light is not capable of keeping Q in the state Q^- under
aerobic conditions. Under anaerobic conditions F increases slowly
up to F_{max} (Fig. 1,B) reflecting photoreduction of Q by the measuring light. This is confirmed by the fact that a pulse of O_2 (bubbling the sample with air) lowers the fluorescence yield temporarily
(Fig.1,B). After creation of anaerobic conditions the actinic light 2
induces a significant decrease of F which is caused by photoreduction of some substance since O_2 added during the illumination results in an increase of F (Fig. 1,B). DCMU (10 μM) almost totally abolishes the photoinduced decrease of F ($-\Delta F$) (Fig. 1,B,
dashed trace). Actinic light 1 (exciting predominantly PS I) does
not have any effect on F of the ACC-1 mutant.

Rate and extent of the photoinduced decrease of F under anaerobic conditions are enhanced upon addition of 1 μM CCCP. Again
the fluorescence decrease is due to a reductive process since the
air bubbling during illumination leads to an increase of F (Fig. 1,B).
The negative ΔF induced by light 2 in the presence of CCCP, is
significantly diminished upon addition of DCMU and it is strongly reactivated by subsequent addition of ascorbate (Fig.1,B) or dithionite.
Addition of $NADP^+$ (3 mM) results in a remarkable (by a factor of
20-30) increase in the rate of the dark relaxation of the negative
photoinduced ΔF (revealing acceleration of the dark oxidation of
Pheo) and lowers the dark level of F (Fig.1,B). Similar effects were seen upon addition of 3 mM benzyl viologen or methyl viologen.

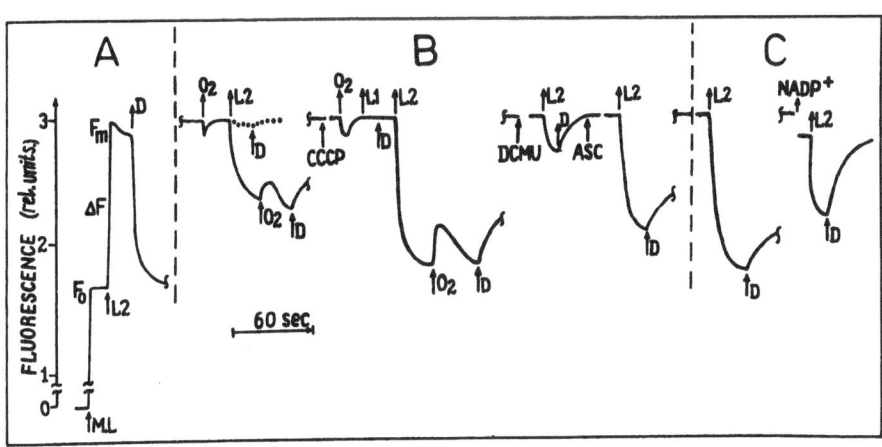

FIGURE 1. Kinetics of photoinduced ΔF in the ACC–1 <u>Chlamydo-</u>
<u>monas</u> mutant, lacking PS 1 and Chl 'b', under aerobic (A) and
anaerobic (B, C) conditions; Chl concn is 6 μg/ml. Designations:
$\uparrow$, the moment of switching on (L, light) and off (D, dark) or addition of agents: ML, the measuring light (λ = 480 nm; approx.
0.15 $J \cdot m^{-2} \cdot s^{-1}$) exciting F of Chl; L_2, actinic light 2 exciting mainly PS II ($\lambda \approx$ 600 nm; 10^2 $J \cdot m^{-2} \cdot s^{-1}$); L_1, actinic light 1 exciting
predominantly PS I ($\lambda >$ 710 nm; 30 $J \cdot m^{-2} \cdot s^{-1}$); GGOC, glucose
(10 mM), glucose oxidase (50 U/ml) and catalase (approx. 1000
U/ml); O_2, a fast air bubbling (1 cm^3) through the sample; CCCP,
carbonyl cyanide–m–chlorophenylhydrazone (1 μM); DCMU, 3–(3,4-
dichlorophenyl–1,1–dimethylurea)(10 μM); asc, sodium ascorbate
(5 mM); NADP+, oxidized NADP ($\sim$3 mM) $\dashv\vdash$, dark period for
10- 30 min. The dashed line shows the effect of light 2 in the presence of DCMU in a separate experiment. C, a separate sample in
the presence of GGOC and 1 μM CCCP.

In all cases, the photoinduced decrease of F under anaerobic
conditions is accompanied by absorbance changes (with the same
kinetics) the spectrum of which is shown in Fig.2 (points). It is
quite similar to the spectrum of photoreduction of Pheo in PS II
particles in the presence of dithionite measured earlier [1] and
shown in Fig.2 by a solid line. Photoinduced ΔA related to reduction of Pheo are inhibited (like ΔF) upon addition of DCMU
(Fig.2, inset), and they are reactivated by ascorbate.
        Thus, photoreduction of Pheo is observed in PS II of the who-
le algae cells in the absence of any exogenic reductants like dithionite when the sole source of electrons for a PS II reaction
center is the water –splitting system. The results provide convin-
cing evidence for the conclusion [1,2] that the Pheo photoreduction is characteristic of functionally active, native reaction centers
of PS II and that the photoaccumulation of Pheo$\bar{\cdot}$ results from donation of electron from the water–oxidizing system to the state

$[P_{680}^{+\bullet} \text{ Pheo}^-]$ Q^- formed in the primary photoreaction. Inhibition of the Pheo photoreduction by DCMU strongly supports the conclusion. As pointed out earlier [2] DCMU blocks electron transfer between Q^- and plastoquinone and makes it impossible for PS II during continuous illumination to reach the state $S_4 [P_{680}\text{Pheo}^-] Q^-$ which immediately produces O_2 and the long-lived state $S_0 [P_{680}\text{Pheo}^-]Q^-$ (S_0–S_4 are the states of the water-oxidizing system [6–8]). The states $S_2 [P_{680}\text{Pheo}^-] Q^-$ and (especially) $S_3 [P_{680}\text{Pheo}^-] Q^-$ which can be formed by light in the presence of DCMU may be unstable due to the possibility of a fast charge recombination between Pheo$^-$ and the states S_2 and S_3. This can result in the decrease of the long-lived spectral changes related to photoaccumulation of Pheo$^-$ in our experiments. Similarly, the well-known effect of acceleration of the decay of S_3 and S_2 states in the presence of CCCP (or other ADRY reagents) [6–8] can be responsible for the enhancement of spectral changes, related to photoreduction of Pheo, upon addition of CCCP (Fig.1,B). In other words, in the presence of CCCP an additional path for formation of the long-lived Pheo$^-$ from the states $S_2 [P_{680} \text{ Pheo}^-] Q^-$ and $S_3 [P_{680}\text{Pheo}^-] Q^-$ is opened.

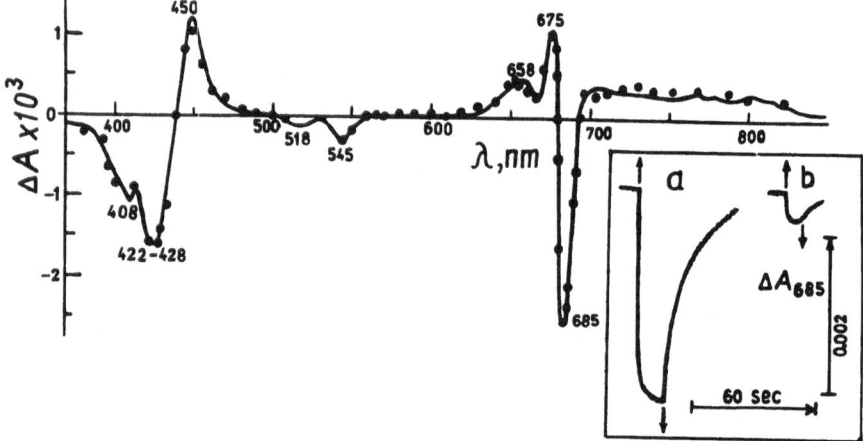

FIGURE 2. The spectrum of light-induced absorbance changes (ΔA) in the ACC–1 <u>Chlamydomonas</u> mutant under anaerobic conditions (see Fig.1, C) in the presence of GGOC, 1 μM CCCP and 3 mM NADP$^+$ (closed circles); the solid line is the light-minus-dark difference absorption spectrum observed at photoreduction of Pheo in pea PS II preparation in the presence of dithionite (taken from Ref.1). The spectra are normalized at 685 nm. Inset: kinetics of ΔA at 685 nm before (a) and after (b) addition of 10 μM DCMU; ↑ and ↓ , the actinic light 2 is on and off, respectively. Chl concn. is 6 μg/mL.

Electron donation from ascorbate or dithionite to the water-splitting system or directly to $P_{680}^{+\bullet}$ (preventing charge recombination between Pheo$^-$ and the oxidized donors) could be responsible for the reactivation of the Pheo photoreduction by these reductants after its inhibition by DCMU. Inhibition of the Pheo photoreduction upon removal of Mn acting in the donor side of PS II and its reactivation by

Mn^{2+} [1] indicates reduction of $P_{680}^{+\bullet}$ (leading to accumulation of the state $[P_{680}Pheo^{-}]$ Q^{-}) through the water–splitting system rather than as a result of direct electron donation to $P_{680}^{+\bullet}$ from dithionite.

Acceleration of dark oxidation of $Pheo^{-}$ upon addition of $NADP^{+}$, benzyl viologen and methyl viologen to the ACC–1 mutant lacking PS I, reveals capability of these typical electron acceptors of PS I to accept electrons from PS II by oxidizing $Pheo^{-}$ directly or through some electron carrier(s). The conclusion is confirmed by the data on lowering of F_{max}, equal to $F_0 + \Delta F$ (Fig.1; see also Refs [2] and [9]) which evidently reflects quenching of the recombination luminescence (ΔF) originating from $[P_{680}^{+\bullet}$ $Pheo^{-\bullet}]$ [1] due to oxidation of $Pheo^{-}$. Our results are consistent with the recently described 'alternative' scheme of electron transport from water to $NADP^{+}$ without involving PS I [10]. However, taking into account the high rate of electron transfer from $Pheo^{-}$ to Q ($\leqslant$ 200 ps) [11] one cannot expect high efficiency of photoreduction of $NADP^{+}$ by PS II when Q is oxidized. It can probably take place (as an alternative electron path) only at high light intensity when Q is accumulated in the reduced state.

Anabaena variabilis and Chlorella pyrenoidosa

Creation of anaerobic conditions in suspension of Anabaena or Chlorella leads to the appearance of effects (Fig.3) which are similar to those observed in the ACC–1 Chlamydomonas mutant. The fluorescence yield is kept near the level $F_{max} = F_0 + \Delta F$ by the measuring light, and illumination by actinic light 2 results in a decrease of F which is also activated by CCCP and has a reductive nature (compare with Fig.1 and its discussion). The fluorescence decrease is diminished or completely eliminated by DCMU and it is reactivated by subsequent addition of ascorbate or dithionite (Fig.3). Similar effects were seen in cyanobacteria Phormidium laminosum and Cynechococcus elongatus and green algae Scenedesmus obliquus and Chlamydomonas reinhardii (a wild type).

The data indicate that the fluorescence decrease induced by light 2 in the algae and cyanobacteria under anaerobic conditions can also be related to the photoreduction of Pheo in PS II (see previous section).

An additional effect (in comparison with the ACC–1 mutant) is observed in the wild types of algae and cyanobacteria, namely, light 1 also induces a fluorescence decrease (Fig.3') which is quite different from that related to photoreduction of Pheo. The effect is also seen in chloroplasts and is due to photooxidation of Q^{-} by PS 1 [2,4,12]. In fact, it is inhibited by DCMU (Fig.3) and is not reactivated by subsequent addition of ascorbate (in contrast to the fluorescence decrease related to photoreduction of Pheo). Besides, switching on light 2 during illumination by light 1 (Fig.3, above) results in a very fast increase of F back to the level F_{max} thus reflecting photoreduction of Q oxidized by light 1. Of particular interest is also the fact that if the illumination by light 2 is continued the fluorescence starts to decrease slowly (due to the Pheo photoreduction described above) resulting in a typical effect of 'induction of fluorescence' (the 'P–S' decrease). The effect is also seen in chloroplasts [2,12]. Similar effect of fluorescence induction with the P–S–decrease is observed in the algae and cyanobacteria including the ACC–1 mutant lacking PS I under aerobic conditions as well as for a few minutes after creation of anaerobic conditions while F in–

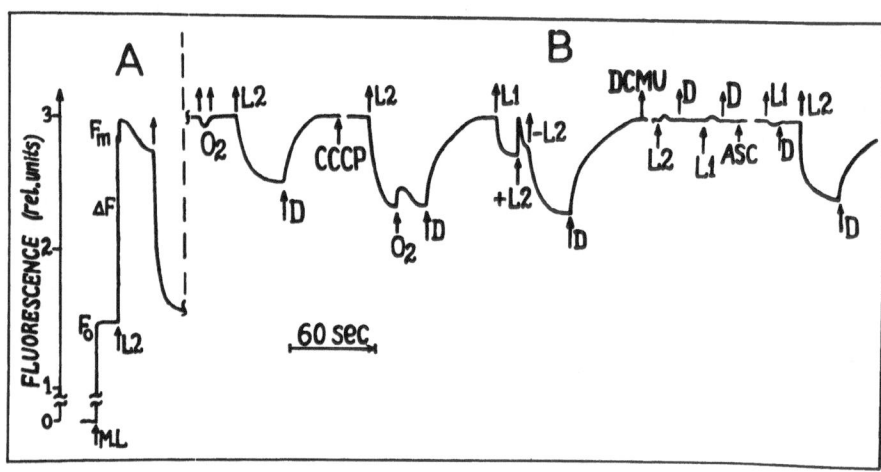

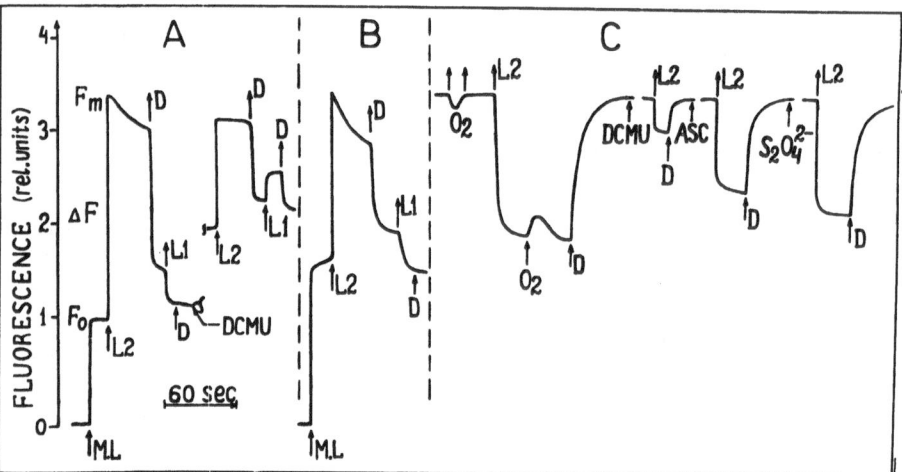

FIGURE 3. Kinetics of photoinduced changes of chlorophyll fluorescence yield in suspension of the whole cells of <u>Anabaena variabilis</u>
(above) and <u>Chlorella pyrenoidosa</u> (below) under aerobic (A) and
anaerobic (B) conditions. Chl concn. is 15 μg/ml. Designations:
+L2 and −L2 is switching light 2 on and off, respectively, during
illumination by light 1, $S_2O_4^{2-}$, dithionite (1 mg/ml).
The other abbreviations as in Fig.1. The measuring light with
  λ = 580 nm, approx. 0.15 $J.m^{-2}.s^{-1}$ was used for the <u>Anabaena</u>
experiments. The dashed trace at the bottom shows the effect of
light 2 and light 1 in the presence of 10 μM DCMU under aerobic
condition in a separate sample.

creases slowly but has not reached yet the level F_{max} in the dark
(Figs 1 and 3). The P–S–fluorescence decrease is inhibited along
with the Pheo photoreduction upon addition of DCMU (Fig.3, below,
dashed line). All these results indicate that photoreduction of Pheo

can contribute to the P–S–decrease during fluorescence induction. In fact, when the water–splitting system is active a strong illumination can induce successive photoreduction of both Q and Pheo thereby producing the long–lived state $[P_{680}Pheo^-] Q^-$.

REFERENCES

1. Klimov VV and Krasnovsky AA (1981) Photosynthetica 4: 592–609
2. Klimov VV, Shuvalov VA and Heber U (1985) Biochim Biophys Acta 809: 345–350
3. Ladygin VG, Allakhverdiev SI, Ananyev GM and Klimov VV (1986) Soviet Plant Physiology (submitted)
4. Duysens LMN and Sweers KE (1963) In: Studies on microalgae and Photosynthetic Bacteria (Miyachi, S, ed.) pp. 353–372, University of Tokyo Press, Tokyo
5. Heber U, Kobayashi Y, Leegood RC and Walker DA (1985) Proc. R. Soc Lond 225: 41–53
6. Etienne AL (1974) Biochim Biophys Acta 333: 320–330
7. Renger G and Eckert HJ (1981) Biochim Biophys Acta 638: 161–171
8. Ghanotakis DF, Yerkes ChI and Babcock GI (1982) Biochim Biophys Acta 682: 21–31
9. Daniel H, Anbudurai PR, Periyannan S, Renganathan M, Bhardwaj R, Kulandaivelu G and Gnanam A (1985) Biochem Biophys Res Commun 126: 1114–1121
10. Arnon DI, Tsujimoto HY and Tang GMS (1981) Proc Natl Acad Sci USA 78: 2942–2946
11. Klimov VV (1984) In: Advances in Photosynthesis Research (Sybesma, C, ed.) Vol.I, pp. 131–138, Martinus Nijhoff/Dr.W.Junk Publishers, the Hague
12. Karapetyan NV and Klimov VV (1973) Soviet Plant Physiol 20: 545–553.

IV.   Electron Transfer in Photosynthesis; Components

Figure 8. Warren L. Butler, Pierre Joliot, Govindjee, and Gernot Renger .... discussions in Japan.

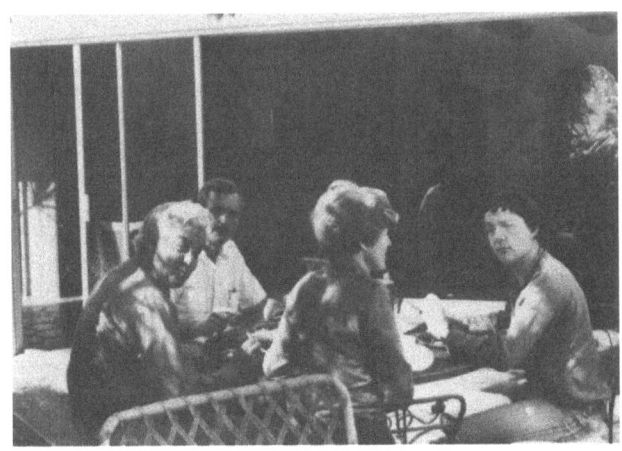

Figure 9. Warren L. Butler, George Hoch, Lila Butler, and Wolfgang Junge .... photo takes at Warren's home in California.

Photosynthesis Research 10: 365–379 (1986)
© Martinus Nijhoff Publishers, Dordrecht

ELECTRON TRANSFER THROUGH PHOTOSYSTEM II ACCEPTORS: INTERACTION WITH ANIONS

GOVINDJEE and J.J. EATON-RYE

Key Words: anions, bicarbonate, electron transfer, photosystem II, quinone

ABSTRACT

We present an overview of anionic interactions with the oxidation-reduction reactions of photosystem II (PSII) acceptors. In section 1, a framework is laid for the electron acceptor side of PSII: the overview begins with a current scheme of the electron transport pathway and of the localization of components in the thylakoid membrane, which is followed by a brief description of the electron acceptor Q or Q_A and the various heterogeneities associated with it. In section 2, we review briefly the nature of the active species of the bicarbonate (HCO_3^-) effect, the location of the site of action of HCO_3^-, and its relationship to interactions with other anions. In section 3, we review data on the anion effects on the reoxidation of Q_A^- and on the various reactions involved in the two-electron gate mechanism of PSII, and provide a hypothesis as to the action of HCO_3^- on the protonation reactions. New data obtained by one of us (G) in collaboration with J.J.S. van Rensen, J.F.H Snel and W. Tonk for HCO_3^--depleted thylakoids, demonstrating the abolition of the binary oscillations contained within the periodicity of 4 observed for proton release, are also reviewed. In section 4, we comment on the measured binding constant of HCO_3^- at the anion binding site. And, in section 5, we review our current concept of the mechanism of the HCO_3^- effect on the electron acceptor side of PSII, and comment on the possible physiological roles for HCO_3^-. Measurements of HCO_3^- reversible anionic inhibition in intact cells of a green alga Scenedesmus are also reviewed.

1. INTRODUCTION

Much of the information regarding the complexities of photosynthesis have been drawn from studies of the variable chlorophyll (Chl) a fluorescence yield [30]. Govindjee et al. [24] and Butler [6] showed that the variable fluorescence yield excited by PSII light could be quenched by simultaneous excitation by PSI light suggesting its relationship to a two photosystem-two light reaction scheme of photosynthesis. Kautsky et al. [44] explained the Chl a fluorescence transient in terms of the oxidation state of a member of the electron transport chain; fluorescence was suggested to be quenched when this component was oxidized by one light reaction, while its photochemical reduction by another light reaction gave rise to an increase in fluorescence. The designation of this acceptor as Q, for "quencher", arose from the work of Duysens [12] and Duysens and Sweers [13] (see Butler [7]). Q may be identified as Q_A, the primary quinone acceptor of PSII, in the electron transfer scheme of photosynthesis shown in Figs. 1 and 2. Figures 1 and 2, should, respectively, serve as a framework for electron transport, and the components, discussed in this book.

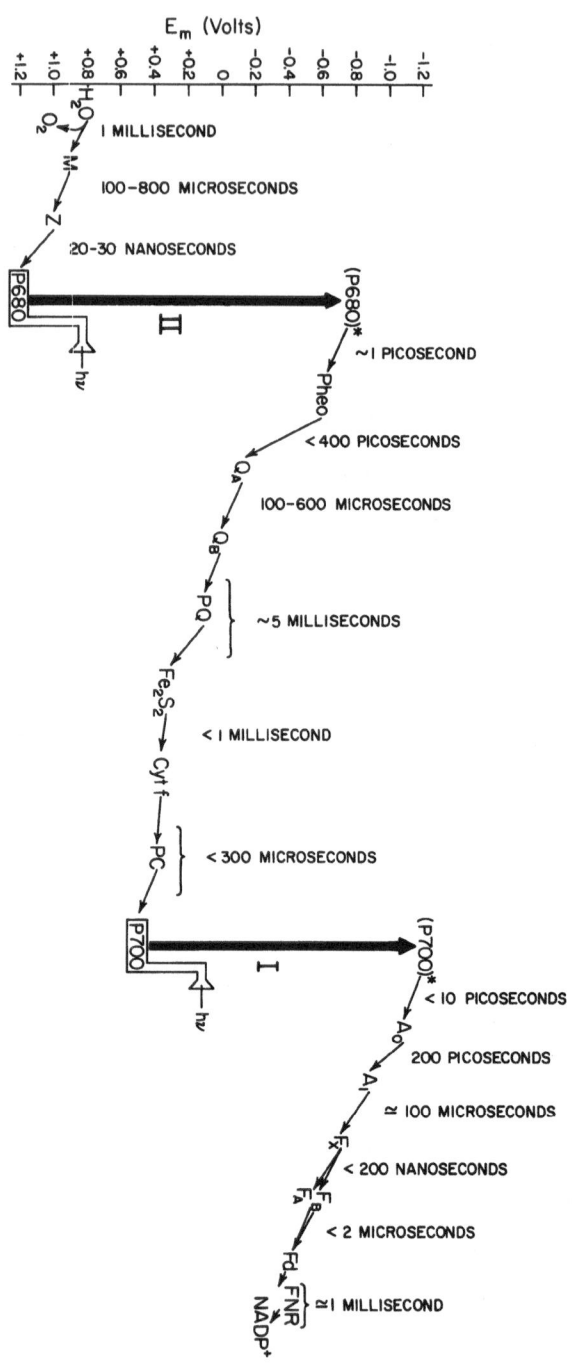

E_m (Volts)

+1.2 +1.0 +0.8 +0.6 +0.4 +0.2 0 -0.2 -0.4 -0.6 -0.8 -1.0 -1.2

H_2O
O_2
M
Z
(P680)
II
hv

(P680)*

1 MILLISECOND

100-800 MICROSECONDS

20-30 NANOSECONDS

~1 PICOSECOND

Pheo

< 400 PICOSECONDS

Q_A

100-600 MICROSECONDS

Q_B

PQ

~5 MILLISECONDS

Fe_2S_2

< 1 MILLISECOND

Cyt f

PC

< 300 MICROSECONDS

(P700)
I
hv

(P700)*

< 10 PICOSECONDS

A_0

200 PICOSECONDS

A_1

≃ 100 MICROSECONDS

F_X

< 200 NANOSECONDS

F_A
F_B

< 2 MICROSECONDS

Fd
FNR
$NADP^+$

≃ 1 MILLISECOND

FIGURE 1.
See legend
on the
following
page.

FIGURE 1. Pathway of noncyclic electron flow from H_2O, the electron donor of photosynthesis, to nicotinamide adenine dinucleotide phosphate ($NADP^+$), the physiological electron acceptor. $E_{m,7}$ on the ordinate stands for mid-point redox potential. Light quanta ($h\nu$) are absorbed in two sets of antenna chlorophyll molecules, the excitation energy is transferred to the reaction center chlorophyll a molecules of photosystem II (P680) and photosystem I (P700) forming (P680)* and (P700)*, and the latter two initiate electron transport. M stands for an all-purpose complex, the "M complex" or the oxygen evolving complex, but it specifically reflects the electron carriers that undergo redox reactions and charge accumulation; Z is the electron donor to P680; Pheo represents pheophytin; Q_A, Q_B and PQ are plastoquinone molecules (see Fig. 3); Fe_2S_2 represents the Rieske iron-sulphur center, Cyt f stands for cytochrome f, PC is plastocyanin; A_0 is suggested to be a chlorophyll molecule, A_1 is possibly a quinone; F_A, F_B, and F_X are thought to be 4Fe-4S centers and FNR is ferredoxin NADP oxidoreductase. Estimated or directly measured times for various reactions are also indicated. In the case of PSII these are taken from [26] and for the PSI from [63]. The values for the intersystem chain are from [31]. In the case of PSI it has also been suggested that A_1 directly reduces F_A and/or F_B in approximately 200 ns while F_X reduction, in approximately 100 µs by A_1, represents a side pathway. For a detailed discussion see [63].

A heterogeneous population of electron acceptors seems to be present in PSII. Q does not represent a single chemical entity [2,85]: (1) Redox potentiometric titrations have revealed two components: Q_H which has an $E_{m,7}$ (mid-point potential at pH 7) of about 0 mV and Q_L which has an $E_{m,7}$ of about -250 mV [9,34]. (2) Parallel measurements on C550 (an absorbance change at 550 nm [16]) and variable Chl a fluorescence following single saturating flashes, in DCMU (3-(3,4-dichlorophenyl)-1,1-dimethylurea)-treated samples, revealed the existence of two Q's: Q_1 and Q_2, where Q_1 was related to all of C550 and to 70% of variable fluorescence yield [37, 39]. Reduction of Q_H and Q_1 is associated with the creation of a membrane potential (ΔA 515), whereas reduction of Q_2 and Q_L is not [11,38]. Furthermore, Q_1 gives a semiquinone signal X-320, whereas Q_2 does not [40]. It appears that Q_1 and Q_H are the same acceptor located on a side different from that of Q_2 and Q_L.

Q_B, the secondary quinone acceptor of PSII (Figs. 1-3), is thought to function in a two-electron gating mechanism [5,83]. Electrons are first transferred from reduced pheophytin ($Pheo^-$) to Q_A, which can only be reduced to the semiquinone form. Q_A^- is then oxidized by Q_B (Fig. 3). After two such events, Q_B is reduced to plastoquinol ($Q_B^{2-}(2H^+)$) which then exchanges with a plastoquinone (PQ) from the plastoquinone pool (PQ pool). Independently, Velthuys [82] and Wraight [93] proposed that the mode of action of a number of PSII herbicides (e.g., DCMU in plants) is to compete with the quinone for the secondary acceptor binding site, the so-called B-site. Following a single actinic flash an equilibrium for an electron is set up between the two quinone acceptors. While Q_B and plastoquinol ($Q_B^{2-}(2H^+)$) are bound loosely at the B-site, Q_B^- is bound tightly and the equilibrium K_E (Fig. 3) is displaced towards $Q_B^-(H^+)$. A value of 20 for this parameter has been estimated at pH 7.6 [60]. In the presence of a non-electron accepting herbicide (I), such as DCMU, $Q_A I$ is produced (Fig. 3); K_O and K_I are the association constants for Q_B and I respectively when

368

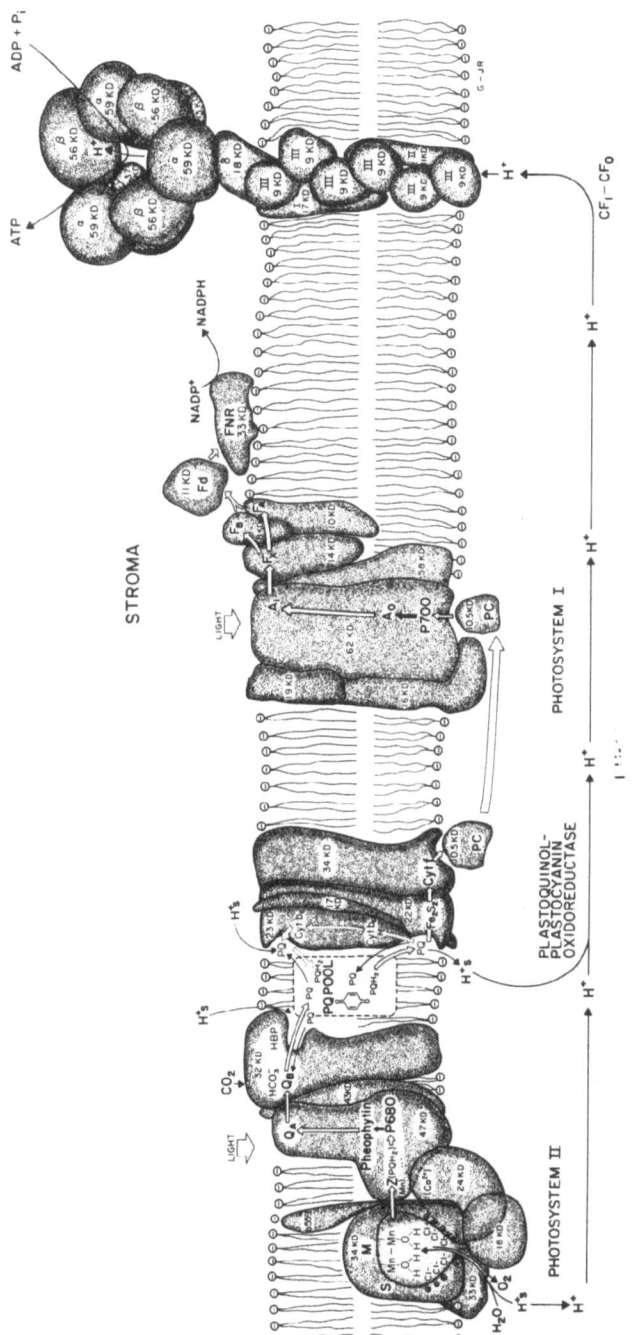

FIGURE 2.
See legend
on the
following
page.

FIGURE 2. A stylized model of the electron transport chain with the light-harvesting pigment-protein complexes omitted. The depiction of PSII is adapted from [26] and the organization of the plastoquinol-plastocyanin oxidoreductase or cytochrome b_6/f complex is based on [52] and [59]. The organization of PSI is adapted from a recent overview given in [59] and the chapter by R. Malkin in this volume. The organization of the H^+-ATPase (CF_1-CF_0) is highly schematic. The hydrophobic CF_0 appears to contain 4-6 copies of the DCCD (N,N'-dicyclohexylcarbodiimide) binding protein or subunit III but CF_0 has not yet been purified [54]. A model for isolated CF_1 has recently been proposed [75]. The subunit stoichiometry shown here is $\alpha_3\beta_3\alpha\gamma\delta\epsilon$ [54].

Q_A is reduced. When K_T' is $\gg K_O'$ centers become stable in the state Q_A^-I. Since Q_A^- is not a quencher of fluorescence, the presence of Q_A^-I may be detected by measurments of the variable Chl $\underline{a}$ fluorescence yield. In the presence of DCMU, however, the formation of centers in the state Q_A^-I appears to be present only in 50-70% of PSII [48-51,92]. This apparent partial displacement of Q_B has been attributed to heterogeneity of PSII electron acceptors rather than equilibrium between the possible states indicated in Fig. 3. Centers which do exhibit electron back-transfer from Q_B^- to Q_A in the presence of DCMU are known as B-type; those accounting for the remainder of the variable fluorescence are described as non-B-type. Lavergne [50] has suggested that non-B-type centers are not connected to the main electron transfer pathway; and further B-type centers possess many characteristics of Q_1 centers while non-B-type centers resemble Q_2 centers [2].

There is an additional complexity. PSIIα and PSIIβ centers are characterized by kinetic components of the steady-state fluorescence induction curve. The Chl $\underline{a}$ transient, in the presence of DCMU, exhibits a fast sigmoidal phase corresponding to PSIIα and a slower exponential phase corresponding to PSIIβ [56,57]. The sigmoidicity of the α phase has been suggested to arise as a consequence of interconnected antennae serving these centers. In this matrix model [36] the α-centers exist in a statistical pigment bed (see also [8]). Energy transfer is allowed between PSIIα units such that an exciton arriving at a closed reaction center is able to visit other centers until it encounters an open trap. The first-order kinetics of PSIIβ centers, by contrast, arise from centers where energy transfer from closed to open centers is not possible. It has been proposed that PSIIα are associated with stacked appressed thylakoid membranes and PSIIβ is present in the stroma lamellae (see e.g., [1]). Studies employing absorbance difference spectroscopy have shown that while α-centers contain Q_1, β-centers contain both Q_1 and Q_2 [55,58].

Recently, a population of PSII centers have been identified that are able to evolve oxygen in the presence of halogenated benzoquinones, artificial electron acceptors for PSII, but are not connected to the main electron transport pathway (T. Graan and D. R. Ort, personal communication). These centers appear to represent about 40% of the total PSII present. The relationship of these centers to other PSII heterogeneities has yet to be characterized.

For further details of PSII the reader is referred to published reviews [2,26,30,76,85].

In the remainder of this overview we shall discuss some of the studies that have explored the effects of anions on the acceptor side of

PSII. Electron transfer at the level of the two-electron gate has been shown to be inhibited by the presence of formate and NO_2^- ([15,25,62]). Whether this effect is due to the removal of bound HCO_3^- or it is a direct inhibitory effect is not yet clear. However, this inhibition is uniquely reversed by the addition of HCO_3^- (see e.g., [21,74]). Furthermore, a wide range of monovalent anions have been shown to be competitive inhibitors of HCO_3^- binding [74; cf. 21]. These findings suggest the existence of an anion binding site on PSII that, when occupied by HCO_3^-, facilitates electron transport into the PQ pool. We are currently investigating the possibility that acetate, formate and NO_2^- inhibit electron flow [74] by displacing HCO_3^-. An alternative approach is that the stimulation of electron transport by HCO_3^- is simply due to the removal of inhibitory anions [68,71].

2. THE BICARBONATE EFFECT

Warburg and Krippahl [91] reported a stimulatory effect of HCO_3^- on the Hill reaction. Originally, it was assumed by Warburg that this effect was on the oxygen evolving mechanism, i.e., he assumed that O_2 was evolved from CO_2. Recent studies have shown [23,84,87] that this effect is on the electron acceptor side of PSII.

Good [21] studied the conditions necessary for HCO_3^--depletion and found that the presence of anions, particularly formate, acetate and chloride, facilitated the depletion process. Since HCO_3^- in solution sets up the following equilibria:

$$CO_2 + H_2O \rightleftharpoons H_2CO_3 \rightleftharpoons H^+ + HCO_3^- \rightleftharpoons 2H^+ + CO_3^{2-} \qquad (1)$$

the nature of the active species involved has been the subject of several studies. The most effective pH to stimulate the Hill reaction in HCO_3^--depleted thylakoids, upon addition of HCO_3^-, was found to be in the pH 6-7 range [46,70]. In confirmation of this, the maximal HCO_3^--restored/HCO_3^--depleted ratio of Hill reaction rates was found to be at pH 6.5 [88]. Furthermore, addition of CO_2 to HCO_3^--depleted samples was found to stimulate Hill activity more readily than addition of HCO_3^- to HCO_3^--depleted samples at 5°C and pH 7.3 [64,65]. Since the pK for the overall reaction ($CO_2 + H_2O \rightleftharpoons H^+ + HCO_3^-$) is 6.4, it was suggested (e.g., [88]) that CO_2 was the species required for diffusion to the active site in HCO_3^--depleted membranes but that HCO_3^- was the active species in restoring the activity. That the active species is indeed HCO_3^- has been recently shown [4,15] by taking advantage of the pH dependence of [CO_2]/[HCO_3^-] ratio at equilibrium. The rate of restored electron transport, in HCO_3^--depleted membranes in the presence of formate, was found to depend on the HCO_3^- concentration when the CO_2 concentration was held constant. This work also demonstrated that H_2CO_3 and CO_3^{2-} have no direct involvement in reversing HCO_3^--depletion.

The location of the HCO_3^- effect in the electron transport chain has been identified through several approaches. Wydrzynski and Govindjee [94] studied the effect of this phenomenon on the Chl a fluorescence induction kinetics and observed an accelerated rise in HCO_3^--depleted samples. This demonstrated that the reoxidation of Q_A^- had been impaired in the depleted samples. Employing specific inhibitors and electron donors and acceptors, which enabled the electron transport chain to be dissected into a number of clearly defined partial reactions, the HCO_3^- effect was located on the electron acceptor side of PSII [14,46]. Competitive binding studies with several PSII herbicides, which bind near Q_B, also support this view [47, 67,73,79,80,89]. We anticipate that a study of the HCO_3^- specific reversal of anionic inhibition will add substantially to our understanding of PSII acceptor side chemistry.

3. ANIONIC INTERACTIONS ON PSII ACCEPTOR SIDE QUINONE CHEMISTRY

Kinetics of Q_A^- reoxidation may be followed by monitoring the decay of Chl $\underline{a}$ variable fluorescence by a double-flash technique [53]. Following a single-turnover actinic flash, a second weak flash, sampling approximately 1% of the centers [35], is given at specified times. The fluorescence yield from the weak analytical flash is a function of $[Q_A^-]$, the relationship being non-linear [18,36]. Adoption of this technique has shown Q_A^- reoxidation to be inhibited identically in samples HCO_3^--depleted in the presence of formate [25,41,43,62], and similar samples even in the presence of atmospheric CO_2 (390 μl/l) [62]. This phenomenon has also been measured by the absorbance change at 320 nm [17,66] and by the 515 nm absorbance change both in thylakoids [41] and in intact chloroplasts [78]. No specific measurements have been made yet to address the differential effects, if any, of this inhibition upon the various PSII heterogenous populations. However, it is evident from the correlation between the fluorescence and absorption measurements that this phenomenon is associated with Q_1 i.e., the HCO_3^- effect is in the major PSII centers.

The extent of the anionic interaction is dependent upon flash number [25]. Using Chl $\underline{a}$ fluorescence, we have measured [62] half-times for Q_A^- reoxidation of 1.2 ms for HCO_3^--depleted and formate incubated samples, and 230 μs for control and HCO_3^--restored thylakoids after a single flash. After the third flash, we obtained half-times of 13 ms for HCO_3^--depleted, 10 ms for formate-incubated, and 360 μs for control and HCO_3^--restored samples. The half-times after flash 2 were intermediate between flash 1 and 3; subsequent flashes yielded results similar to flash 3. The above conclusion was also evident from absorbance changes at 320 nm [17] and 515 nm [78].

The kinetics of Q_A^- reoxidation for flash 3 are expected to resemble those of flash 1 [60] since, following the formation of plastoquinol after the second flash, $Q_B^{2-}(2H^+)$ should readily exchange with a PQ from the PQ pool and Q_A^- should be oxidized by this PQ species. The exchange reactions at the B-site have been determined to occur with a half-time < 2.5 ms (Robinson, H.H. and Crofts, A.R., personal communication). We have found [15] that the inhibition for the third flash in HCO_3^--depleted/anion inhibited centers is large and is the same when the dark time between the second and third flashes is 30 ms or 1 s; this result indicates that the exchange reactions are greatly decreased in the inhibited or HCO_3^--depleted case.

A possible explanation for the above observations is that the binding of inhibitory anions to PSII may alter the association constant, K_0 for Q_B (see Fig. 3). Although there is no direct measure of the value for K_0, a number of methods for estimating a value are available [10]. One method is to analyze the decay kinetics of Q_A^- by monitoring the variable Chl $\underline{a}$ fluorescence after a single flash. Biphasic kinetics are observed for this decay; 60-70% of centers undergo oxidation by a first-order process with a half-time of ~0.15 ms and the remainder by slower processes of indeterminate order [10]. If it is assumed that the centers exhibiting first-order kinetics represent centers in the state $Q_A Q_B$ before the flash,

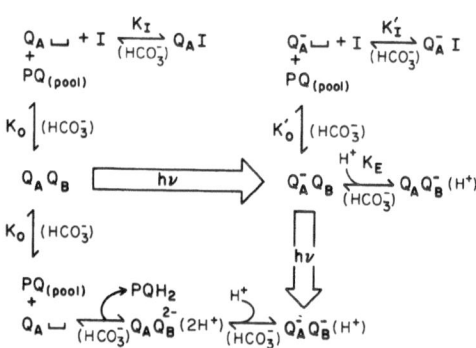

FIGURE 3. Diagramatic presentation of the possible reactions associated with the secondary quinone binding site of the B-site. Photochemical reactions are shown as open arrows; |_____| represents the empty B-site. K_O and K_I are the association constants for plastoquinone and herbicide respectively when Q_A is oxidized and K_O' and K_I' are the association constants when Q_A is reduced. K_E is the equilibrium constant for the sharing of an electron between Q_A and Q_B. The reactions apparently influenced by HCO_3^- are indicated (see [15] and text for details).

a value of 500 M^{-1} for K_O can be calculated [10]. We have analyzed our earlier data for HCO_3^--depleted and formate-incubated thylakoids [62] and found that K_O is reduced to 200 M^{-1} in these samples [15]. A second effect is also evident from this analysis. The half-time of the fast phase is increased approximately 4-fold (i.e., from ~0.2 ms to ~0.8 ms) in these samples [cf. 62]. The mechanism of this second effect cannot be explained from the available data.

In addition to the slowing and reduction of the fast phase of Q_A^- reoxidation, a shift in the equilibrium for the sharing of an electron between Q_A and Q_B (see Fig. 3, K_E) has been reported in the thylakoids that have been HCO_3^--depleted in the presence of formate [90]. A two-fold shift in this equilibrium towards Q_A^- was observed by comparing the rates of the back-reaction with the S_2 state (for a discussion of S-states, see [26]) of the oxygen evolving complex both in the presence and absence of DCMU. In the absence of DCMU, the back-reaction from Q_B^- to S_2 was inhibited two-fold [90].

The equilibrium (see Fig. 3, K_E) for the sharing of an electron between Q_A and Q_B is pH dependent [61]. It has, therefore, been suggested that the presence of a proton in association with the B-site stabilizes the electron on $Q_B^-(H^+)$. We suggest that HCO_3^--depletion inhibits protonation at the B-site. In addition, the fraction of centers decaying through the rapid first-order process after a second flash, has been shown to be proportional to the fraction of centers in which $Q_B^-(H^+)$ is present [61]. Therefore the inhibition on the $Q_B^-(H^+)$ protonation suggested above may also account for the inhibited kinetics of Q_A^- reoxidation observed after the second flash [25,62]. By analogy, the maximal inhibition observed after the third flash may result from Q_B^{2-} not becoming protonated and therefore not able to exchange with the PQ pool. This interpretation suggests that the rate-limiting step introduced by HCO_3^--depletion and/or anion

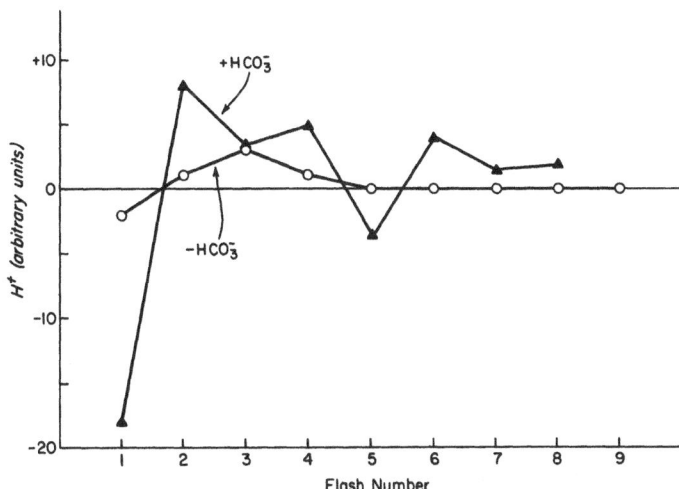

FIGURE 4. The effect of HCO_3^--depletion in the presence of formate on the proton oscillations in uncoupled pea thylakoids measured by a sensitive pH electrode. The electron acceptor was methyl viologen; the sample included methylamine and gramicidin-D to allow the protons to leak out of the thylakoids (from [27]).

inhibition is the rate of protonation of Q_B^{2-}. A role for HCO_3^- in protolytic reactions in PSII has also been proposed as a result of comparative studies with carbonic anhydrase [69,71].

Although an effect of HCO_3^- on protonation at the B-site has not been shown, an effect of HCO_3^--depletion, in the presence of formate, has been shown by Govindjee et al. [27] on proton release at the level of PQH_2 oxidation. Using ferricyanide or methyl viologen as electron acceptor, and an uncoupler of phosphorylation (e.g., methylamine) to bring the protons released into the lumen into the vicinity of a sensitive pH electrode, Fowler [20] had measured an oscillation with a combined period of 4 (protons released from the oxygen evolving complex) and of 2 (protons released from PQH_2 oxidation). This was confirmed by Govindjee et al. [27] for the first time by a pH electrode, although confirmation and extension by other methods have already been made (see e.g., Förster and Junge [19]). When the samples were depleted of HCO_3^-, the binary oscillation was abolished (Fig. 4). This result is consistent with our picture that HCO_3^--depletion blocks electron flow prior to PQH_2 oxidation. Unfortunately, this result does not provide any clue as to any direct effect of anions on the protonation reactions per se.

4. THE BINDING CONSTANT FOR BICARBONATE AT THE ANION SITE

Given the unique ability of HCO_3^- to reverse the anion inhibition of quinone mediated acceptor side electron transfer, studies have been performed to determine its binding constant (K_b). A value of 80 μM has recently been obtained using $H^{14}CO_3^-$ in maize thylakoids with 1 binding site per PSII [72,73]. $H^{14}CO_3^-$ binding has been shown to be competitive with HCO_2^-, NO_2^-, NO_3^-, and $CH_3CO_2^-$ and F^- [74]. This list is almost certainly not exhaustive. NO_2^- is of particular interest. Formate has routinely been employed in HCO_3^--depletion procedures since the work of Good [21]. This reflects, in part, its structural homology with HCO_3^- as well as its specificity. NO_2^-, however, has the same degree of charge delocalization as does HCO_3^-. Blubaugh and Govindjee [4] have discussed the significance of this homology suggesting that the unique behavior of HCO_3^- may result from the hydroxyl group on this anion. A similar suggestion was made earlier by Good [21]. Stemler and Murphy [74] have demonstrated that NO_2^- is an even more effective competitor of $H^{14}CO_3^-$-binding than formate. We have reported [15] that HCO_3^--depletion can also inhibit Q_A^- reoxidation and steady-state electron transport supported by methyl viologen when formate is replaced by NO_2^- in HCO_3^--depletion and reaction media. Jursinic and Stemler [42] have also demonstrated, using identical experimental conditions as employed for the $H^{14}CO_3^-$-binding constant determination, an 80 μM K_m (concentration required to restore half-maximal activity) for Q_A^- reoxidation as monitored by the decay of variable Chl $\underline{a}$ fluorescence following a single actinic flash. These findings therefore appear to confirm that the binding constant measured in $H^{14}CO_3^-$-binding studies is for the binding site at which HCO_3^- facilitates electron transfer through PSII in the presence of inhibitory anions.

However, the magnitude of the HCO_3^--binding constant may have been over-estimated. Since it has been shown that HCO_3^- and various anions are competitive at the 80 μM site [74], it follows a priori that the K_b for HCO_3^- will depend upon the anionic strength used in the experimental conditions. The K_b determination was in fact performed in buffers containing 200 mM NaCl, which is much higher than the [Cl^-] needed for PSII activity [33]. This level of Cl^- has already been established to facilitate HCO_3^--depletion almost certainly by increasing the binding constant. In fact, we have demonstrated [15] that the time course of HCO_3^-depletion is dependent on the [Cl^-]. Thus, the binding constant under native conditions is expected to be smaller than 80 μM.

Furthermore, it is difficult, if not impossible, to be sure that there is only one binding site and only one binding constant. A hint of at least two separate binding sites was presented by Blubaugh and Govindjee [3]. The existence of a tight binding site may have been overlooked since none of the experiments show data on the amount of intrinsic bound HCO_3^- in the sample.

5. MECHANISM OF BICARBONATE ACTION AND POSSIBLE PHYSIOLOGICAL ROLES FOR BICARBONATE

The specificity of HCO_3^- in reversing the inhibition induced by anions on PSII acceptors has lead to speculation regarding an in vivo role for HCO_3^-. The phenomenon is clearly associated with PSII-Q_1-B-type centers and therefore is a characteristic of the principal electron transport pathway. Bound HCO_3^- has also been suggested [47] to produce a conformational change in the 32 kD herbicide/quinone binding protein (Fig. 2), facilitating efficient reduction of Q_B [86], and of exchange of Q_B^{2-} (2H$^+$) with a PQ molecule of the PQ pool (Fig. 3). Indeed, the phenomenon is irrefutably associated with the oxidation of Q_A^- in these centers and strong evidence suggesting a direct involvement on the exchange reactions of the two-electron gate has been collected [15,25,29,62]. Herbicide action has even been proposed to result from the displacement of HCO_3^- from its binding site [80,88].

One physiological role suggested is that HCO_3^- may act as a regulatory anion balancing the production of ATP and reductant (NADPH) needed for CO_2 assimilation [84]. A detailed scheme has been proposed where HCO_3^- protects against inhibitory formate produced in photorespiration [68]. A HCO_3^- effect has been shown in the Hill reaction by intact chloroplasts [78] and by intact cells in the presence of formate. Figure 5 shows measurements of Govindjee et al. [28] on HCO_3^- reversible anionic inhibition of O_2 evolution in intact cells of a green alga Scenedesmus; in these experiments, HCO_3^--depletion of cells was done by first letting the cells perform photosynthesis and use up ambient CO_2, then formate was added to remove bound HCO_3^-, and parabenzoquinone was used to diminish respiration. Furthermore, the parabenzoquinone Hill reaction was measured in order to separate the HCO_3^- effect from that due to the operation of CO_2 fixation. It is clear from Fig. 5 that HCO_3^- was required for the Hill reaction by Scenedesmus cells. The insert in Fig. 5 shows the Chl a fluorescence transient of Scenedesmus cells without the addition of parabenzoquinone. The results, shown here, are similar to those on chloroplasts [86]. Apparently, this suggests that the two have the same basis. This is further supported by the data on other green algae in which electron acceptors beyond ferredoxin-NADP reductase had no effect on the fluorescence transient (see discussions by Govindjee and Satoh [22]). However, we cannot reject the possibility that the absence of CO_2 fixation in CO_2-free cells may also give a faster rising fluorescence transient.

Attempts to see the HCO_3^- effect in the absence of inhibitory anions have met with partial success. These results are reminiscent of the early days of HCO_3^- research when small HCO_3^- effects on the Hill reaction were observed without the use of inhibitory anions [23,91] and large effects with inhibitory anions present [21,23]. In the absence of inhibitory anions, we have observed a fully reversible HCO_3^- effect on Q_A^- reoxidation [14]; this effect is also present in steady-state oxygen evolution and in the Chl a fluorescence induction kinetics [15]. The effect, however, is less dramatic than when inhibitory anions are present. For example, after 3 actinic flashes the kinetics of Q_A^- reoxidation for HCO_3^--depleted thylakoids were found to have a half-time of approximately 2 ms in the absence of formate [14]. This is to be compared with approximately 13 ms in the presence of formate [62]. However, 20 mM Cl^- was present in the formate free case. It is possible that in the absence of HCO_3^- this low $[Cl^-]$ might become inhibitory. Bound HCO_3^- may possibly be necessary in vivo to protect against inhibition from anions such as $CH_3CO_2^-$, NO_2^- and $\overline{Cl^-}$ as suggested for formate [68].

Arguments against a physiological role for HCO_3^- have been based upon the 80 μM binding constant [72]. This claim stems from an estimated in vivo CO_2 concentration of < 5 μM [32] which, it has been suggested, would result in the HCO_3^- binding site being unoccupied. However, Blubaugh and Govindjee [4] have pointed out that while the CO_2 concentration may be quite low in the chloroplast, at pH 8.0, the approximate pH of the stroma, the HCO_3^- concentration may be as high as 220 μM. This is well above the estimated binding constant [4,15]. Furthermore, the binding constant under native conditions may even be much lower than 80 μM.

The last word of this debate has not yet been heard. However HCO_3^- reversible anionic inhibition of PSII is clearly a real phenomenon and has already proved itself an important gateway into the complex reactions of the acceptor side of PSII.

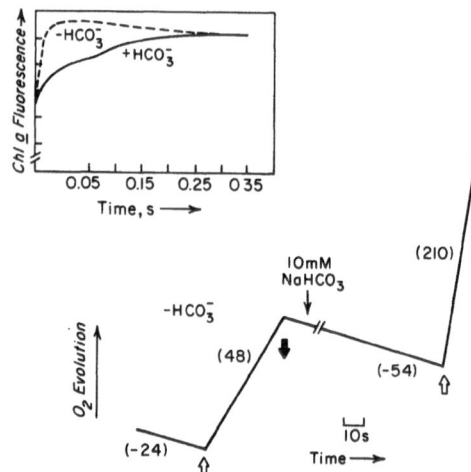

FIGURE 5. The effect of HCO$_3^-$-depletion in the presence of formate on the Hill reaction in intact cells of <u>Scenedesmus</u>. The electron acceptor was parabenzoquinone. The numbers in parentheses indicate O$_2$ exchange in μmoles(mg Chl)$^{-1}$ h^{-1}. Open and closed arrows indicate light on, and off, respectively. The insert shows the effect of HCO$_3^-$-depletion, in the presence of formate but in the absence of parabenzoquinone, on the chlorophyll <u>a</u> fluorescence transient (from [28]).

ACKNOWLEDGEMENT

We thank the National Science Foundation (Supplementary grant to PCM83-06061) for financial support.

REFERENCES

1. Anderson JM and Melis A (1984) Proc Natl Acad Sci USA 80: 745-749
2. Black MT, Brearley TM and Horton P (1984) Photosynthesis Res, in press
3. Blubaugh DJ and Govindjee (1984) Z Naturforsch 39C: 378-381
4. Blubaugh DJ and Govindjee (1986) Biochim Biophys Acta 848: 147-151
5. Bouges-Bocquet B (1973) Biochim Biophys Acta 314: 250-256
6. Butler WL (1962) Biochim Biophys Acta 64: 309-317
7. Butler WL (1966) In: Current Topics in Bioenergetics (Sanadi, DR, ed.) Vol. I, pp. 49-73, Academic Press, New York
8. Butler WL (1980) Proc Natl Acad Sci USA 77: 4697-4701
9. Cramer WA and Butler WL (1969) Biochim Biophys Acta 172: 503-510
10. Crofts AR, Robinson HH and Snozzi M (1984) Advances in Photosynthesis Research 1: 461-468
11. Diner BA and Delosme R (1983) Biochim Biophys Acta 722: 443-451
12. Duysens LNM (1963) Proc Roy Soc B157: 301-313
13. Duysens LNM and Sweers HE (1963) In: Microalgae and Photosynthetic Bacteria (Japanese Society of Plant Physiologists, ed.) pp. 353-372, University of Tokyo Press, Tokyo
14. Eaton-Rye JJ and Govindjee (1984) Photobiochem Photobiophys 8: 279-288

15. Eaton-Rye JJ, Blubaugh DJ and Govindjee (1986) In: Ion Interactions in Energy Transfer Biomembranes (Papageorgiou, GC, Barber, J and Papa S, eds.), pp. 263-278, Plenum Publication Corp., New York

16. Erixon K and Butler WL (1971) Biochim Biophys Acta 234: 381-389

17. Farineau J and Mathis P (1983) In: The Oxygen Evolving System of Photosynthesis (Inoue, Y, Crofts, AR, Govindjee, Murata, N, Renger, G and Satoh, K, eds.) pp. 317-325, Academic Press, New York

18. Forbush B and Kok B (1968) Biochim Biophys Acta 162: 243-353

19. Förster V and Junge W (1985) Photochem Photobiol 41: 183-190

20. Fowler CF (1977) Biochim Biophys Acta 462: 414-421

21. Good NE (1963) Plant Physiol 38: 298-304

22. Govindjee and Satoh K (1986) In: Light Emission by Plants and Bacteria (Govindjee, Amesz, J and Fork, DC, eds.) in press, Academic Press, Orlando

23. Govindjee and van Rensen JJS (1978) Biochim Biophys Acta 505: 183-213

24. Govindjee, Ichimura S, Cederstrand CN and Rabinowitch E (1960) Arch Biochem Biophys 89: 322-323

25. Govindjee, Pulles MPJ, Govindjee R, van Gorkom JH and Duysens LMN (1976) Biochim Biophys Acta 449: 602-605

26. Govindjee, Kambara T and Coleman W (1985) Photochem Photobiol 42: 187-210.

27. Govindjee, van Rensen JJS, Snel JFH and Tonk W (1983) Unpublished observations

28. Govindjee, Snel JFH and van Rensen JJS (1983) Unpublished observations.

29. Govindjee, Nakatani HY, Rutherford AW and Inoue Y (1984) Biochim Biophys Acta 766: 416-423

30. Govindjee, Amesz J and Fork DC (Editors) (1986) Light Emission by Plants and Bacteria. Academic Press, Orlando

31. Haehnel W (1984) Ann Rev Plant Physiol 35: 659-693

32. Hesketh JE, Wooley JT and Peters DB (1982) In: Photosynthesis (Govindjee, ed.) Vol II, pp. 387-418, Academic Press, New York

33. Hind G, Nakatani HY and Izawa S (1969) Biochim Biophys Acta 183: 361-373

34. Horton P and Croze E (1979) Biochim Biophys Acta 545: 188-201

35. Joliot A (1974) In: Proceedings of the Third International Congress on Photosynthesis (Avron, M, ed.) Vol I, pp. 315-322, Elsevier, Amsterdam

36. Joliot P and Joliot A (1964) C R Acad Sci Paris Ser D 258: 4622-4625

37. Joliot P and Joliot A (1973) Biochim Biophys Acta 305: 302-316

38. Joliot P and Joliot A (1979) Biochim Biophys Acta 546: 93-103

39. Joliot P and Joliot A (1981) In: Photosynthesis II, Electron Transport and Photophosphorylation (Akoyunoglou, G, ed.) pp. 885-889, Balaban, Philadelphia.

40. Joliot P and Joliot A (1981) FEBS Letts 134: 155-158

41. Jursinic P and Stemler A (1982) Biochim Biophys Acta 681: 419-428

42. Jursinic P and Stemler A (1986) Photochem Photobiol 43: 205-212

43. Jursinic P, Warden J and Govindjee (1976) Biochim Biophys Acta 681: 419-428

44. Kautsky H, Appel W and Amann H (1960) Biochem Z 332: 277-292

45. Kelly PM and Izawa S (1978) Biochim Biophys Acta 502: 198-210

46. Khanna R, Govindjee and Wydrzynski T (1977) Biochim Biophys Acta 462: 208-214

47. Khanna R, Pfister K, Keresztes A, van Rensen JJS and Govindjee (1981) Biochim Biophys Acta 634: 105-116

48. Lavergne J (1982) Photobiochem Photobiophys 3: 257-271
49. Lavergne J (1982) Photobiochem Photobiophys 3: 273-285
50. Lavergne J (1982) Biochim Biophys Acta 682: 345-353
51. Lavergne J and Etienne AL (1980) Biochim Biophys Acta 593: 136-148
52. Mansfield RW and Anderson JM (1985) Biochim Biophys Acta 809: 435-444
53. Mauzerall D (1972) Proc Nat Acad Sci USA 69: 1358-1362
54. McCarty RE and Nalin CM (1986) In: Encyclopedia of Plant Physiology: Photosynthetic Membranes (Staehlin, A, Arntzen, CJ, eds.), in press, Springer-Verlag, Berlin
55. Melis A and Duysens LNM (1979) Photochem Photobiol 29: 373-392
56. Melis A and Homann PH (1975) Photochem Photobiol 21: 431-437
57. Melis A and Homann PH (1976) Photochem Photobiol 23: 343-350
58. Melis A and Schrieber U (1979) Biochim Biophys Acta 547: 47-57
59. Ort DR (1986) In: Encyclopedia of Plant Physiology: Photosynthetic Membranes (Staehlin, A, and Arntzen, CJ, eds.), in press, Springer-Verlag, Berlin
60. Robinson HH and Crofts AR (1983) FEBS Lett 153: 221-226
61. Robinson HH and Crofts AR (1984) Advances in Photosynthesis Research 1: 447-480
62. Robinson HH, Eaton Rye JJ, van Rensen JJS and Govindjee (1984) Z Naturforsch 30C: 382-385
63. Rutherford AW and Heathcote P (1985) Photosynthesis Res 6: 295-316
64. Sarojini G and Govindjee (1981) In: Photosynthesis II, Electron Transport and Photophosphorylation (Akoyunoglou, G, ed.), pp. 143-149, Balaban, Philadelphia
65. Sarojini G and Govindjee (1981) Biochim Biophys Acta 636: 168-174
66. Siggel U, Khanna R, Renger G and Govindjee (1977) Biochim Biophys Acta 462: 196-207
67. Snel JFH and van Rensen JJS (1983) Physiol Plant 57: 422-427
68. Snel JFH and van Rensen JJS (1984) Plant Physiol 75: 146-150
69. Stemler A (1985) In: Inorganic Carbon Transport in Aquatic Photosynthetic Organisms (Berry, J and Lucas, W, eds.), pp. 377-387, American Society of Plant Physiologists
70. Stemler A and Govindjee (1973) Plant Physiol 52: 119-123
71. Stemler A and Jursinic P (1983) Archiv Biochem Biophys 221: 227-237
72. Stemler A and Murphy JB (1983) Photochem Photobiol 38: 701-707
73. Stemler A and Murphy JB (1984) Plant Physiol 76: 179-182
74. Stemler A and Murphy JB (1985) Plant Physiol 77: 974-977
75. Tiedge H, Schafer G and Mayer F (1983) Eur J Biochem 132: 37-45
76. van Gorkom HJ (1985) Photosynthesis Res 6: 97-112
77. van Rensen JJS (1984) Z Naturfursch 39C: 1021-1023
78. van Rensen JJS and Snel JFH (1985) Photosynthesis Res 6: 231-246
79. van Rensen JJS and Vermaas WFJ (1981) In: Photosynthesis II, Electron Transport and Photophosphorylation (Akoyunoglou, G, ed.), pp. 151-156, Balaban, Philadelphia
80. van Rensen JJS and Vermaas WFJ (1981) Physiol Plant 51: 106-110
81. Reference was deleted.
82. Velthuys BR (1981) FEBS Lett 120: 277-281
83. Velthuys BR and Amesz J (1974) Biochim Biophys Acta 333: 85-94
84. Vermaas WFJ and Govindjee (1981) Proc Indian Natl Sci Acad B47: 581-605
85. Vermaas WFJ and Govindjee (1981) Photochem Photobiol 34: 775-793
86. Vermaas WFJ and Govindjee (1982) Biochim Biophys Acta 680: 202-209

87. Vermaas WFJ and Govindjee (1982) In: Photosynthesis, Development, Carbon Metabolism and Plant Productivity (Govindjee, ed.) Vol II, pp. 541-558, Academic Press, New York
88. Vermaas WFJ and van Rensen JJS (1981) Biochim Biophys Acta 636: 168-174
99. Vermaas WFJ, van Rensen JJS and Govindjee (1982) Biochim Biophys Acta 681: 242-247
90. Vermaas WFJ, Renger G and Dohnt G (1984) Biochim Biophys Acta 764: 194-202
91. Warburg O and Krippahl G (1958) Z Naturforsch 13B: 509-514
92. Wollman F-A (1978) Biochim Biophys Acta 503: 263-273
93. Wraight CA (1981) Israel J Chem 21: 348-354
94. Wydrzynski T and Govindjee (1975) Biochim Biophys Acta 384: 403-408

ADDENDUM

W.F.J. Vermaas and A.W. Rutherford (FEBS Lett 175: 243-247, 1984) have reported that the EPR signal of the iron-quinone in PSII is much larger in HCO_3^--depleted particles. This suggests an interaction of HCO_3^- at the Q_A-Q_B level. M.C.W. Evans (Physiol Veg 23: 563-569, 1985) has looked at this iron-quinone in HCO_3^--depleted PSII samples by EPR and observed two signals with E_m = 50 mV and E_m = -250 mV. On the basis of this and other results Evans has presented a model for electron flow from Q_A to PQ which is different from that presented in this overview. In the Evans' model, two bound semiquinones act in a concerted fashion to reduce a PQ molecule. Further experiments are needed to judge the merits of this proposal.

AUTHORS' ADDRESS

Departments of Plant Biology and of Physiology and Biophysics
University of Illinois
289 Morrill Hall, 505 S. Goodwin Ave.
Urbana, Il. 61801 (U.S.A.)

.

Photosynthesis Research 10: 381–392 (1986)
© *Martinus Nijhoff Publishers, Dordrecht*

INHIBITORS OF PHOTOSYSTEM II AND THE TOPOLOGY OF THE HERBICIDE
AND Q_B BINDING POLYPEPTIDE IN THE THYLAKOID MEMBRANE

A. Trebst and W. Draber

Dept. of Biology, Ruhr-University Bochum, FRG and
Agrochemicals, Chemical Research, Bayer AG, Monheim, FRG

Key words: herbicides, herbicide binding polypeptide, D-1
protein, D-2 protein, photosystem II, membrane topology

Abstract: The folding through the thylakoid membrane of the
D-1 herbicide binding polypeptide and of the homologous D-2
subunit of photosystem II is predicted from comparison of
amino acid sequences and hydropathy index plots with the
folding of the subunits L and M of a bacterial photosystem.
As the functional amino acids involved in Q and Fe binding in
the bacterial photosystem of R. viridis, as indicated by the
X-ray structure, are conserved in the homologous D-1 and D-2
subunits of photosystem II, a detailed topology of the binding
niche of Q_B and of herbicides on photosystem II is proposed.
The model is supported by the observed amino acid changes in
herbicide tolerant plants and algae. These changes are all in
the binding domain on the matrix side of the D-1 polypeptide,
and turn out to be of functional significance in the Q_B binding.
New inhibitors of Q_B function are described. Their chemical
structure, i.e. pyridones, quinolones, chromones and benzo-
diones, contains the features of the phenolic type herbicides.
Their essential elements, π-charges at particular atoms, QSAR
and steric requirements for optimal inhibitory potency are
discussed and compared with the "classical" herbicides of the
urea/triazine type.

Inhibitors, many of them commercial herbicides, were - and
still are - highly instrumental in studies on the function and
the architecture of the photosynthetic apparatus. They were
particularly essential for defining the reactions and the
binding sites of the redox carriers on the acceptor side of
PS II, the role of bound (Q_A) and of exchangeable (Q_B) plastoquinone species and for understanding fluorescence, quenching
and light distribution phenomena (1,2,3). At the same time
photosynthesis studies provided great details on the mode of
action and structure activity relationships of PS II herbicides
(4) that allow rational design and molecular modelling of new
compounds.
The core complex of photosystem II consists of five major
integral, membrane spanning, polypeptide subunits of 47, 44,
34, 32 and 10 kDa molecular weight. There may be more subunits,
not yet clearly identified. The core complex contains the
reaction center chlorophyll P680, several core antenna chlorophylls, pheophytin, Fe, cytochrome b559, two acceptor plastoquinones Q_A and Q_B and a primary electron donor possibly also

a plastoquinone (2,5,6). Together with peripheral polypeptides attached to the donor side the integral peptides also form the binding domain for manganese involved in oxygen evolution. The integral complex binds inhibitors, among them many commercial herbicides, here grouped in two classes of an urea/triazine- or of an phenol-type. Functional studies have indicated that these "photosystem II herbicides" inhibit photosynthetic elec- tron flow between the primary quinone acceptor Q_A and the secondary quinone Q_B possibly by displacing Q_B (see (3)). The binding affinity of the inhibitors (7) and of Q_B to the photo- system II complex is lowered by trypsin treatment of the mem- brane; at the same time Q_A becomes accessible to ferricyanide (8,9). Therefore a herbicide binding protein shield above the quinone acceptors was postulated (9). The subunit of the PS II complex responsible for herbicide binding and for the trypsin sensitive shield was then identified by photoaffinity labeling. Azidotriazine binds to the 32 kDa polypeptide of PS II (10) that is rapidly turning over and is coded for by a photogene psbA in the plastome (11). Also an azidourea (12) - and an azidotriazinone (13) - derivative bind covalently to the 32 kDa polypeptide. Since then it is called the 32 kDa (or D-1) herbicide and Q_B binding polypeptide. An azidonitrophenol, that also inhibits at Q_B, however, binds to higher molecular weight subunits of PS II about 41 kDa in thylakoids (14) - as azidourea does also in addition to labeling the 32 kDa poly- peptide (12) - and 47 kDa in PS II enriched particles.

FIGURE 1. Alignment of the amino acid sequences of the D-1 and D-2 subunit of PS II of spinach (16,17) with the L and M sub- unit of the photosystem of R. capsulata (24) indicating five hydrophobic segments common to each. Conserved amino acids involved in the binding of functional components are under- lined.

The topology of the Q_B and herbicide binding polypeptides

The primary amino acid sequence of polypeptides determines
the secondary folding. This folding may be predicted with the
use of appropriate algorithms for hydrophobic membrane polypeptides (15). All five integral subunits of PS II are coded
for in the plastome (16). The genes have been localized and
sequenced (16,17). From hydropathy index plots of the amino
acid sequence it has been predicted that the 32 kDa polypeptide
would fold through the membrane in hydrophobic ℓ -helices seven
times (18) or perhaps even eight times (19). Therefore this
polypeptide is exposed in hydrophilic sequences on both the
matrix and lumen side of the membrane and it is not just a
shield above the hydrophobic membrane on the matrix side. More
recently the gene psbD coding for the 34 kDa (D-2) polypeptide
of PS II has been sequenced (20-23). Its amino acid sequence
is somewhat homologous and its hydropathy index plot is highly
homologous to that of the 32 kDa polypeptide (20). Furthermore,
as noted first by Youvan et al. (24), there is homology of the
D-1 protein to the L and M subunits of the photosystem of
purple bacteria.

Inspection of the hydropathy index plots and of the amino
acid sequence of these four polypeptides, D-1 and D-2 of PS II
and L and M of the bacterial photosystem, show that the folding
prediction of Rao et al. (18) should be revised. By considering
only those sequences that are common to all four polypeptides
it is predicted that they have five hydrophobic sequences that
are large enough to span the membrane in five hydrophobic ℓ -he-
lices   (25) (Fig. 1). The folding of the D-1 (32 kDa) and D-2
(34 kDa) polypeptides according to this interpretation is
shown in Fig. 2 and 3. The folding prediction gets support by
a) inspecting the bacterial system for more homologies to PS II
in conserved amino acids and by b) considering amino acid
changes in the 32 kDa polypeptide of herbicide tolerant plants
and algae.

FIGURE 2. Pre-
dicted folding
of the amino
acid sequence
of the 32 kDa
D-1 subunit of
PS II of spi-
nach (17) in
five hydropho-
bic helices
spanning the
membrane.
Arrows indi-
cate mutations
in herbicide
tolerant
plants.

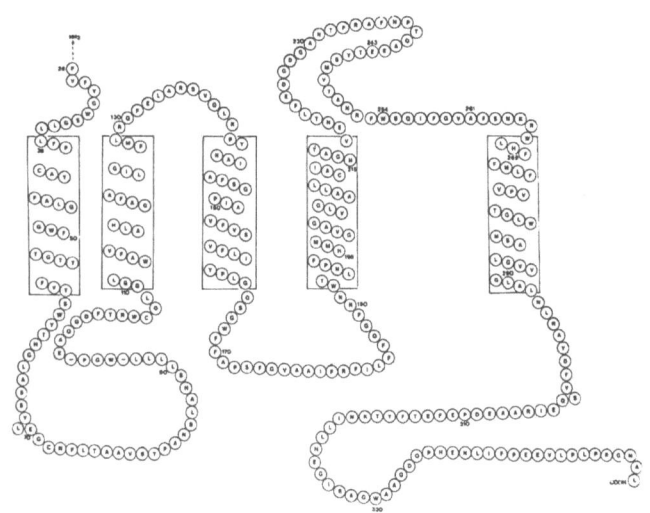

FIGURE 3. Predicted folding of the amino acid sequence of the 34 kDa D-2 subunit of PS II of spinach (21) in five hydrophobic helices spanning the membrane.

In composition and function photosystem II shares many properties with the bacterial photosystem: a reaction center, pheophytin and two quinone acceptors, connected via an iron atom (2,3,26,27). Also PS II herbicides bind to the bacterial reaction center (3,28,29) and in particular to the L subunit (29-31). Now it appears that also certain core polypeptides and their folding through the membrane are similar in PS II and the bacterial photosystem (32,33). Therefore one can use the recent results from the X-ray structure of the bacterial photosystem of R. viridis by Deisenhofer, Michel et al. (32,33) to inspect for amino acids conserved in PS II that are shown to be involved in pigment and redox carrier binding in the bacterial system (32,33). The result is that the architecture of the polypeptides involved in Q_A, Fe and Q_B binding in photosystem II appears to be quite analogous to the bacterial system.

Accordingly in a model for PS II the binding niche for the acceptor quinones Q_A and Q_B is composed of two subunits, one each of the D-1 and the D-2 protein that both span the membrane five times in a highly symmetrical manner. Helices IV and V of each subunit are of particular relevance of quinone and herbicide binding. As already proposed from EXAF studies (34) and now seen in the X-ray structure (32,33) of the bacterial system, histidines are responsible for Fe binding. Four histidines are involved according to the X-ray structure and these are conserved in the D-1/D-2 subunits: his 215 and his 272 in the D-1 and the homologous his 215 and his 269 in the D-2 polypeptide. The menaquinone Q_A in the crystallized bacterial center of R. viridis is close to his 217 (homologous to his 215 in D-2), to a peptide bond of ala 258 (homologous to ala 261 in D-2) and to a trp 250 of the M subunit in a parallel helix (33). Again this tryptophan (trp 254) is conserved in D-2 (see Fig. 1). From the homologous folding of the D-1 (and L) subunit it follows that his 215 and phe 255 are involved in Q_B and possibly also the peptide bond of ala 263 or ser 264 for the second hydrogen bridge to Q_B.

There are also significant differences in the topology of
the Q_A/Q_B binding niche between the bacterial system and photosystem II. The connecting amino acid sequence between helices
IV and V is longer in D-1 compared with L and is longer as in
D-2, whereas this is the opposite in L vs M. The fifth ligand
for the Fe is glu 232 in M subunit. An equivalent amino acid
might be glu 243 in D-2. But this glu is also present in D-1,
but not in L and it is impossible that both subunits provide
the fifth ligand. Particular striking is the difference in the
position of an arg. It is in helix V both in the M and L subunit beyond the his 264 (or 230 respectively) that is involved
in Fe binding, whereas in D-1 and D-2 an arginine is outside
helix V and before his 269 (or 272 respectively).

The predicted folding of the 32 kDa/D-1 protein seems not
to be in agreement with trypsin digestion experiments (35). In
particular according to the predicted folding in Fig. 2 the
carboxylend of D-1 would be inside the lumen space rather than
in the stroma, as assumed so far (35). The predicted folding,
however, is supported by mutant data, because it allows quite
reasonable explanations for those amino acid changes in the
32 kDa polypeptide, identified so far in herbicide tolerant
mutants of higher plants and algae (36-40). All changes are on
that part of the 32 kDa polypeptide that is oriented towards
the matrix side in the folding prediction in Fig. 2, as is to
be expected for inhibitors of Q_B functioning on the matrix
side. Indeed they all appear to have a functional significance
in Q_B binding. As indicated in Fig. 4, the val 219, changed in
triazine tolerance (38), is above the his 215 that is implied
directly in Q_B binding. As val 219 is in the membrane spanning
hydrophobic helix, its precise orientation with respect to the
his 215 can be taken from a helical wheel (see (18)), as in an
$\mathcal{L}$-helix there are 3.6 amino acids in one turn. Possibly increasing the bulky sidechain from valine to isoleucine in the

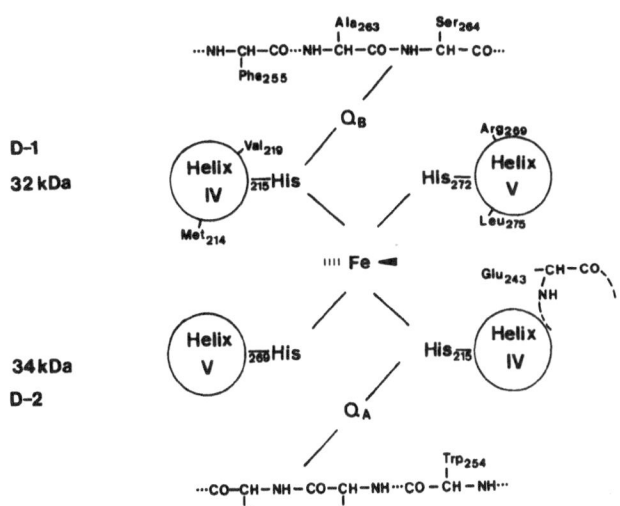

FIGURE 4. Proposed
binding of the $Q_A/$
Q_B (and herbi-
cides) and Fe via
four histidines
of the D-1 and
D-2 subunits of
PS II, following
the homology with
and the X-ray
structure of a
bacterial photo-
system.
The amino acid
changes in herbi-
cide tolerant mu-
tants are indica-
ted (val 219,
phe 255, ser 264,
leu 275).

mutation leads to steric hindrance of binding of an inhibitor towards the histidine below. Particularly supportive of the model of the D-1/D-2 folding is the mutation in phe 255 in DCMU resistance (38). Because the X-ray structure of the bacterial system shows this phe to be on a membrane parallel helix that extends into the hydrophobic space below the Q (33) this phe might interact with Q_B, but also with those herbicides with an aromatic ring. The recently reported leu 275 change (39) is on the fifth helix. Leu 275 is below the his 272 that is involved in Fe binding and it is also facing into the binding niche (Fig. 4). The ser 264 changed in triazine resistance in both higher plants (36) and algae (37,40) might be involved in Q_B binding. Because it is very striking that this ser 264 in D-1 is on a homologous position with that ala 261 of D-2 that is involved in Q_A binding - if following the homology of M and D-2 and the X-ray data. Perhaps the hydroxylgroup of the serine is involved in destabilizing Q_B/H_2 binding and that its exchange to alanine or glycine increases Q_B semiquinone binding so that the inhibitor can no longer displace Q_B, i.e. the system becomes herbicide tolerant. The mutations reported in terbutryn resistance of the bacterial system on an isoleucine next to his in helix V (31) in the L subunit fits the model. The methionine 214, tagged by azidotriazine (41) (but not by azidourea) (12), is somewhat pointing away from the binding niche - thus perhaps indicating some leeway in the exact binding during photoaffinity labeling experiments. An arginine, at the end of a helix conferring a particular dipolmoment, has been suggested to be of importance in herbicide binding by theoretical considerations (42); an arginine with that property might be arg 269 at the end of helix V above his 275 (Fig.4).

The model for the binding niche of Q_A and Q_B in PS II presented here is based on conserved functional amino acids in the Q binding of the subunits of the R. viridis reaction center. It furthermore incorporates amino acid mutations in the D-1 polypeptide of herbicide tolerant plants. This seems to be a strong argument that herbicides and Q_B indeed occupy the same binding niche and displace each other directly (28,43). This might be no contradiction to models proposing allosteric mechanisms and overlapping binding sites for different herbicides (4,44,45). Different sidechains in the binding domain of the D-1 polypeptide are involved in binding of different inhibitors and quinones depending on size and reactivity of special groups. The subtle changes in the effectiveness of herbicides brought about by changes of reactive groups in the dimensions of the binding domain and of lipids surrounding the PS II complex is particularly obvious in herbicide tolerant mutants of plants. The cross resistance pattern to various herbicide analogues can be quite different from one species to another (see (46-52)). Similar subtle changes in certain herbicide binding efficiency, but not in others, brought about by conformational changes in the binding niche by tris treatment are used to differentiate the two groups of PS II inhibitors (53, see below).

The identity of a 34 kDa polypeptide of PS II in Scenedesmus labeled by azidotriazine (54) is not yet clear. It could be the D-1, but perhaps also the D-2 protein. A mutation in that poly-

peptide leads to a blockage of electron flow at the donor side
of PS II (54). This is no contradiction to assuming that the
D-1 protein is changed on the donor side of PS II, as the
azidotriazine binding D-1 protein is exposed on the lumen side
as well and is part of the attachement of the peripheral 33 kDa
polypeptide and Mn to the integral complex, according to the
model presented here.
    There have been reports (55-57) that particularly phenols,
but also DCMU have an additional binding site and inhibitory
effect on the donor side of PS II.However, after trypsin treatment (8,9,58) or LHCP removal (59) DCMU has no longer an effect on the donor side. Therefore the "second site" is likely
just a relay of changes through transmembrane helices onto a
function of the D-1 protein on the donor side.There is a third
binding site for the PS II inhibitors also on the cytochrome
b/f-complex (60) (at the Q_c site) equivalent to the DCMU effect
on the cytochrome b/c_1-complex of yeast (61).
    The description of the Q_A, Q_B binding niche in PS II in
Fig. 4 follows the homology to the bacterial reaction center
of R. viridis (32,33). It is consequent therefore to suggest
that the D-1/D-2 polypeptide also forms the reaction center of
PS II as the L and M subunit of the bacterial system do (27,29).
Indeed the two his for the binding of a reaction center chlorophyll are conserved - his 198 in D-1/D-2. However, so far the
47 kDa subunit of PS II has been assigned the role of the
reaction center polypeptide (see 2,5,6,26). But if one were to
accept the direct involvement of the D-1 and D-2 polypeptide
also in the reaction center of PS II, then further details on
the conformation of the herbicide binding niche can be obtained.
This is because chlorophyll binding to the conserved his 198 in
helix IV on the same helical span of both the D-1 and D-2 and
M and L subunits and of Fe to his 215 exactly 16 amino acids
apart from the his 215 in Fe binding brings constraints. The
strict orientation of the residues on an $\mathcal{L}$-helix requires
that the helices IV of D-1 and D-2 are tilted against each
other in order to accomodate both chlorophyll and Fe binding
between them (note also the 4 gly in the center of helix IV of
D-1 that may bend the helix). This large tilt is indeed directly seen in the X-ray structure of the bacterial system
(32,33).

The chemistry of inhibitors
    It is long known that the "classical" inhibitors of the
acceptor side of photosystem II - triazines, ureas, triazinones,
anilides etc. - all inhibit at the same functional site, at Q_B,
although they belong to quite different chemical groups. They
also displace each other from the membrane (7), which led to
the concept of overlapping binding sites of functionally related, but chemically different compounds (4,44). As already
discussed above, it has been possible to specify such differences even further in cross resistance studies with herbicide
tolerant plants and also in trypsin digestion experiments or
tris treated thylakoids (53). Such experiments further substantiated that there is a second family of PS II inhibitors.
This second "non-classical" type of inhibitors consisted at
first of just nitro- and halogen-substituted phenols like dino-

seb and ioxynil (62,63). Recently, however, quite a number of
phenol analogues have been described as potent PS II inhibitors: pyridones (53), quinolones (64), naphthoquinones (64,65),
benzoquinones (66,67), pyrones (68), a natural occuring chro-
mone (stigmatellin) (69), dioxobenzthiazoles (70) and cyanoacrylates (71) (Fig. 5). They displace each other as well as
the classical inhibitors from the membrane. Similar to the
triazine/urea family they are inactive in trypsin treated membranes. But there are significant differences in functional
and chemical behaviour, some of it already summarized (72).
Atrazine or DCMU tolerant mutants are still equally, if not
more sensitive to the phenols (44,48-51). In Anacystis, however, a diuron mutant has also lost sensitivity of PS II towards hydroxyquinoline (52), a compound that belongs to the
phenol-type inhibitors (64). As the cross resistance in herbicide tolerant mutants also tris treatment may be used to differentiate between subfamilies of PS II inhibitors. Tris
treatment (that removes the three peripheral peptides of PS II
and oxygen evolution (73)) leads to a decrease in inhibitory
potency of the classical inhibitors triazine, triazinone and
ureas, but not of phenol, quinolone and pyridone inhibitors
(53,64).
   Already very early a chemical element common to all herbicides of the classical type (triazine, urea family) has been
proposed - a sp² hybrid bound to N, O or ⇒CH and attached to
a lipophilic substituent(see 4,74,75 and many other reviews).
In QSAR studies of this "classical" group the inhibitory potency follows lipophilic and in some cases electronic parameters (4,74,75). In the second "non-classical" type of inhibitors a common essential element can now also be formulated

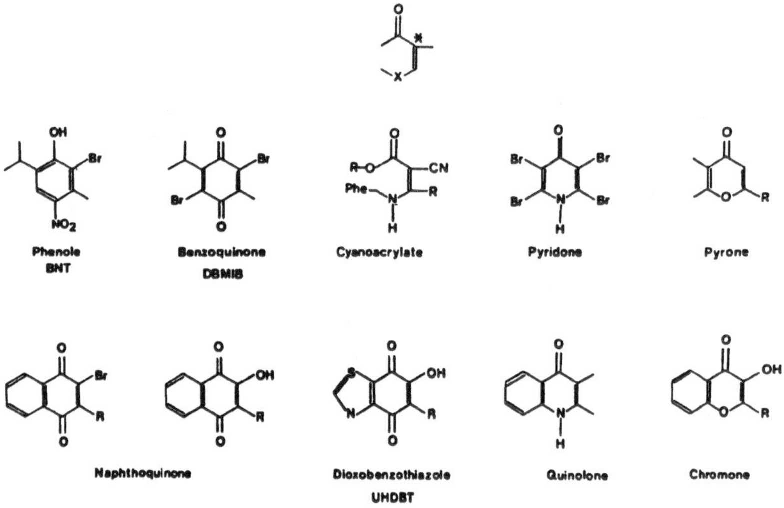

FIGURE 5. The chemical element common to "phenol-type" inhibitors of PS II.
The 𝜋-charge on the atom with an asterisk can be positive or
negative, conferring particular properties to the interaction
with the binding subunit.

(Fig. 5). This element is also observed in the cyanoacrylates, when written in a planar way as in Fig. 5. In QSAR calculations the phenol-type inhibitor group follows steric (4) with little influence of electronic substitution parameters or in a related quinone group redox potential, steric and electronic parameters (65). Draber and Fedtke (76) have used steric parameters to calculate the size of the binding niche for the herbicides: small substituents on one side of a planar system with the lipophilic and larger substituent extending towards the other side.

It may turn out that the charge distribution in the different herbicides may be of particular significance to identify specific amino acid residues involved in the binding of one, but not another herbicide. It was proposed that a positive π-charge at a particular atom is of importance for the "classical" type of inhibitors. It suggests the importance of a particular amino acid in the binding niche that compensates this charge (Fig. 6) (63,77). The "non-classical" type, however, was shown to carry a negative charge at the equivalent atom (63,77). Recently an interesting transition has been observed (64). Upon a certain substitution naphthoquinones change the π-charge at atom 2 (see Fig. 5) from a negative to a positive value (63). This brings about a change in binding characteristics. Those compounds with a negative charge (phenols, pyridones, quinolones) are different from the diuron family in tris treated thylakoids (they do not loose inhibitory potency), whereas those that have a positive charge (hydroxynaphthoquinones) lost inhibitory potency and behaved like diuron (53,63). It indicates that in those compounds with a negative charge another amino acid residue in the binding site compensates for that charge than in those which have a positive π-charge.

The present state of knowledge of the steric conditions for both the target and the inhibitors for optimal interaction makes molecular modelling possible and allows a more rational design of new inhibitors. The model provides reasonable suggestions for site specific mutagenesis to obtain herbicide tolerance without loss of photosynthetic efficiency.

ACKNOWLEDGEMENT

Work at Bochum was supported by Deutsche Forschungsgemeinschaft and Fonds der chemischen Industrie.

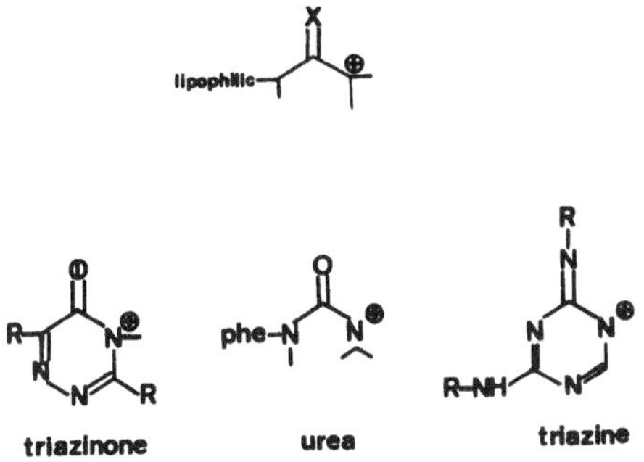

FIGURE 6. A positive π-charge on a particular nitrogen is common to the urea/triazine type photosystem II herbicides.

References

1. Butler WL and Kitajima M (1975) Biochim. Biophys. Acta 376: 116-125
2. Govindjee, Kambara T and Coleman W (1985) Photochem. Photobiol. 42: 187-210
3. Stein RR, Castellvi AL, Bogacz JP and Wraight CA (1984) J. Cell Biochem. 24: 243-259
4. Trebst A and Draber W (1979) In: Advances in Pesticide Science (Geissbühler H, ed.), Part 2, pp. 223-234, Pergamon Press, Oxford, New York
5. Critchley C (1985) Biochim. Biophys. Acta 811: 33-46
6. Satoh K (1985) Photochem. Photobiol. 42: 845-853
7. Tischer W and Strotmann H (1977) Biochim. Biophys. Acta 460: 113-125
8. Regitz G and Ohad I (1976) J. Biol. Chem. 251: 247-252
9. Renger G (1976) Biochim. Biophys. Acta 635: 236-248
10. Pfister K, Steinback KE, Gardner G and Arntzen CJ (1981) Proc. Natl. Acad. Sci. USA 78: 981-985
11. Mattoo AK, Pick U, Hoffmann-Falk H and Edelman M (1981) Proc. Natl. Acad. Sci. USA 78: 1572-1576
12. Boschetti A, Tellenbach M and Gerber A (1985) Biochim. Biophys. Acta 810: 12-19
13. Oettmeier W, Masson, K, Soll HJ and Draber W (1984) Biochim. Biophys. Acta 767: 590-595
14. Oettmeier W, Masson K and Johanningmeier U (1980) FEBS Lett. 118: 267-270
15. Eisenberg D (1984) Ann. Rev. Biochem. 53: 593-623
16. Herrmann RG, Westhoff P, Alt J, Tittgen J and Nelson N (1985) In: Molecular Form and Function of the Plant Genome (van Vloten-Doting L, Groot GSP and Hall TC, eds.), pp. 233-256, Plenum Publishing Corporation
17. Zurawski G, Bohnert HJ, Whitfeld PR and Bottomley W (1982) Proc. Natl. Acad. Sci. USA 79: 7699-7703
18. Rao JKM, Hargrave PA and Argos P (1983) FEBS Lett. 156: 165-169
19. Kyle DJ (1985) Photochem. Photobiol. 41: 107-116
20. Rochaix JD, Dron M, Rahire M and Malnoe P (1984) Plant Molec. Biol. 3: 363-370
21. Alt H, Morris J, Westhoff P and Herrmann RG (1984) Current Genetics 8: 597-606
22. Holschuh K, Bottomley W and Whitfeld PR (1984) Nuc. Acids Res. 12: 8819-8834
23. Rasmussen OF, Bookjans G, Stumman BM and Hennigsen KW (1984) Plant Molec. Biol. 3: 191-199
24. Youvan DC, Bylina EJ, Alberti M, Begusch H and Hearst JE (1984) Cell 37: 949-957
25. Trebst A (1986) Z. Naturforsch. 40c: 237-241
26. Zimmermann JL and Rutherford AW (1985) Physiol. Veg. 23: 425-434
27. Debus RJ, Feher G and Okamura MY (1985) Biochemistry 24: 2488-2500
28. Wraight CA (1981) Israel J. Chem. 21: 348-354
29. Okamura MY (1984) In: Biosynthesis of the Photosynthetic Apparatus: Molecular Biology, Development and Regulation, pp. 381-390, Alan R. Liss Inc.

30. De Vitry C and Diner B (1984) FEBS Lett. 167: 327-331
31. Brown AR, Gilbert CW, Guy R and Arntzen CJ (1984) Proc. Natl. Acad. Sci. USA 81: 6310-6314
32. Deisenhofer J, Epp O, Miki K, Huber R and Michel H (1984) J. Mol. Biol. 180: 385-398
33. Deisenhofer J, Epp O, Miki K, Huber R and Michel H (1985) Nature 318: 618-624
34. Bunker G, Stern EA, Blankenship RE and Parson WW (1982) Biophys. J. 37: 539-551
35. Marder JB, Goloubinoff P and Edelman M (1984) J. Biol. Chem. 259: 3900-3908
36. Hirschberg J and McIntosh L (1983) Science 222: 1346-1348
37. Erickson JM, Rahire M, Bennoun P, Delepelaire P, Diner P and Rochaix JD (1984) Proc. Natl. Acad. Sci. USA 81: 3617-3621
38. Erickson JM, Rahire M, Rochaix JD and Mets L (1985) Science 228: 204-207
39. Erickson JM and Rochaix JD (1985) In: Abstracts. First International Congress of Plant Molecular Biology (Galau GA, ed.), p. 54/OR-25-02, The University of Georgia Center for Continuing Education for the Int. Soc. for Plant Mol. Biol., Athens
40. Golden SS and Haselkorn R (1985) Science 229: 1104-1107
41. Steinback KE and Wolber PK (1985) In: Molecular Biology of the Photosynthetic Apparatus (Arntzen CJ, Bogorad L, Bonitz S and Steinback KE, eds.), Cold Spring Harbour Meeting, in press
42. Shipman LL (1981) J. Theor. Biol. 90: 123-148
43. Velthuys BR (1981) FEBS Lett. 126: 277-281
44. Pfister K and Arntzen CJ (1979) Z. Naturforsch. 34c: 996-1009
45. Vermaas WFJ, Renger G and Arntzen CJ (1984) Z. Naturforsch. 39c: 368-373
46. Gressel J (1985) In: Weed Physiology, Vol. II: Herbicide Physiology (Duke SO, ed.), pp. 159-189, CRC Press , Boca Raton, FL
47. van Rensen JJS (1985): In: Weed Science Advances, Vol. 1 (Turner RG, ed.), Butterworth, London, in press
48. Thiel A and Böger P (1984) Pest. Biochem. Physiol. 22: 232-242
49. Oettmeier W, Masson K, Fedtke C, Konze J and Schmidt RR (1982) Pest. Biochem. Physiol. 18: 357-367
50. Galloway RE and Mets LJ (1984) Plant Physiol. 74: 469-474
51. Pucheu N, Oettmeier W, Heisterkamp U, Masson K and Wildner GF (1984) Z. Naturforsch. 39c: 437-439
52. Golden SS and Sherman LA (1984) Biochim. Biophys. Acta 764: 239-246
53. Trebst A, Depka B, Ridley SM and Hawkins AF (1985) Z. Naturforsch. 40c: 391-399
54. Metz JG, Bricker TM and Seibert M (1985) FEBS Lett. 185: 191-196.
55. Renger G (1973) Biochim. Biophys. Acta 314: 113-116
56. van Assche JC and Carles PM (1982) In: Biochemical Responses Induced by Herbicides. ACS Symposium Series 181 (Moreland DE, St.John JB and Hess FD, eds.), pp. 1-21, American Chemical Society, Washington D.C.

57. Carpentier R, Fuerst P, Nakatani HY and Arntzen CJ (1985) Biochim. Biophys. Acta 808: 293-299
58. Völker M, Ono T, Inoue Y and Renger G (1985) Biochim. Biophys. Acta 806: 25-34
59. Ikeuchi M, Yuasa M and Inoue Y (1985) FEBS Lett. 185: 316-322
60. Hartung A and Trebst A (1985) Physiol. Veg. 23: 635-648
61. Colson AM, TheVan L, Convent B, Briquet M and Goffeau A (1977) Eur. J. Biochem. 74: 521-526
62. van Rensen JJS, van der Vet M and van Vliet WPA (1977) Photochem. Photobiol. 25: 579-583
63. Trebst A, Draber W and Donner WT (1983) In: IUPAC Pesticide Chemistry. Human Welfare and the Environment (Miyamoto et al., eds.), pp. 85-90, Pergamon Press, Oxford
64. Draber W, Pittel H, Dittgens K, Knops HJ, Trebst A and Wietoska H (1986) Z. Naturforsch., in press
65. Oettmeier W, Dierig C and Masson K (1986) Quant. Struct. - Act. Relat., in press
66. Bauer K and Köcher H (1979) Z. Naturforsch. 34c: 961-963
67. Oettmeier W, Reimer S and Link K (1978) Z. Naturforsch. 33c: 695-703
68. Kuwabara M, Yoshida S, Takahashi N and Fujita Y (1980) Plant & Cell Physiol. 21: 745-753
69. Oettmeier W, Godde D, Kunze B and Höfle G (1981) Biochim. Biophys. Acta 807: 216-219
70. Oettmeier W, Masson K and Godde D (1981) Z. Naturforsch. 36c: 272-275
71. Phillips J and Huppatz J (1984) Z. Naturforsch. 39c: 335-337.
72. Oettmeier W and Trebst A (1983) In: The Oxygen Evolving System of Photosynthesis (Inoue Y, Crofts AR, Govindjee, Murata N, Renger G and Satoh K, eds.), pp. 411-420, Academic Press Japan, Tokyo
73. Murata N and Miyao M (1985) Trends Biochem. Sci. 10: 122-124
74. Büchel KH (1972) Pestic. Sci. 3: 89-110
75. Kakkis E, Palmire VC, Strong CD, Bertsch W, Hansch C and Schirmer U (1984) J. Agric. Food Chem. 32: 133-144
76. Draber W and Fedtke C (1979) In: Advances in Pesticide Science, Part 3 (Geissbühler H, ed.), pp. 475-486, Pergamon Press, Oxford, New York
77. Trebst A, Donner W and Draber W (1984) Z. Naturforsch. 39c: 405-411

Photosynthesis Research 10: 393–403 (1986)
© *Martinus Nijhoff Publishers, Dordrecht*

ON THE STRUCTURE AND FUNCTION OF CYTOCHROME b-559

W. A. CRAMER*, S. M. THEG, and W. R. WIDGER

ABSTRACT. A summary of biochemical, biophysical, and molecular biological data is presented which led to the identification of two different polypeptides (α and β, MW = 9.16 and 4.27 kDa) in the cytochrome b-559 protein. The presence of a single His residue on each polypeptide, and the conclusion from spectroscopy that the heme coordination must be bis-histidine led to an obligatory requirement for coordination of a single heme through a heme cross-linked dimer. This structure does not have a precedent among soluble or membrane bound cytochromes. The possible participation of the cytochrome in the pathway of photoactivation is discussed.

I. *Introduction*

   Cytochrome b-559 is an intrinsic polypeptide of photosystem II (1), and its M_r 10,000 polypeptide is recognized in SDS-PAGE gels of PSII particles (e.g., ref. 2). The location of the cytochrome genes (3), as well as their nucleotide sequence in the spinach plastid chromosome (4), is known. The presence of a cytochrome b-559 polypeptide in the minimum PSII complement naturally suggested a function associated with the water splitting reactions. The following such functions have been proposed: (i) A redox function for cytochrome b-559 in water splitting (5,6), as well (ii) as a function in which the reduced cytochrome might function as an H^+ acceptor of the water splitting reactions (7,8); (iii) a cycle around PSII (9-10), based initially on the discovery of b-559 photooxidation by PSII at $77°K$ (12); consistent with such a cycle are the proposals (iv) that the cytochrome may function as a donor to the PSII donor, Z^+, or to ADRY-like compounds interacting with Z^+, inferred from effects of ADRY reagents on cytochrome turnover (13-15) and Z^+ reduction (16), or (v) in a cycle that functions in photoheterotrophic but not autotrophic algae (17).

   Regarding structure, recent advances in understanding of the amino acid composition, heme coordination, nucleotide and amino acid sequence, and polypeptide composition of cytochrome b-559 have relied on a combination of biochemical, biophysical, and molecular biological approaches (3,4,18-20). These data have led to conclusions concerning the arrangement of the cytochrome in the membrane (21, see below):

II. *Studies on the Structure of Cytochrome b-559.*

   The initial step in the purification, the extraction of membranes with 2% Triton X-100 and 4 M urea, was developed by Wasserman and coworkers (22,23), who reported that the purified cytochrome was made of small electrophoretically similar subunits of M_r 5,600 with three different NH_2-termini, in a complex oligomeric (octameric) lipoprotein of 110,000 molecular weight. However, the absence of histidine in the amino acid composition (23) was extremely unlikely for a heme protein, and the heterogeneity in the NH_2-termini was also an indication of lack of preparation purity. In retrospect, the small polypeptide thought to be a b-559 subunit was probably a heterogeneous mixture of peptides resulting from background proteolysis, since the amino acid composition was quite different from that later determined for pure cytochrome b-559 (4,18). Furthermore, the molecular weight of the lipoprotein is probably not ~100,000, and this value is probably only a mean of values for the large heterogeneous non-specific aggregates (MW = 100-300,000) of the cytochrome that are obtained in aqueous solution (24).

It was difficult to reproducibly obtain a high yield of cytochrome in the initial extraction until we systematically unstacked the thylakoids beforehand, suggesting that the extracted cytochrome b-559 arises from the appressed membrane region. Because of the small size of the cytochrome and a tendency toward proteolysis, all procedures were carried out in the presence of a cocktail of protease inhibitors (18).

Starting from 600 mg of chlorophyll, 5-10 mg of the cytochrome could be purified after three chromatography steps, as judged by a predominant M_r 10,000 band on SDS-PAGE gels (18). A similar M_r value for the cytochrome was obtained starting with PSII particles from maize (19). The gels also showed other minor bands, particularly one near M_r 8,000 and a weakly staining M_r 6,000 band. Because the M_r 8,000 band was found to be more pronounced in the absence of proteolysis inhibitors, it and the 6 kDa band were initially assumed to be proteolysis products of the dominant M_r 10,000 polypeptide. It was realized at this time, however, that the small amount of the M_r 6,000 band might be a result of a weak affinity for Coomassie stain, since (a) the amino acid composition of the M_r 6,000 component was somewhat similar to the M_r 10,000 with one His residue in each (8), and (b) it showed a significant 280 nm absorbance relative to the M_r 10,000 band in the HPLC elution profile (Fig. 1). Therein will lie a story.

The M_r 10,000 band could be purified on reverse phase HPLC (Fig. 1) as a single band on an overloaded gel, and the purity was confirmed by an NH_2-terminal amino acid sequence of twenty-seven residues obtained from it (18). Monospecific polyclonal antibody to the HPLC-purified polypeptide was used to locate the gene for this cytochrome b-559 polypeptide in the spinach plastid chromosome by hybrid selection-translation using a library of restricted DNA fragments and immunoprecipitation (3).

A reading frame on the spinach plastid chromosome could be located that corresponded exactly to that determined for the twenty-seven residues of the NH_2-terminus of the M_r 10,000 polypeptide, and whose nucleotide sequence corresponded, after post-translational processing of the NH_2-terminal residue, to an 82 amino acid polypeptide with MW = 9,162 (4). The presence of a single histidine residue is an important aspect of the sequence, since at just about this time spectroscopic

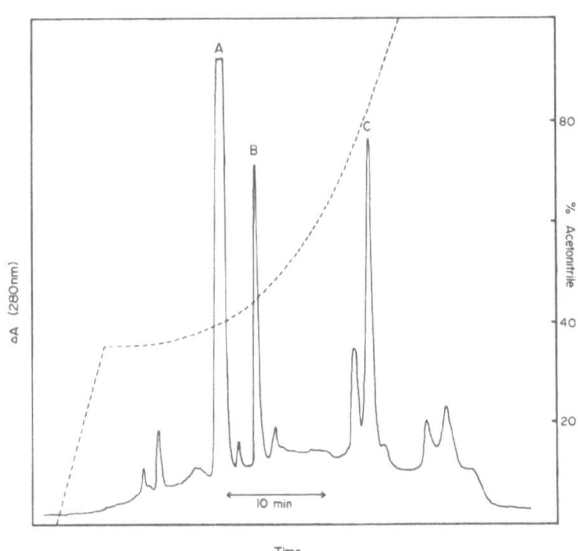

Fig. 1. Separation by reverse phase HPLC of the purified polypeptide components of cytochrome b-559 (13). The elution bands A-C contain free heme, the 4.27 kDa, and the 9.16 kDa polypeptides, respectively.

measurements carried out with the purified cytochrome showed that the heme coordination must be bis-histidine (20). The absence of methionine as the sixth ligand was proven by the lack of a 695 nm absorption band of the oxidized cytochrome (Fig. 2), the absence of lysine in the protein (4), and EPR and Raman spectral data (20).

    Continuation of the nucleotide sequence in Herrmann's laboratory showed that the TAG stop codon of this reading frame overlapped a ribosome binding sequence, GGAGG, and that an initiator codon for another reading frame encoding a 4.27 kDa polypeptide was present nine nucleotides downstream from the terminal G codon (4). Proximity of the downstream reading frame suggested possible relevance to the b-559 protein. This idea was strengthened by alignment of the sequences which showed that the polypeptide products of each gene, psbE and psbF, contain a single His residue positioned five residues from an Arg residue on the NH_2-terminal side of a 25-26 residue non-polar domain. The corresponding position of the Arg residue defining the end of the hydrophobic domain and the homology between identical (solid box) and like (dashed box) residues in the hydrophobic domain is shown (Table I).

Table I. Comparison of psbE and psbF genes.

NH_2- R Y W V I H S I T I P S L F I A G W L F V S T G L A Y D psbE

NH_2- R W L A I H G L A V P T V S F L G S I S A M Q F I Q R psbF

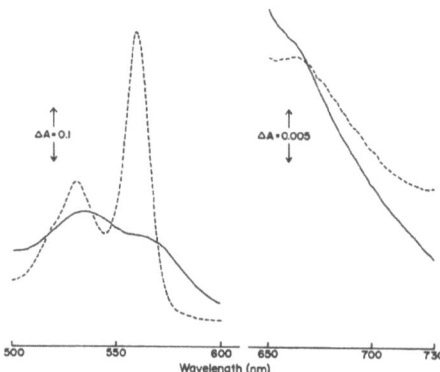

Fig. 2. Comparison of the visible light absorbance properties of purified cyt b-559 in the α-band (left) and 695 nm (right) region, the latter diagnostic of a methionine-ferric heme interaction that generates an absorption band with ϵ_{mM} = 800 M^{-1} cm^{-1} (20). Note the 20-fold scale expansion for the right-hand panel; (——) oxidized with ferricyanide; (---) reduced by dithionite (ref. 20). From the α-band spectra the ϵ_{mM} values are: peak (559.5nm) minus trough (577nm) of difference spectrum, 21.5; reduced minus oxidized at 559.5 and 552 nm, 17.5 and 5.0, respectively. The isobestic points are at 570, 548, and 538 nm.

Given the DNA sequence data, we were led to ask whether the small polypeptide was present in the SDS-PAGE gels. As stated above, the band running near M_r 6,000 had originally been thought to be a proteolysis product of the M_r 10,000 band. The initial amino acid composition data were consistent with this designation since only a single histidine residue was found (18). An amount of the small polypeptide sufficient for NH2-terminal sequencing was separated and isolated by the reverse phase HPLC. Comparison of the sequences showed that the M_r 6,000 polypeptide was the product of the downstream gene (21):

Measured
NH2 ... Thr-Ile-(Asp)-X-(Thr)-Tyr-Pro-Ile-Phe- ... COOH
Predicted from Nucleotide Sequence
NH2 ... -Thr-Ile-Asp-Arg-Thr-Tyr-Pro-Ile-Phe- ... COOH

The predicted amino acids were also present in positions 3 and 5, but one other residue was found in these cycles. The residue in position 4 could not be determined. The smaller b-559 polypeptide, like the larger one, appears to have lost the N-terminal amino acid by post-translational processing, so that the molecular weights are 9,162 and 4,268, respectively. Establishment of the identity of the b-559 M_r 6,000 polypeptide with the protein product of the downstream gene allowed the stoichiometry of the 9.16 and 4.27 kDa polypeptides to be

determined through the respective areas under the peaks in the HPLC
elution profile normalized to tryptophan content. The stoichiometry is
close to 1:1, implying that the polypeptides are present in equal molar
ratio in the b-559 protein (21). The 1:1 stoichiometry of the 9.16 kDa
(α): 4.27 kDa (β) polypeptides and the bis-histidine coordination
resulting from polypeptides containing only a single histidine imply
that the simplest model for coordination of a single b-559 heme is a
heme cross-linked ($\alpha\beta$) dimer. Assuming that the 25-26 residue non-
polar domains on each polypeptide span the membrane in an α-helical
conformation, the two histidine residues on α and β would be positioned
five residues into the hydrophobic phase from Arg residues positioned
at the polar interface (Fig. 3). This heme cross-linked dimeric
structure does not have a precedent among soluble or membrane-bound
cytochromes.

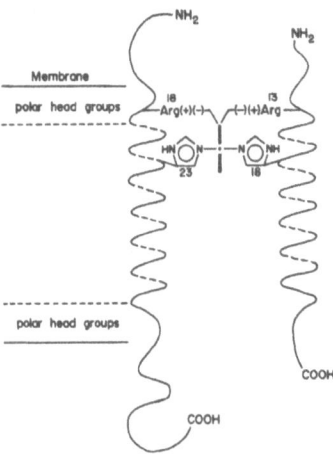

Fig. 3. Model for folding of a monoheme unit of cytochrome b-559 in
the hydrophobic core of the thylakoid membrane (16).

Without any information on the arrangement of the cytochrome b-559
protein relative to other PSII proteins, the nature of the packing of
the heme is not known. The well-known lability of this cytochrome in
situ (1) may be a consequence of this structure. Recent studies of the
orientation of the heme plane of ferric-cytochrome b-559 show it to be
perpendicular to the plane of the membrane (25), consistent with the
model shown in Fig. 3. Upon formation of a high-spin cytochrome after
aging of the membranes, however, the heme plane tilts by 45° (25).
Evidently, coordination of at least one of the His residues is labile
with respect to this treatment, and may also be correlated with the
shift from high to low midpoint potential of the cytochrome (1,26).

III. *Structural Models for a Two Heme Cytochrome b-559.*

The choice of models for the arrangement of the b-559 polypeptide
in the membrane becomes less obvious if the protein contains two hemes.
There is a substantial amount of evidence indicating the presence of
two b-559 hemes per P680 reaction center positioned at two different
environments in the membrane. The stoichiometry of one or two hemes

per P680 depends, however, upon the chlorophyll/P680 stoichiometry (27,28). The b-559 content in different PSII particle preparations can be one or two per P680 (29-32). Two different heme environments are indicated by the biphasic oxidation by ferricyanide of two high potential hemes (33). This provides an explanation for the observations that one b-559 heme can be photo-oxidized at 77°K (34), and that one heme can be photoreduced by PSII in 100 msec at room temperature (35), although the question of whether these hemes are distinct has not been answered. The simplest models for coordination of two hemes in one cytochrome b-559 protein would involve a tetramer of polypeptides to coordinate the two hemes (Fig. 4).

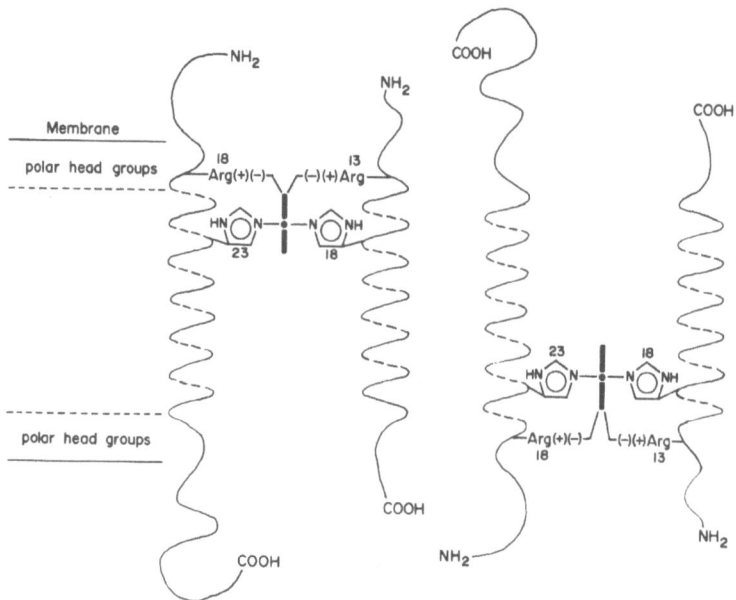

Fig. 4. Tetrameric polypeptide model for arrangement of cytochrome b-559 containing two hemes in the hydrophobic core of the thylakoid membrane, assuming two heme cross-linked heterodimers.

The two models that can be constructed for a cytochrome b-559 polypeptide tetramer would consist of two heme cross-linked dimers, two homodimers $(\alpha)_2$ $(\beta)_2$, or heterodimers $(\alpha\beta)_2$ [Fig. 4], with the hemes on opposite sides of the hydrophobic phase. Both of these models for cytochrome b-559 invoke the precedent of the model proposed for the trans-membrane arrangement of the hemes in cytochrome b of the mitochondrial and chromatophore b-c1 complexes and cytochrome b6 of the chloroplast b6-f complex (36-38).

An argument against a possible $(\alpha)_2(\beta)_2$ homodimer model is that the two dimers differ greatly in charge, since the α and β subunits have a net charge at neutral pH of +2 and -3, respectively. The two homodimers would then differ by 10 charge units, as well as by about 10 kDa in molecular weight, and should have been separable on the detergent-containing DEAE columns used for purification of the protein. An argument against the heterodimer model is that there is no a priori basis for two identical (αβ) units to orient oppositely in the membrane during its synthesis. An answer to the latter point is that the b-559 polypeptide could be assembled in the membrane as a tetrameric unit. With either model, the predicted heme content, one heme per 14,000 molecular weight, is close to the measured value of ~1/17,000 (18). The two different protein and lipid environments proposed for the two cytochrome hemes, one close to the stromal, and the other close to the lumenal interface, might also explain the greater lability toward conversion to lower potential (not reducible by hydroquinone) of approximately half the cytochrome b-559 population (1). No indication has been obtained thus far in the purification work for the presence of a distinct low potential b-559 protein (discussed in ref. 18).

IV. *The Question of Function.*

Independent of the question of homo- vs. hetero-dimers, the model of Fig. 4 suggests that the two hemes would span most of the membrane dielectric, implying that this cytochrome could participate in a trans-membrane pathway. The obvious possibility for such a pathway would be a cycle around photosystem II. The lack of evidence for such a cycle in chloroplasts with unimpaired water splitting (1) is consistent with rapid reduction of P680 by other intermediates in the water splitting pathway (e.g., ref. 39). Furthermore, competent electron transfer in PSII with water as the donor can occur in the absence of high potential (hydroquinone-reducible) cytochrome b-559 (1,40-42), thereby arguing against hypotheses for b-559 function that would obligatorily involve a high potential b-559 in water splitting, the most recent such proposal concerning a function in proton-linked PSII electron transport (7). Thus, we know of no viable hypotheses linking cytochrome b-559 to electron transport in PSII while the water splitting enzyme is operational. Because of these data, and the fact that the amplitude of the photooxidation of cytochrome b-559 by PSII is increased when water splitting is blocked or impaired (e.g., 12) one is led to look for the function of the cytochrome in chloroplasts with impaired PSII activity.

V. *A Proposal for the Function of Cytochrome b-559 in Chloroplast Development and in Response to Stress.*

In this section we would like to propose a role for cytochrome b-559 as a mediator in the (re)assembly of the water splitting enzyme, either in developing chloroplasts or in thylakoids recovering from stress-induced damage.

It has been known for some time that light is required for the assembly of the components of the oxygen evolving complex into a functional water splitting enzyme (for a review, see 43). This process, known as photoactivation, has been demonstrated in both algae (green and blue-green) and in higher plants, and occurs when the oxygen evolving complex is in a non-functional, but "ready to assemble" state. Such a situation is found, for instance, in algae and chloroplasts of plants grown without Mn^{2+} (44), or washed with NH_2OH (45) or TRIS (46) and then supplied with externally added Mn^{2+}, and in chloroplasts of seedlings grown in darkness (47,48) or intermittent light (49,50).

It is clear that photoactivation is a PSII-dependent process. Both its action spectrum and absorption cross-section are the same as those of PSII, it is sensitive to DCMU, and it has been observed in a PSI-less mutant (43). The rate of photoactivation is also proportional to the number of inactive PSII units (43), suggesting that the oxidants produced in reaction centers not connected to functional oxygen evolving complexes play a critical role in the process.

As discussed above, while cytochrome $\underline{b}$-559 has never been shown to compete effectively with the oxygen evolving complex for oxidants produced at the PSII reaction center, it is readily photooxidized by PSII when the water splitting enzyme is inhibited (e.g.,12). In developing chloroplasts, its photooxidation by PSII can be observed at the onset of assembly of the photochemical apparatus (51). It was, in fact, noted in ref. 43 that "the terminal step in formation of the capacity for O_2 evolution may be the photoactivation of Mn." It therefore seems plausible that the cytochrome could play a role in funneling the oxidizing potential of P680 in inhibited chains to the thylakoid component responsible for the final assembly of the oxygen evolving complex, and specifically to the oxidation of Mn^{2+} (Fig. 5). Consideration of the mid-point potentials of Mn-containing superoxide dismutases that range between +180 and +320 mV (52) indicates that the middle-high potential forms (Em > 200 mV) of cytochrome $\underline{b}$-559 could be capable of oxidizing complexed Mn. In addition, the requirement for low light intensities for photoactivation (4% of saturating intensity for photosynthesis (43)) may be a result of the relatively slow redox reaction times (100 ms, 36) of cytochrome $\underline{b}$559. Finally, although a number of seemingly unrelated compounds (uncouplers, DCCD) act through unknown mechanisms to inhibit photoactivation in certain cases (54), one can understand the inhibition by hydrazine and hydroquinone (43) as being due to their competition with cytochrome $\underline{b}$-559 for the PSII oxidant. Indeed, our hypothesis predicts that competent PSII electron donors should be able to inhibit photoactivation. [The idea that cytochrome $\underline{b}$-559 might be involved in the pathway of photoactivation was also independently put forward by Prof. G. Cheniae (personal communication)].

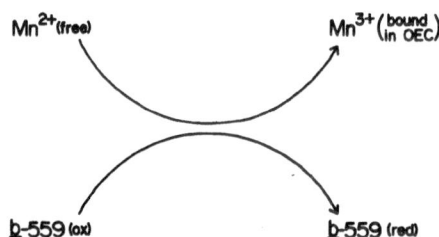

Fig. 5. Proposed role for cytochrome $\underline{b}$-559 in the oxidation of Mn^{+2} to Mn^{+3} during the process of photoactivation or repair of stress damage to PSII.

An important aspect of this hypothesis is that cytochrome $\underline{b}$-559 would be important not only for chloroplast development, but also as part of a stress-response system for PSII, mediating repair to damaged oxygen evolving complexes. These enzymatic complexes are among the most labile in the electron transport chain, and have been shown to be damaged by heat (55), chilling (56,57), and water stress (O. Canaani, personal communication), all conditions potentially encountered by plants in their natural habitats. Thus, in spite of the lack of a demonstrated role of cytochrome $\underline{b}$-559 in chloroplasts with a normally functioning water splitting complex, it would be understandable that this cytochrome is maintained throughout the life of the chloroplast and not degraded after the completion of development.

A number of experiments can be suggested to test this working hypothesis. First, one could examine the process of photoactivation in plants or algae in which the gene for cytochrome $\underline{b}$-559 is altered or deleted. Such a deletion has been accomplished in the Synechocystis 6803 cyanobacterium (H. Pakrasi, personal communication). Alternatively, one could investigate the effects of an antibody to the cytochrome on photoactivation of Mn-extracted PSII membrane preparations. The use of PSII preparations may be essential in this case, as there appears to be an accessibility problem with the antibody added to typical thylakoid preparations (unpublished data). In either experiment, the model predicts that photoactivation should be severely hampered. Second, cytochrome $\underline{b}$-559 could be monitored during a train of photoactivating flashes; the model predicts that the cytochrome should undergo continuous oxidation-reduction cycles, the reduction should be faster in the presence of added Mn^{2+}, and the rate and/or extent of oxidation by PSII should decrease as the recovery proceeds. Experiments such as these could determine whether the model proposed above brings us closer to uncovering the elusive function of cytochrome $\underline{b}$-559 in chloroplasts.

ACKNOWLEDGMENT

W. A. C. is indebted to Warren for a great deal, including an introduction to the problem discussed in this article. Research supported by the National Science Foundation and the Indiana Corporation for Science and Technology.

REFERENCES

1. Cramer WA and Whitmarsh J (1977) Ann Rev Plant Physiol 28:133-172
2. Tang X-S and Katoh S (1985) FEBS Lett 179:60-64
3. Westhoff P, Alt J, Widger WR, Cramer WA, and Herrmann RG (1985) Plant Molec Biol 4:103-110
4. Herrmann RG, Alt J, Schiller B, Widger WR and Cramer WA (1984) FEBS Lett 176:239-244
5. Lundegårdh H (1961) Proc Natl Acad Sci USA 53:703-710
6. Cox RP and Bendall DS (1972) Biochim Biophys Acta 283:124-135
7. Butler WL (1978) FEBS Lett 95:19-25
8. Matsuda H and Butler WL (1983) Biochim Biophys Acta 724:123-127
9. Boardman NK, Anderson JM and Hiller RG (1971) Biochim Biophys Acta 234:126-136

10. Cramer WA and Böhme H (1972) Biochim Biophys Acta 256:358-369
11. Heber U, Kirk MR and Boardman NK (1979) Biochim Biophys Acta 546:292-306
12. Knaff DB and Arnon, DI (1969) Proc Natl Acad Sci USA 63:956-962
13. Cramer WA, and Butler WL (1967) Biochim Biophys Acta 143:332-339
14. Ben-Hayyim G (1974) Eur J Biochem 41:191-196
15. Velthuys BR (1981) FEBS Lett 126:272-276
16. Yerkes CJ and Crofts AR (1984) In Sybesma C, ed. Adv Photosyn Res I, pp. 489-492. Martinus Nijhoff/Dr. W. Junk, The Hague.
17. Mende D (1980) Plant Sci Lett 17:215-220
18. Widger WR, Cramer WA, Hermodson M, Meyer D, and Gullifor M (1984) J Biol Chem 259:3870-3876
19. Metz JG, Ulmer G, Bricker TM and Miles D (1983) Biochim Biophys Acta 725:203-209
20. Babcock GT, Widger WR, Cramer WA, Oertling WT and Metz JG (1985) Biochemistry 24:3638-3644
21. Widger WR, Cramer WA, Hermodson M, and Herrmann RG (1985) FEBS Lett. 191:186-190.
22. Garewal HS and Wasserman AR (1974) Biochemistry 13:4063-4071
23. Garewal HS and Wasserman AR (1974) Biochemistry 13:4072-4079
24. Widger WR and Cramer WA (1985) Unpublished data
25. Rutherford AW (1985) Biochim Biophys Acta 807:189-201
26. Wada K and Arnon DI (1971) Proc Nat Acad Sci USA 68:3064-3068
27. Anderson JM and Melis A (1983) Proc Natl Acad Sci USA 80:745-749
28. Whitmarsh J and Ort DR (1984) Arch Biochem Biophys 231:378-389
29. Sandusky PO, Selvius De Roo CL, Hicks DB, Yocum CF, Ghanotakis DF and Babcock GT (1983) In Inoue Y, Crofts AR, Govindjee, Murata N, Renger G and Satoh K, eds. The Oxygen Evolving System of Photosynthesis, pp 189-199. Academic Press, Tokyo.
30. Bricker TM, Metz JG, Miles D and Sherman LA (1983) Biochim Biophys Acta 724:447-455
31. Yamamoto Y, Tabata K, Isogai Y, Nishimura M, Okayama S, Matsuura K, and Itoh S (1984) Biochim Biophys Acta 767:493-500
32. Murata N, Miyao M, Omata T, Matsunami H and Kuwabara T (1984) Biochim Biophys Acta 765:363-369
33. Selak ME, Koch-Whitmarsh BE and Whitmarsh J (1984) In Sybesma C, ed. Adv Photosyn Res, pp 493-496. Martinus Nijhoff/Dr. W. Junk, The Hague.
34. Vermeglio A and Mathis P (1975) Avron M, ed. Proc 3rd Int Cong Photosyn, pp 323-334. Elsevier, Amsterdam.
35. Whitmarsh J and Cramer WA (1977) Biochim Biophys Acta 460:280-289
36. Widger WR, Cramer WA, Herrmann RG and Trebst A (1984) Proc Natl Acad Sci USA 81:674-678
37. Cramer WA, Widger WR, Herrmann RG and Trebst A (1985) Trends Biochem Sci 10:125-129
38. Cramer WA, Widger WR, Black MT and Girvin ME (1986) Topics in Photosynthesis (J Barber, ed), vol. 8, in press
39. Van Best JA and Mathis P (1978) Biochim Biophys Acta 503:178-188
40. Cramer WA, Whitmarsh J and Widger WR (1981) in Photosynthesis: Electron Transport and Phosphorylation (Akoyunoglou, G, ed) pp 509-522. Balban Int Sci Serv, Philadelphia
41. Briantais J-M, Vernotte C, Miyao M, Murata N, and Picaud M (1985) Biochim Biophys Acta 808:348-351

42. Ghanotakis DF, Yocum CF and Babcock GT (1985) Photosyn Res. In Press
43. Radmer R and Cheniae GM (1977) In Barber J, ed. Primary Processes of Photosynthesis, pp 303-348. Amsterdam:Elsevier/North Holland Biomedical Press
44. Cheniae GM and Martin IF (1971) Biochim Biophys Acta 253:167-181
45. Cheniae GM and Martin IF (1972) Plant Physiol 50:87-94
46. Yamashita T, Inoue Y, Kobayashi Y and Shibata K (1978) Plant and Cell Physiol 19:895-900
47. Oku T and Tomita G (1976) Physiol Plant 38:181-185
48. Oku T and Tomita G (1980) Physiol Plant 48:99-103
49. Ono T and Inoue Y (1982) Plant Physiol 69:1418-1422
50. Ono T and Inoue Y (1983) Biochim Biophys Acta 723:191-201
51. Baker NR and Butler WL (1976) Plant Physiol 58:526-529.
52. Lawrence GD and Sawyer DT (1978) Coord Chem Rev 27:173-193
53. Cheniae GM and Martin IF (1969) Plant Physiol 44:351-360
54. Yamashita, T and Tomita C (1975) Plant Cell Physiol 16:283-296.
55. Diner BA and Joliot P (1977) In Trebst A and Avron M, eds. Encyclopedia of Plant Physiology, Vol 5, pp 187-205. Berlin:Springer-Verlag
56. Margulies MM (1972) Biochim Biophys Acta 267:96-103
57. Smillie RM and Nott R (1979) Plant Physiol 63:796-801

Authors' address:
    Department of Biological Sciences
    Lilly Hall of Life Sciences, Purdue University,
    West Lafayette, Indiana 47907 (U.S.A.)

Photosynthesis Research 10: 405–413 (1986)
© *Martinus Nijhoff Publishers, Dordrecht*

REVERSE ELECTRON FLOW IN CHLOROPLASTS

YOSEPHA SHAHAK AND MORDHAY AVRON
Biochemistry Department, Weizmann Institute of Science, Rehovot 76100, Israel

I. ABSTRACT

Energy dependent reverse electron flow reactions in isolated thylakoids provide a unique tool to study, in the dark, the coupling between the ATP synthase, proton transport and the electron transfer system. Appropriate experimental conditions have been established to follow experimentally the following reactions:
1. ATP driven proton uptake into the inner-thylakoid space, which requires preactivation of the ATP synthase.
2. ATP driven reverse electron transport, which involves proton transport as an intermediate, and results in the reduction of Q_A by an externally added electron donor.
3. ATP driven luminescence, which requires the presence of an oxidized partner on the water side of photosystem II, and involves electron transport from Q_B to Q_A.
4. ΔpH driven reverse electron flow, which does not require the participation of the ATP synthase, and uses reduced intermediates between the two photosystems as electron donors for the reduction of Q_A.
5. ΔpH driven luminescence which again uses reduced intermdiates between the two photosystems as electron donors for Q_A reduction, and requires the presence of an oxidized partner on the water side of photosystem II.
Several of these reactions have been shown to occur in intact chloroplasts and may provide an important regulatory mechansim *in vivo*.

II. INTRODUCTION

Energy conservation in the photosynthetic systems is effected via several steps in series. Light energy absorbed by the pigment systems drives electron flow along the electron transport chain, which in turn drives proton transport across the membrane, and ATP formation by the ATP synthase [1,2]. It is well accepted today that these are the fundamental steps of energy conservation in the chloroplast. Nevertheless the molecular mechanisms of the two proton pumps, the electron transport system and the ATP synthase complex, are still a subject of intensive research.

Different experimental approaches have been taken to explore this system. These include redox changes of the different components under continuous illumination as well as under single turnover flashes, the use of specific inhibitors, artificial electron donors and acceptors, chemical and genetic modifications of the components, isolation and characterization of single carriers or whole complexes and reconstitution into model systems. Reverse electron flow, which will be reviewed in this communication, is one more approach in this category. This reaction provides a unique tool to study in the dark the coupling between electron transport, proton transport and the ATP synthase in the thylakoid membrane. In this approach, the two pumps are reversed: the ATP synthase is triggered to hydrolyse ATP and build up a proton gradient which, in turn, drives the coupled electron flow backwards, up the hill with respect to redox potential. Energy driven reverse electron flows have been observed also in chromatophores and mitochondria. Baltscheffsky has shown that in both systems cytochrome b becomes reduced and c becomes oxidized during the hydrolysis of either ATP or pyrophosphate [3]. Only the work done in chloroplasts will be reviewed here.

III. REVERSAL OF PROTON PUMPING BY THE ATP SYNTHASE

The chloroplast ATP synthase, unlike the mitochondrial or chromatophore enzyme, is latent in the dark. ATP hydrolysis can be activated by preillumination of the chloroplasts. Two major factors are required for light activation of the ATPase: a proton electrochemical gradient $(\Delta\mu H^+)^1$ and a thiol reductant. Thus, for light- activation of the enzyme in broken and washed (type C according to the nomenclature of Hall [4]) chloroplasts, an electron carrier and DTT are added to the light-activation stage (see 5, 6 for review). Light activation also occurs under physiological conditions, since it was demonstrated both in intact (type A) chloroplasts [7, 8] and in whole leaves [9, 10]. Here, no additions are required. The physiological reductant is probably thioredoxin [11-14]. The activation process seems to involve reduction of a regulatory disulfide bond located in the γ subunit of CF_1 [15].

ATP hydrolysis by the light activated enzyme is the reversal of ATP synthesis. During ATP hydrolysis, protons are pumped electrogenically into the thylakoid lumen leading to the formation of $\Delta\mu H^+$ in the dark. This has been indicated by the following findings: (i) ATPase activity is stimulated by uncouplers [16]; (ii) ATP hydrolysis is accompanied by the uptake of protons from the medium [17], leading to the formation of a proton concentration gradient (ΔpH) across the thylakoid membrane [18]; (iii) Using the membrane potential probe oxonol VI,the electrogenic nature of the ATP dependent proton transport was demonstrated [19-21]. $\Delta\Psi$ formation was further supported by the detection of an ATP induced carotenoid electrochromic shift in the dark after light activation [22].

IV. ATP INDUCED Q REDUCTION

IV.1. Background

Early attempts to demonstrate that ATP hydrolysis can lead to reverse electron flow met with the experimental difficulty of devising a measureable system. Neither an external electron donor which could be followed spectroscopically, nor an acceptor which would not be directly reduced by the donor were readily available. The choice of Q, the native acceptor of PSII, as the acceptor solved the problem. Q is a quencher of PS II chlorophyll fluorescence in its oxidized, but not in the reduced form (reviewed in [23]). Therefore, chlorophyll

---

1 **Abbreviations:** CF_0 (CF_1), chloroplast coupling factor 0 (1); DTT, dithiothreitol; DBMIB, 2,5-dibromothymoquinone; DNP-INT, 2,4-dinitrophenylether of 2-iodo-4-nitrothymol; DCMU, 3-(3,4-dichlorophenyl)-1,1 dimethylurea; PSII (I), Photosystem II (I); Q (or Q_A, quencher, the primary quinone acceptor of PSII; Q_B the secondary quinone acceptor; PQ, plastoquinone; PMS, phenazinemethosulfate; $\Delta\mu H^+$ (ΔpH, $\Delta\Psi$), transmembrane electrochemical (chemical, electro) proton gradient.

a fluorescence yield was used to monitor the redox state of Q under different conditions. Furthermore, Q is not reduced by most externally added mild reductants. Upon illumination of chloroplasts, a typical curve of chlorophyll fluorescence induction is observed. Kinetically, the curve is composed of a rapid rise to a level designated Fo, followed by a slow rise, (variable fluorescence, Fv) which reaches its maximal level (Fmax) [23]. Fo corresponds to a fully oxidized Q and Fmax to a fully reduced Q. If light intensity is sufficiently low, the fluorescence level can be set at close to the Fo level and then dark reduction of Q can be easily monitored.

This approach was successfully used by Rienits *et al.* to demonstrate that the ATPase can indeed drive Q reduction by reverse electron flow in the dark [24]. The procedure includes activation of the ATPase of type C chloroplasts by strong preillumination in the presence of DTT and PMS, a dark interval to allow the decay of the light induced chlorophyll fluorescence level to close to the Fo level, followed by injection of ATP. Detailed procedures and instrumentation for this reaction, as well as the other reverse reactions to be discussed below, have been described [25].

IV.2. Dependence on the H^+-ATPase

The addition of ATP to preactivated chloroplasts induces the reduction of Q. Under optimal conditions, fluorescence is increased to about one third of Fv. The reaction depends upon the preactivation of the ATPase, it is inhibited if a too long dark interval is introduced between the activation and ATP addition, it is specific for ATP, requires Mg^{2+} and is inhibited by uncouplers and by energy transfer inhibitors such as Dio 9 or phlorizin [24, 26]. All these facts indicate that ATP hydrolysis by the H^+-ATPase is the initial driving force for the dark reduction of Q.

IV.3. What part of the electron transport chain is involved?

Several observations relate to this question: (1) The reaction requires an electron donor, but it can be of a rather high redox potential (e.g. benzohydroquinone, E'o=+260 mv) [26]. (2) ATP induced cytochrome f oxidation has been measured, concomitant with Q reduction [26]. However, because of interfering light scattering changes, the observed absorbance changes were not as clear cut as would be desired. (3) Q reduction is inhibited by DBMIB (which has to be prereduced to avoid its debromination by DTT [27]) and by DNP INT (Shahak and Avron, unpublished). (4) In chloroplasts isolated from a mutant of *Lemna perpusilla* (number 1073) which is defective in its Rieske protein [28, 29], ATP induced Q reduction is blocked, but the ATPase is fully active [27]. (5) DCMU inhibits the reaction [26]. These data suggest that in ATP induced Q reduction, electrons are transfered from the reductant (DTT) *via* cytochrome f-plastocyanin, through the b_6f complex and up to Q_A.

IV.4. Reverse electron flow in intact chloroplasts

Schreiber has found that ATP induced reverse electron flow can also take place in type A chloroplasts. The addition of ATP to chloroplasts which were preilluminated while intact (in the absence of DTT) and then osmotically shocked (to rupture the envelope which is impermeable to nucleotides), induces reduction of Q. The latter is transient [30], probably due to the lack of an electron donor. Moreover, the reaction was demonstrated in chloroplasts which were kept intact also during the assay period. In this case the reaction was initiated by dihydroxyacetone phosphate, under conditions where it increases the stromal ATP pool [30]. The electron donor in this system has not been defined.

IV.5. **What form of energy links ATP hydrolysis and reverse electron flow?**

Based on the chemiosmotic theory, the intermdiary energy transducer during ATP-induced Q reduction should be the proton electrochemical gradient. Thus, under no conditions should Q get reduced during this reaction, before the formation of ΔpH. Avron and Schreiber have measured simultaneously the time course of ATP induced ΔpH formation and Q reduction and the results suggested agreement with the above prediction [31].

A more detailed study of the kinetics of ATP-induced Q reduction showed two distinct reduction phases: a fast initial one followed by a slow one [32]. The fast phase differs in many respects from the slow phase: (1) it does not require the addition of an electron donor; (2) it is not inhibited by uncouplers (3) hydroxylamine enhances the fast, but not the slow phase; (4) the ratio of fast to slow phase is favored by a high degree of chloroplast integrity. Thus, the ratio is markedly higher in type D as compared with type C chloroplasts; (5) electron transfer inhibitors of the b_6f complex do not inhibit the fast phase; (6) even more intriguing is the finding that the fast phase is quite insensitive to DCMU [33].

It would seem that the slow phase, which is essentially the only one measured in the earlier investigations, does indeed reflect a chemiosmotic type of reverse electron flow in which $\Delta\mu$H$^+$ is the intermediary energy form. Shcreiber suggested that the fast phase reflects a more direct coupling between the ATPase and a redox component in the vicinity of PSII, which is not mediated by $\Delta\mu$H$^+$ [32]. Since the fast phase is insensitive to DCMU, it was speculated that the direct coupling occurs between the ATPase and non-B-type PSII reaction centers located on the margins of the grana stacks [33]. We feel, however, that the origin of the fast phase of ATP-induced Q reduction is not clear. (i) If Q reduction is involved, where do the electrons come from? Despite being a steady state reaction, it does not require an external donor. (ii) The fact that the fast phase is mostly lost in broken and washed chloroplasts may suggest that it requires peripheral factors. What are they and how are they involved? (iii) The suggested lack of involvement of $\Delta\Psi$ requires more evidence. Alternatively, we suggest that the fast phase may reflect a non-reductive effect of the ATPase, or a product of its activity, on the chlorophyll fluorescence yield.

V. ATP INDUCED LIGHT EMISSION

V.1. **Background**

ATP induced reverse electron flow does not end at the Q site but can, under proper conditions, induce the emission of light by PSII. Chlorophyll a luminescence results from a recombination of the first stable oxidized and reduced products of light excitation of PSII, Z$^+$ and Q$^-$ In addition to the natural delayed light emission, a variety of treatments such as temperature jump, external electric pulse, injection of salt, organic solvents, DCMU, dithionite etc. (see [34] for review), induce luminescence. These treatments either increase the supply of electrons at Q, reduce the required activation energy for the recombination reaction, or provide an additional source of activation energy.

In order to measure ATP induced chlorophyll luminescence, the chloroplasts have to be preconditioned so that the ATPase is active and PSII luminescence is limited by the amount of Q$^-$ Schreiber and Avron have developed such a procedure [35]. It includes preillumination of type C chloroplasts in the presence of DTT and MgCl$_2$ to activate the ATPase, followed by 2 min dark to let the native luminescence decay and most of the acceptor pool to become oxidized (the presence of Pi maintains an active ATPase). Then a short flash of light again forms Z$^+$ and Q$^-$, followed by a few seconds in the dark to oxidize Q⁻ (the dark life time of Z^+ is longer), and finally by injection of ATP which gives rise to a burst of light.

V.2. Properties

Since the reaction requires preactivation of the ATPase and is inhibited by the specific ATPase inhibitor, tentoxin, ATP hydrolysis by the ATPase is clearly the driving reaction [35]. ATP-induced luminescence is inhibited by DCMU, indicating that the reaction is indeed a reverse electron flow. DCMU by itself stimulates luminescence, probably due to the induction of a rapid reduction of Q_A by Q_B [36]. DCMU induced luminescence is much smaller than ATP-induced luminescence, but if the injection of DCMU precedes that of ATP, ATP induced luminescence is inhibited [37].

No requirement of ATP-induced luminescence for an external electron donor could be observed. Moreover, PMS, which is required as an electron mediator for the major phase of ATP-Q reduction, inhibits ATP-induced luminescence [37]. The inhibition might result from the chemical reduction of Z^+ by the PMS+DTT couple.

Nigericin and valinomycin have very little effect when added separately (in the presence of K^+). However, the combination of both, which dissipates both ΔpH and $\Delta\Psi$, is fully inhibitory [37]. Gramicidin, which also dissipates both parameters, inhibits the reaction as well [35]. ATP probably stimulates luminescence by a two fold mechanism: (i) increasing the amount of Q⁻ and (ii) creating $\Delta\mu H^+$ which favors the charge recombination [37]. Being a transient reaction it might involve only a short part of the electron transport chain, namely between the PQ pool and PSII. Unfortunately, the effect of b_6f inhibitors on ATP-induced luminescence has not been reported so far.

ATP-induced luminescence has been observed in osmotically shocked chloroplasts which had been light activated in their intact state and also in intact chloroplasts with dihydroxyacetone phosphate as the ATP inducer [30].

VI. ACID-BASE-INDUCED Q REDUCTION

VI.1. Background

If, as the chemiosmotic hypothesis suggests (see [38]), ATP-induced reverse electron flow must be mediated *via* a transmembrane proton gradient, then an artificial ΔpH should by itself drive reverse electron flow. Indeed, upon acid-base transition of chloroplasts the reduction of Q can be observed.

Acid-base-induced, unlike ATP-induced, Q reduction is a transient reaction. The injection of base to chloroplasts which have been preincubated in acidic medium, gives rise to enhancement in chlorophyll *a* fluorescence which reaches a maximal level (typically up to about one third of Fv), followed by a slow (1-2 min) decrease back to the low fluorescence level [39]. In the fluorescence rise two phases, a fast and a slow one, can be distinguished [27, 40].

V.2. Optimal conditions

The pH optimum for the acid stage is 5-6, and for the base stage pH 9-9.5. Unlike acid-base phosphorylation [38], Q reduction does not show a preferential requirement for a specific acidic or basic buffer [39], indicating that ΔpH is required for a very short time in order to drive the reaction. This is further supported by the lack of inhibition by uncouplers [41]. Still, preincubation of the chloroplasts in the acid stage for several minutes is required, indicating that it is the transmembranal pH gradient which drives the reaction.

VI.3. Electron donor

Acid-base-induced Q reduction does not require an external electron donor. The source of electrons for the reaction is the reduced intermediates between the two photosystems. The evidence for this comes from the following results: (i) Preillumination by far red light greatly inhibits acid-base induced Q readuction. (ii) A PSI acceptor present during far red preillumination stimulates the inhibitory effect, while a PSI donor prevents the inhibition. (iii) Short red illumination following the far red period, restores the activity [42]. (iv) Oxidants (such as ferricyanide) present in the acid stage, completely inhibit the reaction. (v) Uncouplers do not inhibit the reaction, except those which are known to be ADRY agents, like FCCP and SF_{6847}. The inhibitory effect of the latter probably results from recycling of electrons from the Q_A-PQ area to Z^+ [43].

VI.4. The decay phase

Parallel measurement of the decay of the fluorescence rise (oxidation of Q) and of ΔpH after acid-base transition showed no correlation between the two [27]. The latter but not the former was sensitive to uncouplers, to the nature of the acidic buffer, the osmolarity of the medium and the temperature, within a defined range. On the other hand, the decay of Q^- but not of ΔpH, was retarded by reductants and facilitated by oxidants [40]. Thus, ΔpH is required to drivea pulse of electrons backwards to Q. However, the life time of the reduced Q thus formed does not depend anymore on ΔpH, but rather on the redox capability of the naturally available carriers.

VI.5. Electron carriers involved

Acid-base-induced, like ATP-induced Q reduction is fully inhibited by DCMU, indicating that Q_B is involved in the reaction. However, inhibitors of the b_6f complex do not inhibit the acid-base reaction. In chloroplasts isolated from the Rieske protein defective *Lemna* mutant 1073, the extent of acid-base-induced Q reduction was found to be slightly bigger and the decay slower, in comparison with wild type chloroplast [27]. The reaction is also not inhibited by DNP-INT (Shahak and Avron, unpublished). When added in its reduced form, DBMIB does not prevent acid-base-induced reverse electron flow [41,42]. It can, therefore, be concluded that the acid-base reverse reaction does not include the b_6f complex.

Our present knowledge of the electron transport system indicates only two carriers between Q and the b_6f complex: Q_B and the PQ pool. The question arose whether the latter is involved in the reverse reaction, or ΔpH just induces a transient change in the equilibrium between Q_A and Q_B, giving rise to Q_A reduction by the Q_B semiquinone. The effects of red and far red preillumination described above, suggest that the pool is involved. Indeed, Hardt found a good kinetic correlation between the redox state of the pool in the dark or during far red preillumination and the extent of acid-base-induced Q reduction [40].

VII. ACID-BASE INDUCED LUMINESCENCE

VII.1. Background

In analogy to ATP driven reverse electron flow, the acid-base induced reaction can also give rise to the emission of light by PSII. Here again the system has to be preconditioned so that Z is positively charged, the PQ pool reduced and Q^- limiting. The procedure developed includes preincubation of chloroplasts (type C) at pH 5.3-6.0, illumination for 20 sec to reduce the PQ pool, dark incubation for 8 sec to allow the photoreduced Q to become reoxidized and then the injection of base (to pH 9.0-9.5), during which luminescence is measured. The dark interval can be longer (e.g. 40 sec) but then a short flash (to reform Z^+) has to precede the injection of the base [43].

VII. 2. Acid-base luminescence involves reverse electron flow

The luminescence emitted by chloroplasts upon acid-base trasition has been extensively studied since its discovery by Mayne and Clayton [44]. Kraan et al. [45] suggested that the acid-base luminescence under their conditions does not result from Q reduction by reverse electron flow. Using the procedures described above we concluded that the reaction does involve reverse electron flow, as it is inhibited by DCMU [43]. Furthermore, it is inhibited by far red illumination (in presence of PSI acceptors) given between the red illumination and the injection of base. DBMIB prevents the far red light effect, since it inhibits PQ oxidation by PSI [41].

The major difference between the technique employed by the earlier investigators and by us, rests in the the different acidic pH used. We have found that the sensitivity of acid-base luminescence to DCMU depends very much on the pH of the acid stage: when the pH was higher than 5.5 it was fully inhibited by DCMU, while below 5 it was largely DCMU insensitive [41]. In most of the earlier reports, the acid stage pH was about 4. It is therefore proposed that acid-base luminescence may occur *via* three routes: (i) a structural change which stimulates the recombination reaction; (ii) protonation-deprotonation of Z and Q located on opposite sides of the membrane; (iii) reverse electron flow from the PQ pool. Only the latter should be inhibited by DCMU. The relative contribution of each process depends upon the reaction conditions. When the PQ pool is prereduced and the acidic pH is above 5.5, acid-base luminescence results mostly from reverse electron flow.

VII.3. Comparison of acid-base-reverse-electron-flow Q reduction and luminescence

The two acid-base reverse reactions are similar with respect to pH optima, lack of a specific requirement for buffers, dependence on reduced PQ pool, inhibition by DCMU and the lack of inhibition by inhibitors of the b_6f complex. Differences between the two reactions reflect the dependence of the luminescence, but not fluorescence, on Z^+ and the probability for charge recombination. The difference was demonstrated in the following experiments [27,43,46]: (1) Tris treatment of chloroplasts inhibits acid-base Q reduction indirectly, due to oxidation of the PQ pool during preillumination. The reaction is restored by the addition of a reductant such as ascorbate plus p- phenylenediamine. The reductant does not restore acid-base luminescence after Tris treatment since it keeps Z reduced as well. (2) Acid-base Q reduction is not sensitive to uncouplers, except for the ADRY agents. Inhibition by the latter group can be restored by a reductant. Acid-base luminescence is sensitive to all uncouplers, with the ADRY agents being most efficient. Inhibition by uncouplers is not prevented by reductants. (3) K^+ diffusion gradients applied simultaneously with ΔpH, affect the acid-base luminescence, but not Q reduction. A K^+ gradient which charges the membrane positively inside enhances acid-base luminescence while an oppositely oriented potential suppresses the reaction. These effects relate, most probably, to enhancement or suppression of the Z^+Q^- recombination reaction.

VIII. OTHER REVERSE ELECTRON FLOW REACTIONS

In addition to the energy dependent reverse electron flow discussed so far, Arnon and Chain reported on Q reduction by NADPH in the dark. They followed the reduction of cytochrome b_{559} and the component C_{550} (supposedly reflecting Q) by NADPH plus ferredoxin [47]. Mills et al. [48] showed that under weak illumination the addition of NADPH induced an increase in chlorophyll *a* fluorescence, up to half the maximal level of Fv. The reaction required ferredoxin and Mg^{2+}; was inhibited by inhibitors of ferredoxin-NADP reductase (FNR) and by antimycin A; but was uncoupler insensitive. The following sequence was, therefore, proposed:

$$NADPH \rightarrow FNR \rightarrow ferredoxin \rightarrow cytochrome\ b_6 \rightarrow PQ\ pool \rightarrow Q_B \rightarrow Q_A$$

However, only part of the NADPH dependent chlorophyll fluorescence increase was due to the dark reduction of Q. The rest resulted from photoreduction of Q by the measuring beam due to inhibition of Q reoxidation, since a major part of the PQ pool was reduced by NADPH in the dark [49]. It should be of interest to study the effect of DCMU and different inhibitors of the b_6f complex on the reaction.

The NADPH dependent reverse reaction may have an important regulatory role *in vivo* [47-49], in controlling the relative activity of cyclic *vs* linear electron transport, and thus the ratio of ATP/NADPH.

IX. REFERENCES

1. Mitchell P (1979) Eur J Biochem 95, 1-20
2. Ferguson SJ (1985) Biochim Biophys Acta 811, 47-95
3. Baltscheffsky M (1968) in: Regulatory Functions of Biological Membranes (Jarnefelt J, ed.) pp 277-286, Elsevier, Amsterdam
4. Hall DO (1972) Nature New Biol 235, 125-126
5. Bakker-Grunwald T (1977) in: Encyclopedia of Plant Physiology - Photosynthesis I (Trebst, A and Avron, M, eds.) vol 5 pp 369-373, Springer-Verlag, Berlin
6. Shavit N (1980) Ann Rev Biochem 49, 111-138
7. Inoue Y, Kobayashi Y, Shibata K and Heber U (1978) Biochim Biophys Acta 504, 142-152
8. Mills JD and Hind G (1979) Biochim Biophys Acta 547, 455-462
9. Morita S, Itoh S and Nishimura M (1982) Biochim Biophys Acta 679, 125-130
10. Vallejos RH, Arana JL and Ravizzini RA (1983) J Biol Chem 258, 7317-7321
11. Buchanan BB (1980) Ann Rev Plant Physiol 31, 341-374
12. Mills JD, Mitchell P and Schurmann P (1980) FEBS Lett 112, 173-177
13. Mills JD, Mitchell P and Schurmann P (1981) in: Photosynthesis. Proc 5th Intern Congr Photosynthesis (Akoyunoglou, G, ed.) vol. 2, pp. 839-848, Balaban International Science, Philadelphia
14. Shahak Y (1982) Plant Physiol 70, 87-91
15. Ketcham SR, Davenport JW and McCarty RE (1984) J Biol Chem 259, 7286-7293
16. Carmeli C (1969) Biochim Biophys Acta 189, 256-266
17. Carmeli C and Lifshitz Y (1972) Biochim Biophys Acta 267, 89-95
18. Bakker-Grunwald T and Van Dam K (1973) Biochim Biophys Acta 292, 808-814.
19. Galmiche JM and Girault G (1980) FEBS Lett 118, 72-76
20. Schuurmans JJ, Casey RP and Kraayenhof R (1978) FEBS Lett 94, 405-409
21. Admon A, Shahak Y and Avron M (1982) Biochim Biophys Acta 681, 405-411
22. Schreiber U and Rienits KG (1982) FEBS Lett 141, 287-291
23. Papageorgiou G (1975) in: Bioenergetics of Photosynthesis (Govindjee, ed.) pp 319-371, Academic Press, New York
24. Rienits KG, Hardt H and Avron M (1973) FEBS Lett 33, 28-32
25. Shahak Y and Avron M (1980) in: Methods in Enzymol (San Pietro A, ed.) vol. 69C, pp 630-641, Academic Press, N.Y.
26. Rienits KG, Hardt H and Avron M (1974) Eur J Biochem 43, 291-298

27. Shahak Y (1978) Ph.D. Thesis, Weizmann Institute of Science, Rehovot
28. Shahak Y, Posner HB and Avron M (1976) Plant Physiol 57, 577-579
29. Malkin R and Posner HB (1978) Biochim Biophys Acta 501, 552-554
30. Schreiber U (1980) FEBS Lett 122, 121-124
31. Avron M and Schreiber U (1977) FEBS Lett 77, 1-6
32. Schreiber U (1984) Biochim Biophys Acta 767, 70-79
33. Schreiber U (1984) Biochim Biophys Acta 767, 80-86
34. Malkin S (1977) in: Encyclopedia of Plant Physiology - Photosynthesis I (Trebst A and Avron M, eds.) vol 5, pp 473-491, Springer-Verlag, Berlin
35. Schreiber U and Avron M (1979) FEBS Lett 82, 159-162
36. Velthuys BR and Amesz J (1974) Biochim Biophys Acta 333, 85-94
37. Avron M and Schreiber U (1979) Biochim Biophys Acta 546, 448-454
38. Jagendorf AT (1977) in Encyclopedia of Plant Physiology - Photosynthesis I (Trebst A and Avron M, eds.) pp 307-337, Springer Verlog, Berlin
39. Shahak Y, Hardt H and Avron M (1975) FEBS Lett 54, 151-154
40. Hardt H (1981) Biochim Biophys Acta 635, 631-644
41. Shahak Y, Siderer Y and Avron M (1977) in: Bioenergetics of Membranes (Packer L, Papageorgiou GC and Trebst A, eds.) pp 405-414, Elsevier, Amsterdam
42. Shahak Y, Pick U and Avron M (1976) in: Enzymes, Electron Transport Systems. Proc 10th FEBS Meeting (Desnuelle P and Michelson AM, eds.) vol. 40 pp 305-314, Elsevier, Amsterdam
43. Shahak Y, Siderer Y and Avron M (1977) in: Photosynthetic Organelles, Structure and Function (Miyachi S, Katoh S, Fujita Y and Shibata K, eds.) pp 115-127, Japanese Society for Plant Physiology, Tokyo
44. Mayne BC and Clayton RK (1966) Proc Natl Acad Sci USA 55, 494-497
45. Kraan GPB, Amesz J, Velthuys BR and Steemers RG (1970) Biochim Biophys Acta 223, 129-145
46. Avron M, Admon A and Shahak Y (1981) in: Energy Coupling in Photosynthesis (Selman BR and Selman-Reimer S, eds.) pp 15-23, Elsevier, New York
47. Arnon DI and Chain RK (1975) Proc Natl Acad Sci USA 72, 4961-4965
48. Mills JD, Crowther D, Slovacek RE, Hind G and McCarty RE (1979) Biochim Biophys Acta 547, 127-137
49. Mills JD, Mitchell P and Barber J (1979) Photobiochem Photobiophys 1, 3-9

Photosynthesis Research 10: 415–422 (1986)
© *Martinus Nijhoff Publishers, Dordrecht*

MULTIDISCIPLINARY RESEARCH IN PHOTOSYNTHESIS: A CASE HISTORY
BASED ON THE GREEN ALGA *Chlamydomonas*

R.K. TOGASAKI[+] and J. WHITMARSH[*]

[+]Department of Biology, Indiana University, Bloomington Indiana.
[*]Department of Plant Biology, University of Illinois, Urbana, Illinois.

ABSTRACT

This article examines the contribution of a unicellular green alga Chlamydomonas
to progress in photosynthetic research. The objective is to focus on the aspects of
Chlamydomonas that have provided an advantage over other photosynthetic
organisms in investigating photosynthesis. To do this we discuss several examples
that demonstrate the progress from a genetic study to a multidisciplinary approach
that probes higher levels of complexity within the organism. These examples
include the function and molecular regulation of electron transport components
between photosystem II and photosystem I, the molecular genetics of the herbicide
binding protein of photosystem II, and several different studies that have derived
from a search for rubisco (ribulose-1,5-bisphosphate carboxylase/oxygenase) mutants
in Chlamydomonas, including chloroplast ribosome function, the regulation of the
large subunit of rubisco, and the interaction between photosynthetic electron
transport and carbon metabolism.

INTRODUCTION

Microalgae have served a vital role in photosynthetic research since the
introduction of Chlorella into the field by Warburg in the 1920s (1,2). In these early
studies Chlorella provided a system that was easy to manipulate while maintaining
photosynthetic activity in vivo. The outstanding example is the determination of
the flash-induced oxygen yield in Chlorella by Emerson and Arnold that led them to
develop the concept of the photosynthetic unit (3). More recently, due primarily to
the creation of genetically characterized mutants that are impaired in their
photosynthetic capacity, the green alga Chlamydomonas has added to the utility of
microalgae in studying photosynthesis. One of the key advantages of
Chlamydomonas lies in the application of multiple research tools, including genetics,
physiology, microbiology, cell biology, photobiology, and molecluar biology, to a
single organism. As a consequence research in one discipline can take advantage of
rapid technological advances in allied fields, thereby increasing a database derived
from a single organism. The long range goal is to understand photosynthesis in the
context of the interrelationship of the various biological phenomena occuring in a
single organism.

The aim of this article is to expose the reader to the potential of Chlamydomonas
as a tool for studying photosythesis by presenting a few examples that reveal its
unique advantages over that of higher plants or photosynthetic bacteria, while
pointing out the significant limitations. In this regard, among the photosynthesis
researchers Waren Butler was one of the earliest to appreciate the research
potential of the Chlamydomonas mutant systems. As a biophysicist collaborating
with biologists he published three early papers based on Chlamydomonas mutants,

one dealing with the analysis of electron transport (4), and the other two dealing with low fluorescence mutants with defects on the oxidizing side of photosystem II (5,6). While these early investigations were not pursued by Warren, he was ahead of his time in the novel approach he adopted and did not hesitate to tackle one of the most formidable topics in photosynthesis research, i.e., water oxidation by photosystem II.

CHLAMYDOMONAS

Chlamydomonas reinhardtii is a unicellular, biflagellated green alga, 4 to 6 microns in diameter, with a single, cup-shaped chloroplast occupying approximately half of the cell volume. The most commonly used strains, 137c and Gr21 descend from a strain isolated in 1935 by G. Smith from soil on the University of Massachusetts campus. Chlamydomonas is easily cultured and has mating types that permit genetic analysis. It can be cultured photoautotraphically with carbon dioxide as the sole carbon source, or hetrotrophically with acetate as the sole carbon source. Hence, mutants defective in either photosynthesis or respiration can be selected and maintained by the alternate energy transduction system. The majority of known mutants are now available from the Chlamydomonas National Culture Collection at Duke University (7), and the historical background and techniques for handling the alga are available in the Chlamydomonas Handbook (8).

ELECTRON TRANSPORT BETWEEN PHOTOSYSTEM II AND PHOTOSYSTEM I

In the 1960s, following the discovery that photosynthetic electron transport in plants requires two photosystems operating in series (see e.g., 9), a considerable amount of research focused on identifying the proteins involved in electron transfer from water to NADP and determining their sequence in the electron transport chain. While progress in this effort depended on numerous experimental techniques applied to a variety of organisms, the genetic tractability of Chlamydomonas enabled a unique and strikingly elegant contribution. Levine and coworkers created an array of photosynthetic mutants of Chlamydomonas that appeared to lack a single, obligatory component of the electron transport chain. Spectroscopic and biophysical characterization of these mutants provided the database for determining the identity and sequence of the electron carriers (see Figure 1 in Govindjee and Eaton-Rye, this volume). An indication of the power of this approach is that Gorman and Levine (10) were successful in predicting the existence of an electron carrier they called M (now known as the Rieske FeS protein) operating between plastoquinone and cytochrome f that was not spectroscopically detected in chloroplasts until the next decade (11,12). In some cases interpretations based on correlating the loss of electron transport capacity with the deletion of a redox component proved to be an oversimplification and led to incorrect conclusions, for example, the suggestion that cytochrome b-559 was a necessary component in electron transport between the two photosystems (13). However, the overall record is impressive, not only because of the success of the genetic approach in elucidating the electron transport chain, but also because some of the same mutants used in the early studies are currently being used to probe more deeply into the photosynthetic process, for example, electron transfer within a protein complex. The following discussion describes the development of research into the process of electron transfer in Chlamydomonas between plastoquinone and P700 via cytochrome f, plastocyanin and cytochrome c-552 as it progressed from a genetic approach coupled to biophyscial and biochemical measurements to current experiments designed to elucidate the mechanism of regulation of the genes for the intermediate electron carriers.

An early controversy in electron transport centered on the sequence of the carriers cytochrome f and plastocaynin. Kok, Rurainski, and Harmon (14), and Kok and Rurainski (15) argued that the two components operated in parallel, while Fork and Urbach (16) argued that the sequence of electron transfer was from plastocyanin to cytochrome f to P700. In 1966 Gorman and Levine (17,18) described two mutants of Chlamydomonas, one that lacked cytochrome f and one that lacked plastocyanin that supported a third model. In the mutant without plastocyanin they showed that cytochrome f could be reduced by photosystem II, but could not be oxidized by photosystem I. In addition, in the mutant that lacked cytochrome f they observed electron transport through photosystem I from reduced DPIP (2,6 dichlorophenolindophenol) to NADP, while in the mutant that lacked plastocyanin the reaction was negligible. These observations led them to suggest that the sequence of electron transfer is from cytochrome f to plastocyanin to P700, a conclusion that has been verified since by numerous experiments (19).

The next significant advancement in this area was the demonstration by Wood (20) that the component that Gorman and Levine (21) assumed to be cytochrome f in Chlamydomonas was in fact two different c-type cytochromes. One cytochrome was membrane bound and both spectrally and biochemically analogous to higher plant cytochrome f, while the other cytochrome (cytochrome c-552) was a low molecular weight, soluble protein that had no analogous component in higher plants. Wood showed that cytochrome c-552 and plastocyanin appeared to serve the same function, and that the synthesis of the two proteins was regulated by the presence of copper in the growth medium. When sufficient copper is available the cell makes plastocyanin, but when copper is limited the cell makes cytochrome c-552 (22). The demonstration by Wood that plastocyanin and cytochrome c-552 are interchangeable proteins enabled the next step in this area, namely to move beyond the physiological question of the function of these electron carriers to investigate how their synthesis is regulated. Recently Merchant and Bogorad (23) have attempted to determine the signal that regulates the accumulation of plastocyanin and cytochrome c-552, and the mechanism of signal transduction. They have concluded that copper regulates plastocyanin synthesis at the level of the stable protein and that the accumulation of cytocrhome c-552 is regulated at the level of stable mRNA.

MOLECULAR GENETICS OF HERBICIDE RESISTANCE

Several herbicides, e.g., DCMU (3 (3,4) Dichlorophenyldimethyl urea), Atrazine, Bromacyl, inhibit photosynthetic electron transport on the reducing side of photosystem II by displacing the secondary quinone acceptor Q_B (see Fig. 3 in Govindjee and Eaton-Rye, this volume). Recent work on Atrazine resistant higher plants identified an intrinsic thylakoid membrane protein (psba) as the binding site for atrazine (24). Non-mendelian inheritance of resistant traits to this class of herbicides have been found in both higher plants and Chlamydomonas (25-28), implicating chloroplast gene involvement. The Chlamydomonas gene for this protein (psba) from different mutants has been isolated, cloned, and sequenced, revealing three altered positions in its nucleotide sequence that confer altered response to herbicides (29,30). A change from valine to isoleucine at position 219 resulted in a large increase in DCMU resistance while Atrazine resistance increased only slightly. A change from phenylalanine to tyrosine at position 255 resulted in a decrease in DCMU resistance, but a large increase in Atrazine resistance. A change from serine to alanine at position 264 resulted in a ten-fold increase in DCMU resistance, and a 100-fold increase in Atrazine resistance. In higher plants a change at the same position from serine to glycine resulted in a 1000-fold increase in atrazine resistance without altering DCMU resistance (30). It has been observed in higher plants that mutants that have altered sensitivity to herbicides have lower rates of light-driven electron transport. This is not the case in Chlamydomonas where changes at positions 219 and 255 did not change electron transport capacity, while the change at position 264 resulted in a lower rate (30).

Analysis of photoinhibition, the loss of electron transport capacity under high intensity illumination, in Chlamydomonas by Kyle et al. (31) shows that high intensity illumination destroys the psba protein function and that the loss can be prevented by addition of Atrazine. Protection by Atrazine was absent in an Atrazine-resistant mutant. Overall, these results show that the phenotype of the psb protein in Chlamydomonas can be altered rather flexibly by specific mutation of the structural gene, making the Chlamydomonas system a versatile tool to elucidate electron transport, herbicide resistance, and photoinhibition.

MUTATIONAL ANALYSIS OF RUBISCO

Rubisco (ribulose-1,5-bisphosphate carboxylase) was one of the earliest and remains one of the most elusive objects of mutational analysis in Chlamydomonas research. Attempts to isolate mutants affecting this enzyme have led to unforeseen yet productive results in several areas.

Chloroplast ribosome function. In 1969 Levine and Togasaki (32) reported a mendelian mutant (ac20, later renamed ac20cr1) that appeared to lack rubisco activity. Subsequently, Togasaki and Levine (33) showed that when ac20cr1 is cultured either photoheterotrophically, or on acetate supplmented medium in the light, rubisco synthesis was severely limited, but that upon transfer of the cells to minimal medium in the light, the synthesis of rubisco resumed after a long lag period. Concurrently, Goodenough and Levine (34), studying ultrastructure, showed that the level of chloroplast ribosomes in ac20cr1 is greatly reduced under photoheterotrophic condition, and that the level increases when the cells are transferred to minimal medium in the light in parallel to the increase in the rate of rubisco synthesis. Thus, the concurrent genetic, physiological, biochemical and ultrastructural studies led to the demonstration of the chloroplast ribosome dependence of rubisco synthesis. Subsequent investigations of chloroplast ribosome mutants have enabled the delineation of the molecular genetics of the Chlamydomonas chloroplast genome (8). Recently, the ac20cr1 mutant has been used in the analysis of the interaction between cytoplasmic and chloroplast protein synthesis.

Dobberstein et al. in 1977 (35) demonstrated the in vitro synthesis and processing of the precursor for the small subunit of rubisco in Chlamydomonas reinhardtii, concluding that the cytoplasmically synthesized precursor is imported into the chloroplast and processed prior to integration into the rubisco holoenzyme. Givan (36) showed that in chloroplast ribosome limited cells there is no detectable free pool of cytoplasmically synthesized small subunit. Surzycki et al. (37) showed that other cytoplasmically synthesized chloroplast proteins are present in the chloroplast ribosome limited cells.

The coordination of the synthesis of the large and small subunits was studied by Mishkin and Schmidt (38). In cells with limited chloroplast ribosome, they showed that the level of translatable mRNA for the small subunit is similiar to that of the wild-type control, and concluded that the level of the large subunit does not affect transcription of the small subunit gene. Subsequently, they showed, through pulse-chase experiments (39) in the same chloroplast ribosome limited mutant, that the small subunit is actively synthesized and imported into the chloroplast and that the excess small subunit not incorporated into the holoenzyme is rapidly degraded. Wild-type cells treated with chloramphenicol, an inhibitor of chloroplast ribosome function, showed the same results. These observations led them to conclude that the precise stoichiometry between the two subunits is due to the selective degradation of excess protein, and not due to the regulation of protein synthesis or precursor processing.

Rubisco Large Subunit Mutants. In 1980 Spreitzer and Metz (40) described the first non-mendelian structural mutant of rubisco. The holoenzyme exhibited no enzymatic activity and an altered isoelectric point. Dron et al. (41) demonstrated, by sequencing the chloroplast gene for the large subunit, that the above was indeed a structural mutant. This was the first example of the physical mapping of non-mendelian mutation in the chloroplast DNA of Chlamydomonas. Progress in this field was advanced by the introduction of a rapid screening method, by Spreitzer and Ogren (42), for chloroplast rubisco mutants based on the unique ability of chloroplast genes to recombine in Chlamydomonas. Spreitzer and Ogren found two new chloroplast rubisco mutants that lacked the holoenzyme as well as the individual large and small subunits. This elegant application of Chlamydomonas genetics to a specific enzyme elucidated the structure and function of rubisco at the molecular level. Subsequent reports by Spreitzer et al. (43) demonstrated that these two mutants contain nonsense mutations close to the 3' and 5' ends of their large-subunit gene, respectively, causing premature termination of translation and production of unstable, truncated large subunit. In agreement with Schmidt and Mishkin (39), Spreitzer et al. did not observe the accumulation of excess small subunit in these mutants. One mutant (18-5b) synthesizes a protein that is 25 amino acids shorter than the wild-type protein, but does not accumulate holoenzyme , suggesting the missing segment is necessary for the holoenzyme stability.

Interaction between photosynthetic electron transport and carbon metabolism. Traditionally, thylakoid membrane bound photosynthetic electron transport and photophosphorylation systems have been regarded primarily as the source of NADPH and ATP for stromal metabolic reactions. Recently, evidence is emerging linking photosynthetic electron transport not only to chloroplast carbon metabolism, but also to mitochondrial and cytoplasmic reactions. Chlamydomonas research has played a major role in this development. In 1970, Moll and Levine (44) described a mendelian mutant (F60) that lacked ribulose-5-P kinase activity, but appeared normal in all other photosynthetic parameters studied, including light dependent generation of ATP and NADPH. This mutant provided the first genetic evidence for photosynthetic CO_2 assimilation through RuBP carboxylation in Chlamydomonas, and prior to 1980, was the only stable Calvin cycle mutant available (40,45). During this period the mutant F60 was used in the analysis of algal adaptation to anaerobic conditions. In 1970 Healey (46) reported the stimulation of anaerobic photohydrogen evolution in Chlamydomonas moewusi which is insensitive to DCMU but is inhibited by addition of fluroacetate. He suggested the anaerobically functioning citric acid cycle would generate NADH which in turn would feed electrons into photosystem I beyond the DCMU block. Under anaerobic conditions, photosystem I serves as the electron acceptor for reducing equivalents from respiratory and glycolytic reactions. Under these conditions photosystem I reduces hydrogen ions, producing molecular hydrogen. In 1982 Bennoun (47) reported an electron transport pathway to O_2 in the thylakoid membranes of Chlamydomonas reinhardtii and Chlorella pyrenoidosa that he called chlororespiration. This O_2 uptake pathway shares plastoquinone with the photosynthetic electron transport system, and serves to dissipate the reducing power generated by starch breakdown.

Bamberger et al. (48) used F60 to study the CO_2 and H_2 evolution under anaerobic conditions. The absence of CO_2 fixation in this mutant made possible the analysis of gas exchange in the light and dark, and their relation to starch degradation. Gfeller and Gibbs (49) extended this work to the analysis of fermentative products in dark and light. In the dark glycolytically generated reducing power was dissipated with ethanol as the major electron sink, and H_2 as a minor sink. In the light ethanol is no longer produced, and electrons are primarily consumed by hydrogen evolution. Gfeller and Gibbs postulated that NADPH can donate electrons at a site in the photosynthetic electron transport chain beyond the site of DCMU inhibition. Gfeller and Gibbs (50) showed that in both F60 and wild-type cells, light inhibition of ethanol production and starch breakdown was reversed by the plastoquinone analogue dibromothymoquinone, but not by DCMU. They concluded that the reducing power generated by the breakdown of starch is dissipated in the light through hydrogen evolution mediated by photosystem I and plastoquinone. In the dark, reducing power is dissipated through multiple pathways, including the plastoquinone mediated respiratory pathway, and mitochondrial respiration. The discovery of a membrane bound NADH-plastoquinone oxidoreductase (51), and the demonstration of NADH dependent H_2 photoevolution by a cell free Chlamydomonas preparation (52,53), has allowed the elucidation of the biochemcial mechanism underlying the interrelationship of electron transport and other metabolic reactions.

CONCLUDING REMARKS

Recent advances in transformation systems for Chlamydomonas should greatly facilitate the future molecular genetic analysis (54,55). The availability of Calvin Cycle mutant (40,42,44,45) and photosynthetically competent intact chloroplasts (56-59) open new research opportunities. In higher plants, systems such as Arabidopsis are emerging as a powerful tool for molecular genetic analysis of problems in plant biology, including photosynthesis (60); however, higher plants lack the microbiological advantage of Chlamydomonas. Cyanobacteria share much in common with green plant system including oxygen evolution and recent advances in the molecular genetic capability in this field (61) make them very useful for photosynthetic research; however, since cyanobacteria are prokaryotic, the eukaryotic Chlamydomonas provides a more applicable model for higher plants.

REFERENCES

1. Warburg O (1919) Biochem Z 100: 230-262
2. Warburg O (1920) Biochem Z 103: 188-213
3. Emerson R and Arnold W (1932) J General Physiol 16: 191-205
4. Levine RP, Gorman DS, Avron M and Butler WL (1966) Brookhaven Symp Biol 19: 143-148
5. Epel BL and Butler WL (1972) Biophys J 12: 922-929
6. Epel BL, Butler WL and Levine RP (1972) Biochim Biophys Acta 275: 395-400
7. Harris EH (1984) Plant Molec Biol Reporter 2: 29-41
8. Harris EH (1986) A Chlamydomonas Handbook. Academic Press. In press
9. Hill R and Bendall F (1960) Nature 186: 136-137
10. Levine RP and Gorman DS (1966) Plant Physiol 41: 1293-1300
11. Malkin R and Aparicio PJ (1975) Biochem Biophys Res Comm 63: 1157-1160
12. Whitmarsh J and Cramer WA (1979) Proc Natl Acad Sci USA 76: 4417-4420

13. Levine RP (1969) Ann Rev Plant Physiol 20: 523-540
14. Kok B, Rurainski HJ and Harmon EA (1964) Plant Physiol 39: 513-520
15. Kok B and Rurainski HJ (1965) Biochim Biophys Acta 94: 588-590
16. Fork DC and Urbach W (1965) Proc Natl Acad Sci USA 53: 1307-1315
17. Gorman DS and Levine RP (1966) Plant Physiol 41: 1648-1656
18. Gorman DS and Levine RP (1966) Proc Natl Acad Sci USA 54: 1665-1669
19. Whitmarsh J (1986) In: Encyc. Plant Physiol. Photosynthetic Membranes, pp. 508-527. Eds. A Staehelin and C. Arntzen. Springer-Verlag, Heidelberg
20. Wood PM (1977) Eur J Biochem 72: 605-612
21. Gorman DS and Levine RP (1966) Plant Physiol 41: 1643-1647
22. Wood PM (1978) Eur J Biochem 87: 9-19
23. Merchant S and Bogorad L (1986) Molec Cell Biol 6: 462-469
24. Pfister K, Steinback KE, Gardner G and Arntzen CJ (1981) Proc Natl Acad Sci USA 78: 981-985
25. Galloway RE and Mets L (1982) Plant Physiol 70: 1673-1677
26. Tollenbach M, Gerber A and Boschetti A (1983) FEBS Lett 158: 147-150
27. Arntzen CJ, Pfister K and Steinback KE (1982) In: Herbicide Resistance in Plants (Lebaron, HM and Gressel, J, eds.), pp. 185-214, Wiley, New York
28. Galloway RE and Mets LJ (1984) Plant Physiol 74: 469-474
29. Erickson JM, Rahire M, Bennoun P, Delepelaire P, Diner B and Rochaix JD (1984) Proc Natl Acad Sci USA 81: 3617-3621
30. Erickson JM, Rahire M, Rochaix JP and Mets L (1985) Science 23: 204-207
31. Kyle KJ, Ohad I and Arntzen CJ (1984) Proc Natl Acad Sci USA 81: 4070-4074
32. Levine RP and Togasaki RK (1965) Proc Natl Acad Sci USA 53: 987-990
33. Togasaki RK and Levine RP (1970) J Cell Biol 44: 531-539
34. Goodenough UW and Levine RP (1970) J Cell Biol 44: 547-562
35. Dobberstein B, Blobel, G and Chua N-H (1977) Proc Natl Acad Sci USA 74: 1082-1085
36. Givan AL (1979) Planta 144: 271-276
37. Surzycki SJ, Goodenough UW, Levine RP and Armstrong JJ (1970) In: Control of Organelle Development, Symp Soc Exp Biol, XXIV, pp. 13-37, University Press, Cambridge
38. Mishkind ML and Schmidt GW (1983) Plant Physiol 72: 847-854
39. Schmidt GW and Mishkind ML (1983) Proc Natl Acad Sci USA 80: 2632-2636
40. Spreitzer RJ and Mets L (1980) Nature 285: 114-115
41. Dron M, Rahire M, Rochaix, J-D and Mets L (1983) Plasmid 9: 321-324
42. Spreitzer RJ and Ogren WL (1983) Proc Natl Acad Sci USA 80: 6293-6397
43. Spreitzer RJ, Goldschmidt-Clermont M, Rahire M and Rochaix J-D (1985) Proc Natl Acad Sci USA 82: 5460-5464
44. Moll B and Levine RP (1970) Plant Physiol 46: 576-580
45. Salvucci ME and Ogren WL (1985) Planta 165: 340-347
46. Healey FP (1970) Plant Physiol 45: 153-159
47. Bennoun P (1982) Proc Natl Acad Sci 79: 4352-4356
48. Bamberger ES, King D, Erbes DL and Gibbs M (1982) Plant Physiol 69: 1268-1273
49. Gfeller RP and Gibbs M (1984) Plant Physiol 75: 212-218
50. Gfeller RP and Gibbs M (1985) Plant Physiol 77: 509-511
51. Godde D (1982) Plant Cell Physiol 25: 531-539
52. Ben-Amotz A and Gibbs M (1975) Biochem Biophys Res Commun 64: 355-359
53. Godde D and Trebst A (1980) Arch Microbiol 127: 245-252
54. Rochaix J-D and van Dillewijn J (1982) Nature 296: 70-72
55. Hasnain SE, Manavathu EK and Leung W-C (1985) Molec Cell Biol 5: 3647-3650
56. Klein U, Chen C, Gibbs M and Platt-Aloia KA (1983) Plant Physiol 72: 481-487
57. Klein U, Chen C and Gibbs M (1983) Plant Physiol 72: 488-491
58. Belknap WR (1983) Plant Physiol 72: 1130-1132
59. Mendiola-Morgenthaler L, Leu S and Boschetti A (1985) Plant Sci 38: 33-39

60. Somerville CR (1984) Oxford Surveys Plant Molec & Cell Biol 1: 103-131
61. Wolk CP, Vonshak A, Kehoe P and Elhai J (1984) Proc. Natl. Acad. Sci. 81:1561-1565

Photosynthesis Research 10: 423–429 (1986)
© *Martinus Nijhoff Publishers, Dordrecht*

ENDOR CHARACTERIZATION AND D$_2$O EXCHANGE IN THE Z$^+$/D$^+$ RADICAL IN PHOTOSYSTEM II

T.K. CHANDRASHEKAR, P.J. O'MALLEY, I. RODRIGUEZ and G.T. BABCOCK

ABSTRACT
The early suggestion by Lozier and Butler (Photochem. Photobiol. 17, 133-137 (1973)) that EPR Signal II arises from radicals associated with the water-splitting process in PSII has been confirmed and extended over the intervening years. Recent work has identified the Signal II radicals, D$^+$ and Z$^+$, with plastosemiquinone cation species. In the experiments presented here we have used ENDOR spectroscopy and D$_2$O/H$_2$O exchange to characterize these paramagnets in more detail. The ENDOR matrix region, which arises from protons which interact weakly with the unpaired electron spin, is well-resolved at 4 K and at least seven resonances are apparent. A number of hyperfine couplings in the 3-8 MHz range are observed and are suggested to arise from methyl or hydroxyl protons which occur as substituents on the plastosemiquinone cation ring or from amino acid protons hydrogen-bonded to the 1,4-hydroxyl groups. Orientation selection experiments are consistent with these possibilities. D$_2$O/H$_2$O exchange shows that the D$^+$/Z$^+$ site is accessible to solvent. However, the exchange occurs slowly and is not complete even after 72 hours which suggests that the free radicals are functionally isolated from solvent water.

1. INTRODUCTION

In a paper published in 1973, Lozier and Butler made the important observation that the so-called Signal II, a stable free radical which had been associated with PSII, was rapidly abolished in the presence of lipophilic anions (1). Since these anions are also able to destabilize the higher S-states in the oxygen-evolving complex (OEC) (2), Lozier and Butler implicated the free radical species in reactions which occur on the oxidizing side of P680. In this paper, then, the authors provided both important tools with which to probe the redox chemistry of the Signal II species and a testable, and ultimately correct, hypothesis for its location in PSII.

Several groups followed the lead suggested by Lozier and Butler and a number of papers appeared subsequently which clarified the role of Signal II-like free radicals in PSII redox reactions (3-11). Two of these paramagnets occur per P680, one of which is the classical stable Signal II species, the other of which is only observed under time-resolved conditions. The former radical is usually referred to as D$^+$ and does not appear to be involved in main chain electron transport; its EPR spectrum is now designated as Signal IIs. The less stable radical has been identified as the Z species which mediates electron transfer between P680$^+$ and the OEC. Under O$_2$-evolving conditions, electron transfer proceeds through Z with submillisecond kinetics and EPR Signal II$_{vf}$ arises from the transient free radical, Z$^+$. The functional connection between Z and the OEC is labile and, when disrupted, Signal IIf is used to describe the EPR spectrum of the now less rapidly reduced Z^+ intermediate.

Although Z^+ and D^+ are functionally distinct, their EPR properties, as well as their orientation in the membrane (12), are essentially identical; only in terms of microwave power saturation have differences been observed (10,13). Weaver originally postulated a quinone radical origin for Signal II (14) and Kohl and Wood provided strong support for this suggestion in their plastoquinone extraction/reconstitution experiments (15). Hales and Das Gupta used linewidth arguments to propose a plasto-quinone anion radical interacting with a diamagnetic metal cation as the identity of the radical (16). Semiquinone anions, however, are unlikely to have the high redox potential required by the Z^+ species. Quinone cation radicals, on the other hand, are strongly oxidizing and we used this plus model compound, ENDOR and oriented thylakoid membrane data to propose that the D^+/Z^+ species is a plastosemiquinone cation (17-19). Results from time-resolved optical spectroscopy are consistent with this proposal (20,21).

In the studies reported here, we used ENDOR spectroscopy to explore the structure and environment of the D^+/Z^+ radical. ENDOR on immobilized radicals is especially appropriate for this as both isotropic and aniso-tropic magnetic parameters for hyperfine coupled protons may be extracted from the spectra (22-24). Moreover, the technique detects hydrogen-bonded protons directly and is able to provide insight into the solvent accessibility of a radical site. The latter two aspects of ENDOR are particularly useful in light of the hydrogen-bonded structure postulated for D^+/Z^+ and of the enigmatic stability of the D^+ free radical. Thus far, most of our work has been done on model quinones as the general phenomenology and applicability of ENDOR to *in vivo* systems is not well-developed. Nonetheless, we have investigated the Signal II species in some detail and the status of this work is reported below.

2. MATERIALS AND METHODS

Oxygen-evolving PSII particles were isolated from spinach, and tris-washed when required, by using procedures based on those in (12). D_2O/H_2O exchange was carried out by resuspending tris-washed PSII parti-cles in a buffered (pD = 6.0) D_2O solution and incubating in the dark at 4°C for three days. The buffer solution (MES, 50 mM; NaCl, 10 mM) was changed twice during the course of the exchange. Brief periods of room light (2-3 min) were given at 6 hour intervals during the incubation. We have found that freeze-thaw cycles accelerate the exchange process and this was done three times for the samples reported here. The Signal II lineshape in an H_2O control was unchanged following this protocol; pre-liminary data indicate that additional free radicals are generated in substantial amounts only after longer periods of dark, cold incubation. EPR and ENDOR spectra were recorded on samples which contained 3-6 mg Chl/ml as described previously (19,22-24).

3. RESULTS AND DISCUSSION

Figure 1 shows the EPR and ENDOR spectra of the D^+/Z^+ free radical in PSII particles recorded at liquid helium temperature. In the high frequency ENDOR region the characteristic axial resonances of a methyl group are apparent as couplings a and b and have been assigned to the methyl at the plastoquinone ring 2 position as discussed in detail else-where (19). In recent work, Brok et al. proposed alternative assignments for these resonances by suggesting that they arise from the two methylene protons of the isoprenoid chain and one of the hydroxyl protons in PQH_2^+

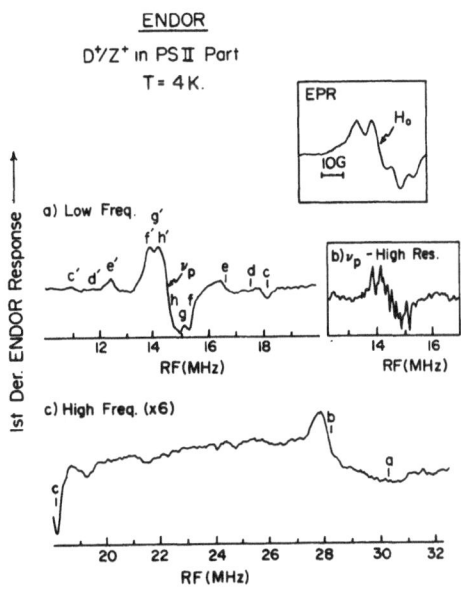

ENDOR
D^+/Z^+ in PS II Part
T = 4 K.

a) Low Freq.

b) v_p - High Res.

c) High Freq. (x6)

FIGURE 1. EPR and ENDOR spectra of Signal II in O_2-evolving PSII particles at 4.0 K. The low frequency spectrum, (a), was recorded under the following conditions: 2.5 mW microwave power, 100 W rf power at 10 MHz, 150 kHz fm depth, 1 s time constant, 1000 s scan; the higher resolution matrix region, inset (b), was recorded under similar conditions except that the fm depth was decreased to 50 kHz; for higher frequency spectrum, (c), the fm depth was 150 kHz and 11 scans were averaged.

(25). We are examining this possibility in oriented PSII membranes and will report our results elsewhere.

In the low frequency ENDOR spectrum, complex structure, particularly in the matrix region around the free proton frequency ($v_H \cong 14.5$ MHz at X-band in our spectrometer) is observed and at least seven resonances are apparent in the high resolution spectrum shown in the inset. Lubitz, Feher and coworkers have noted similar highly resolved matrix region spectra in their ENDOR studies of the Q_A^-/Q_B^- acceptor complex in photosynthetic bacteria (26). The *in vivo* data contrast markedly with our experience with the matrix region in model compounds (e.g. (24)) and can be attributed to the fact that in the protein binding site the radical is in a well-defined, highly structured environment. Such a situation will produce specific dipole–dipole interactions between the radical and amino acid protons in the binding site and lead to the resolved matrix spectrum. In the models, on the other hand, the solvent environment is likely to be disordered and the dipole–dipole interactions will average to produce the poorly resolved spectra we observe. Because the matrix in the D^+/Z^+ radical is produced primarily by relatively weak interactions between the radical and nearby (within ~6 Å) amino acid protons, some of which are likely to be exchangeable, the solvent accessibility of the site may be monitored (see below).

In addition to the matrix, there are at least three other resonances in the low frequency region. Couplings c–c' (7.3 MHz) and d–d' (5.9 MHz) are positive on the low frequency side of v_H and negative on the high frequency side (see also Figure 2), whereas both e and e' (3.9 MHz) are derivative shaped. These couplings are small relative to the a,b resonances (31.3 and 27.1 MHz, respectively) and support our conclusion that the latter couplings determine the partially resolved lineshape of the Z^+/D^+ EPR spectrum (19).

In powder ENDOR spectra, turning points (i.e., resonances) are expected at frequencies which correspond to principal components at the hyperfine tensor. The lineshapes of the c and d resonances are similar to those we

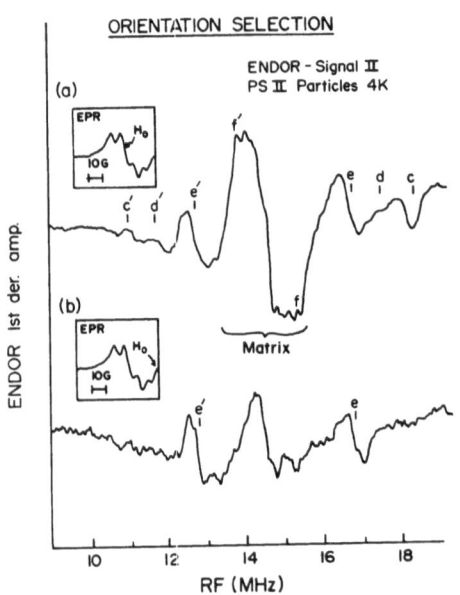

FIGURE 2. *Orientation selection in the ENDOR spectrum of Signal II; the field setting relative to the EPR spectrum is shown in the insets for (a) and (b). ENDOR conditions as in Figure 1; 10 scans were averaged in recording spectrum b.*

observe for the $A_\parallel$ component of methyl protons and also of hydrogen-bonded protons (23,24). This suggests that c and d arise either from the PQH_2^+ methyl at the 3 position, from protons hydrogen-bonded to the 1,4 -OH groups or, possibly, from the -OH protons themselves. In this inter-pretation, the perpendicular components of these tensors would occur as the e-e' resonances which we have observed to have complex substructure in higher resolution spectra. The α-proton at the 6 position in PQH_2^+ is a less likely source for either the c or the d resonances as we have found that this class of protons has only small rhombicities in aromatic radicals, particularly, as is the case here, for situations in which adjacent ring carbons carry substantial unpaired electron spin density ((24), O'Malley and Babcock, unpublished).

For both methyl and hydrogen-bonded protons in quinone radicals the principal axis for the $A_\parallel$ tensor component is expected to lie in the plane of the ring (19,22-24) and a test of this assignment is the orien-tation selection experiment suggested by the Signal II g-anisotropy recently measured by Brok et al. (25). The g_z component (2.0023) is sufficiently different from g_y and g_x (2.0044 and 2.0076, respectively) that recording an ENDOR spectrum with the magnetic field set to the high field side of the EPR spectrum should effectively select only z-oriented molecules. As the direction of g_z is out-of-plane, we expect to lose any in-plane hyperfine tensor components in the ENDOR spectrum. The results of this experiment are shown in Figure 2. Spectrum 2a was recorded at the EPR zero crossing; 2b is the spectrum recorded at the high field side. Resonance c is clearly absent in the latter case (the situation with respect to d is less clear owing to its lower intensity) in agree-ment with the prediction above. Interestingly, there are also major changes with orientation selection in the matrix and the peaks in the region labeled f-f' are absent when z-oriented molecules are selected. This approach should be extremely useful in the deconvolution of the rich but complicated spectra of *in vivo* quinone radicals.

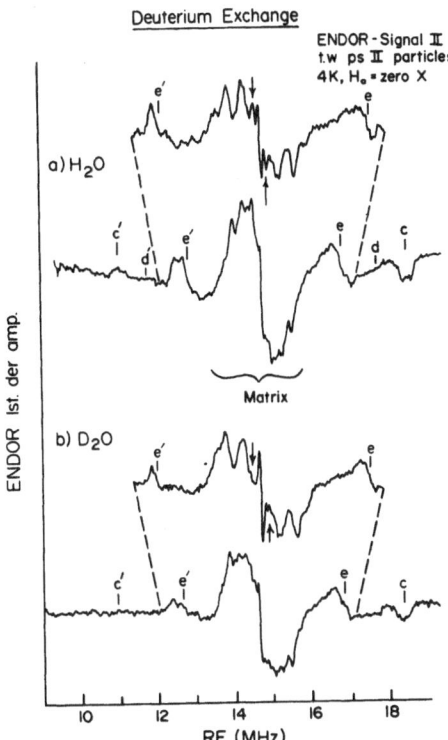

Deuterium Exchange

ENDOR - Signal II
t.w ps II particles
4K, H$_e$ = zero X

a) H$_2$O

Matrix

b) D$_2$O

ENDOR 1st. der. amp.

RF (MHz)

FIGURE 3. The effect of D_2O/H_2O exchange on the ENDOR spectrum of Signal II. D_2O exchange was carried out as described in Methods; ENDOR conditions as in Figure 1. For the expanded regions in (a) and (b) the fm depth was decreased to 50 kHz.

PQH_2^+ hydroxyl protons and hydrogen-bonded protons in the binding site should exchange with solvent water, if the site is accessible, whereas –CH_3 protons on the ring are not expected to exchange under mild conditions. Thus a means by which to distinguish the possible origins of the c-e resonances discussed above (as well as resonances in the matrix) is by D_2O/H_2O exchange (23,24,26). This experiment on the D^+/Z^+ radical is also of interest in terms of the surprising stability of the Signal IIs species: PQH_2^+ is expected to deprotonate rapidly and only if it is well-shielded from solvent is it expected to exhibit the stability of the Signal IIs species. Hales and Das Gupta have already noted that D_2O/H_2O exchange does not alter the lineshape of Signal II (16). We confirmed this in exchange experiments which lasted less than 1 day, i.e., there is no significant change in the EPR nor in the ENDOR when exchange is carried out for 24 hours or less. After three days of exchange following the protocol described in Methods we recorded the spectra shown in Figure 3. Spectrum 3a is the H$_2$O control, 3b is the D$_2$O exchanged sample. The resonance labeled with the arrow in 3a is clearly absent in 3b, moreover, the c' peak is also missing in 3b although its high frequency counterpart, c̄, shows little intensity decrease relative to the H$_2$O control. If exchange is carried out for a two week period resonances c-e continue to decrease in intensity although under these conditions other radicals are also apparent in the EPR spectrum and we don't consider the longer time exchange experiments reliable yet. The results in Figure 3 allow us to make several conclusions. First, the D^+/Z^+ binding sites are well-shielded from solvent and show little evidence of exchange over a

24 hour period. This observation is consistent with the stability of D^+. It also suggests that the proton which is released as the Z species is oxidized (27,28) is most likely a Bohr proton, as also appears to be the situation with quinone acceptors and the proton uptake which accompanies their reduction (29). Second, the site is slowly accessible over long periods, but after three days only the most weakly coupled (and hence probably fairly distant >3-4 Å) protons are unambiguously exchanged. Finally, there does appear to be evidence for hydrogen-bonded protons in the D^+/Z^+ site although this conclusion must be considered tentative.

In conclusion, the ENDOR experiments on the D^+/Z^+ reported above indicate that this technique provides high resolution data on both the radical and on protons from amino acids in its immediate vicinity. Spectral assignments are facilitated by the orientation selection and D_2O exchange techniques described here and the use of oriented thylakoids and PSII particles (19,30) should provide an additional technique by which to assign resonances to specific protons.

REFERENCES

1. Lozier RH and Butler WL (1973) Photochem Photobiol 17:133-137
2. Renger G, Bouges-Bocquet B and Delosme R (1973) Biochim Biophys Acta 292:796-807
3. Babcock GT and Sauer K (1973) Biochim Biophys Acta 325:483-503
4. Esser AF (1974) Photochem Photobiol 20:167-172
5. Warden JT and Bolton JR (1974) Photochem Photobiol 20:245-250
6. Velthuys BR and Visser JWM (1975) FEBS Lett 55:109-112
7. Babcock GT and Sauer K (1975) Biochim Biophys Acta 376:329-344
8. Blankenship RE, Babcock GT, Warden JT and Sauer K (1975) FEBS Lett 51:287-293
9. Babcock GT, Blankenship RE and Sauer K (1976) FEBS Lett 61:286-289
10. Warden JT, Blankenship RE and Sauer K (1976) Biochim Biophys Acta 423:462-478
11. Boussac A and Etienne AL (1982) Biochem Biophys Res Commun 109:1200-1205
12. Berthold DA, Babcock GT and Yocum CF (1981) FEBS Lett 134:231-234
13. Yocum CF, Yerkes CT, Blankenship RE, Sharp RR and Babcock GT (1981) Proc Nat Acad Sci USA 78:7507-7511
14. Weaver EC (1962) Arch Biochem Biophys 99:193-196
15. Kohl DH and Wood PM (1969) Plant Physiol 44:1439-1445
16. Hales BJ and Das Gupta A (1981) Biochim Biophys Acta 637:303-311
17. Ghanotakis DF, O'Malley PJ, Babcock GT and Yocum CF (1983) In: The Oxygen-Evolving System of Plant Photosynthesis (Inoue, Y et al., eds.) pp. 91-101, Academic Press, Tokyo
18. O'Malley PJ and Babcock GT (1984) Biochem Biophys Acta 765:370-379
19. O'Malley PJ, Babcock GT and Prince RC (1984) Biochem Biophys Acta 766:283-288
20. Dekker JP, van Gorkom HJ, Brok M and Ouwehand L (1984) Biochem Biophys Acta 764:301-309
21. Diner BA and DeVitry C (1984) In: Advances in Photosynthesis Research (Sybesma, C, ed.) Vol. I, pp. 407-412, Martinus Nijhoff/Dr. W. Junk Publishers, The Hague
22. O'Malley PJ and Babcock GT (1984) J Chem Phys 80:3912-3913
23. O'Malley PJ, Chandrashekar TK and Babcock GT (1985) In: Antenna and Reaction Centers of Photosynthetic Bacteria (Michele-Beyerle, ME, ed.) Vol. 42, pp. 339-344, Springer-Verlag, Berlin

24. O'Malley PJ and Babcock GT (1985) J Am Chem Soc, submitted
25. Brok M, Ebskamp FCR and Hoff AJ (1985) Biochem Biophys Acta, in
     press
26. Lubitz W, Abresch EC, Debus RJ, Isaacsson RA, Okomura MY and Feher G
     (1985) Biochem Biophys Acta, in press.
27. Renger G and Voelker M (1982) FEBS Lett 149:203-207
28. Förster V and Junge W (1985) Photochem Photobiol 41:183-190
29. Wraight CA (1982) In: Functions in Quinones in Energy-Conserving
     Systems (Trumpower, B, ed.) pp. 181-197, Academic Press, New York
30. Rutherford AW (1983) Biochem Biophys Acta 807:189-201

ACKNOWLEDGEMENTS

We thank Prof. C.F. Yocum and Dr. D.F. Ghanotakis for helpful discussions.
This research was supported by the McKnight Foundation (TKC is a
McKnight postdoctoral fellow) and by the Photosynthesis Program of the
Competitive Research Grants Office of the U.S. Department of Agriculture.

Authors' address: Department of Chemistry, Michigan State University,
                   E. Lansing, MI 48824-1322 USA.

Photosynthesis Research 10: 431–436 (1986)
© Martinus Nijhoff Publishers, Dordrecht

pH DEPENDENT STABILIZATION OF $S_2Q_A^-$ AND $S_2Q_B^-$ CHARGE PAIRS STUDIED BY THERMO-LUMINESCENCE

IMRE VASS[*] and YORINAO INOUE

Solar Energy Research Group, The Institute of Physical and Chemical Research (RIKEN), Hirosawa, Wako, Saitama 351-01, Japan

Key words: Photosystem II, pH dependence, S_2 state, quinone acceptors, thermoluminescence.

1. ABSTRACT

The pH dependence of emission peak temperature and decay time of thermoluminescence arising from $S_2Q_B^-$ and $S_2Q_A^-$ recombinations demonstrates that a stabilization of $S_2Q_B^-$ occurs at low pH whereas stabilization of $S_2Q_A^-$ occurs at high pH. Based on comparative analysis of thermoluminescence parameters of the two types of recombination, we suggest that in the pH range between 5.3 and 7.5, $E_m(S_2/S_1)$ and $E_m(Q_A/Q_A^-)$ are constant, but $E_m(Q_B/Q_B^-)$ gradually increases with decreasing pH, while in the pH range between 7.5 and 8.5, an unusual change occurs on $S_2Q_A^-$ charge pair, which is interpreted as either a decrease in $E_m(S_2/S_1)$ or an increase in $E_m(Q_A/Q_A^-)$.

2. INTRODUCTION

Protonation processes are known to be involved in the functioning of the primary (Q_A) and secondary (Q_B) quinone acceptors of PSII. Based on the equilibrium redox titrations performed at different values of pH, Knaff [1] reported that Q_A/Q_A^- shows a pH dependent midpoint redox potential (E_m) below the pK of Q_A^- (8.9) but a pH independent constant E_m above it. Recent evidence suggests that on the time scale of electron transport within PSII in vivo, only Q_B^- (or a neighbouring group) is protonated but Q_A^- is not; i.e., Q_A/Q_A^- exhibits a pH independent "working redox potential" throughout the physiological pH ranges at the constant level seen above the pK [2]. The different protonation properties of Q_A and Q_B are probably due to the different surroundings of the two species. Q_A is considered to be deeply buried in a proteinaceous environment which acts as shielding barrier [3], while Q_B can easily be detached from its protein binding site in its oxidized or fully reduced form. As for the E_m of the S-state transitions of the water-oxidizing enzyme, a pH dependence is expected when the transition involves proton release, whereas no pH dependence will be expected for the transition which does not involve proton release [4].

The study of charge recombination between the electrons stabilized on Q_A or Q_B and the positive charges stored on water oxidizing enzyme (S-states) provides information about the pH induced changes in the redox potential of these counterparts [2,5,6,8]. Thermoluminescence is an effective method in probing the above type of back reactions [7-9], reflecting the redox span between the recombining charge pairs [10,11].

In the present study the pH dependence of thermoluminescence

[*]On leave from Department of Theoretical Physics, József Attila University, Szeged, Hungary.

components arising from $S_2Q_A^-$ and $S_2Q_B^-$ recombinations were investigated. The results show that protonation of Q_B^- increases $E_m(Q_B/Q_B^-)$ below pH 7.5, whereas above pH 7.5, either a decrease in $E_m(S_2/S_1)$ or an increase in $E_m(Q_A/Q_A^-)$ occurs. Alternative interpretations for the unusual high pH effects are presented.

3. MATERIALS AND METHODS

Thylakoids were isolated daily from market spinach by standard isolation techniques [12], resuspended at high concentration (3-5 mg Chl/ml) in 50 mM N-2-hydroxyethylpiperazine-N'-2'-ethanesulfonic acid (HEPES) (pH 7.5), 5 mM MgCl$_2$, 10 mM NaCl, 0.4 M sorbitol and stored on ice in darkness until use. For thermoluminescence measurements, thylakoids were diluted to 300 μg Chl/ml with one of the following buffers with the same additions as indicated above: 50 mM of L-glutamic acid (pH 5.0-6.0), 4-morpholine-ethanesulfonic acid (MES) (pH 6.0-7.0), HEPES (pH 7.0-7.8), glycylglycine (pH 7.8-9.0) or glycine (pH 9.0-10.0). The actual pH of the suspensions were checked after dilution of the thylakoid stock suspension with respective buffers. Above pH 7.8, 1 μM gramicidin D was included in the medium to assure quick pH equilibration inside and outside the thylakoids. For proteolytic treatment, thylakoids were digested with bovine pancreas trypsin (Boehringer-Mannheim) at room temperature in darkness for 2-10 min at a Chl concentration of 300 μg/ml with an about 1/1 (w/w) trypsin to Chl ratio. The digestion period was changed depending on the pH of the medium.

Thermoluminescence measurements were done as described in [7] with a heating rate of 0.7°C/s. Samples were illuminated with a single Xenon flash (4 μs, 2J) at 2°C for the B band and at -5°C for the Q band. The measurements on dithionite-treated samples were done in a cuvette-like sample holder as described in [7].

For the measurement of thermoluminescence decay course, the thylakoids were illuminated with a single flash at 25°C, incubated in darkness for varying time, and then quickly cooled down. The light-sum of thermoluminescence was determined by calculating the area under the glow peak as a function of the dark incubation period before cooling.

4. RESULTS

The main thermoluminescence bands observed after a single flash illumination of thylakoids are the B band and Q band which arise from recombinations of $S_2Q_B^-$ and $S_2Q_A^-$ charge pairs in the absence and presence of DCMU, respectively [7,9]. Figure 1 depicts these glow curves measured at three typical pHs. The peak temperature of the B band was found at 32°C at pH 7.5, but was shifted to a higher temperature (42°C) at pH 5.3, and also slightly upward to 35°C at pH 8.4 (curves a,b and c). These results are consistent with the earlier observations by Rutheford et al. [8] and Demeter et al. [11] and also with our results in [13]. The upward shift of the peak temperature of the B band at pH 5.3 is due to protonation of Q_B quinone [8,11].

The effect of pH on the Q band is very different from that on the B band. The peak temperature of the Q band in the presence of DCMU was found at 11°C at pH 7.5, and did not show any change at pH 5.3, but showed a remarkable upward shift to 28°C at pH 8.4 (curves d,e and f). This unusual upward shift of the peak temperature was not specific to the Q band induced by DCMU; i.e., after treatment with ioxynil, a phenolic type herbicide which interrupts the electron transport from Q_A to Q_B, the Q band shows a peak temperature far lower (by 20°C) than in the presence of

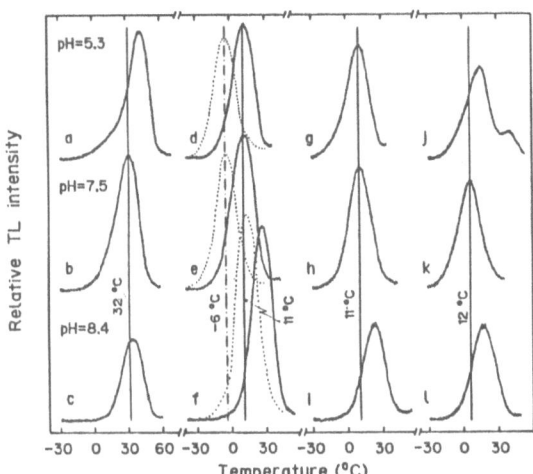

FIGURE 1. Effect of pH on single flash-induced glow curves of spinach thylakoids. B band ($S_2Q_B^-$ recombination) and Q band ($S_2Q_A^-$ recombination) were measured at three typical pHs of acidic (5.3), neutral (7.5) and alkaline (8.4). Q band was induced either by the additions of 5 μM DCMU (solid curves d,e,f), 150 μM ioxynil (dotted curves d,e,f), 0.5 mM sodium dithionite (g,h,i) or by mild trypsination (j,k,l). B band was measured with no additives (a,b,c). Glycerol (30%) was included in sample solutions to avoid liquid solid phase transition at 0°C. Vertical bars indicate the peak temperatures at neutral pH.

DCMU [14], but its response to alkaline pHs is almost the same as that of the DCMU-induced Q band (dotted curves d,e and f).

The same behavior was observed for the other types of Q bands which are induced by redox poising or by enzymatic poisoning of Q_B. As previously reported in [7], an appropriate concentration (0.5 mM) of sodium dithionite preferentially reduces Q_B (but not Q_A), leading to generation of $S_2Q_A^-$ pairs after one flash instead of $S_2Q_B^-$. Similarly, a mild tryptic digestion of thylakoids preferentially inactivates the Q_B protein, which also leads to appearance of Q band [14] by inhibiting the oxidation of Q_A^- by Q_B [3,15,16]. The Q bands brought about by these treatments also exhibit unusual upward shift of peak temperature at pH 8.4 (curves h,i and k,l). It is, however, noteworthy that the Q band induced by tryptic digestion shows an appreciable upward shift at acidic pHs as well (curve j). This is due to protonation of Q_A, as will be discussed later.

Fig. 1 also shows that the B-band intensity is appreciably decreased at high pH, while the intensity of the Q band in the presence of DCMU is much enhanced. A similar enhancement of the Q band at high pH is observed for the Q band induced by ioxynil as well. In contrast, the Q-band amplitude of dithionite treated or trypsinized samples show only slight increase. In these samples, the Q-band intensity is dependent on treatment conditions such as dithionite concentration or trypsin dose. This is

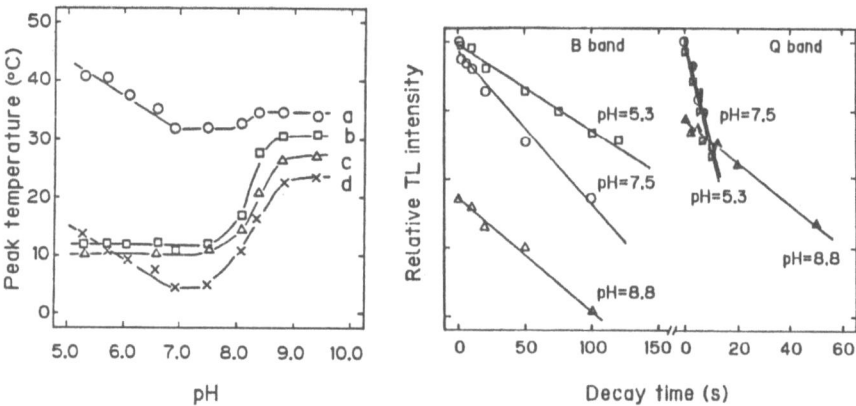

FIGURE 2. pH dependence of thermoluminescence emission temperature. Peak temperatures of the glow-curves measured as in Fig. 1 were plotted as a function of pH. a, B-band; b, DCMU-induced Q band; c, dithionite-induced Q band; d, trypsination-induced Q band.

FIGURE 3. Decay course of thermoluminescence intensity at different pHs. Samples were illuminated with a single saturating flash at 25°C and cooled to 77K after varying dark incubation period. The thermoluminescence light-sum was plotted against the incubation period. B band: pH 5.3 (□); pH 7.5 (o); pH 8.8 (△). Q band (DCMU-induced): pH 5.3 (■); pH 7.5 (●); pH 8.8 (▲).

presumably because these treatments inactivate some of the centers.

More detailed plots of peak temperature versus pH are shown in Fig. 2. The B band (curve a) shows a constant peak position (32°C) between pH 7.0 and 8.0 and a clear upward shift in the acidic region (pH < 7.0). A slight but reproducible increase is also observed in the alkaline region (pH > 8.0). In contrast, the Q band in the presence of DCMU (curve b) exhibits quite a different pH dependence: the peak position keeps a constant temperature (11–12°C) below pH 7.5 but suddenly increases to about 30°C between pH 7.5–8.5. A very close pH course is obtained for the Q band induced by dithionite (curve c). After mild trypsin digestion, the Q band (curve d) shows a similar response at neutral and high pH to that of the DCMU band, but in the acidic region the pH dependence is similar to that of the B band.

The thermoluminescence peak emission temperature approximately corresponds to the stabilization energy of the recombining charge pair [10]. Thus, the upward shift of B-band peak position observed at low pH and that of the Q band at high pH are interpreted as indicating the increased stabilization of $S_2Q_B^-$ at acidic and $S_2Q_A^-$ at alkaline pHs. The pH-induced stabilization was more directly confirmed by measuring the decay course of thermoluminescence intensity (Fig. 3). The half-decay time of the B-band measured at 25°C was about 32s at pH 7.5, increased markedly to about 60s at pH 5.3, and also increased slightly to about 40s at pH 8.8. In contrast, the half-decay time of the Q-band (with DCMU) kept a constant value (2–3s) at pH 5.3–7.5, but remarkably increased to about 20s at pH 8.8. The pH dependent stabilization of $S_2Q_A^-$ and $S_2Q_B^-$ is also supported by the calculated values of the activation free energy for the Q band and B band (will be published elsewhere).

5. DISCUSSION

On the time scale of thermoluminescence measurements the effect of proton motive force on energetic parameters is negligible [10], so that the free energy of separated charge pairs must be stored in the form of redox potential difference. Our thermoluminescence data can be interpreted as reflecting the pH-induced changes in midpoint redox potential of the charge carriers participating in the emission of B and Q bands, and can be summarized as follows:

1. The redox spans of $S_2Q_A^-$ and $S_2Q_B^-$ charge pairs show contrary response to pH changes: $S_2Q_B^-$ is stabilized at acidic pHs, while $S_2Q_A^-$ at alkaline pHs.

2. Below pH 7.5, and assuming that $E_m(S_2/S_1)$ is pH independent in accordance with the generally accepted view that S_1 to S_2 transition does not involve proton release [5,17-20], our data are interpreted as indicating that $E_m(Q_B/Q_B^-)$ increases with decreasing pH (5.3-7.5), while $E_m(Q_A/Q_A^-)$ keeps a constant value. This is consistent with the earlier results by fluorescence [2] and thermoluminescence [8,11] measurements, suggesting the protonation-induced shift of the redox potential of Q_B/Q_B^- couple, and also with the constant "working potential" for Q_A/Q_A^- [1,2].

The pH dependence of $E_m(Q_A/Q_A^-)$ observed after mild trypsin digestion of thylakoids indicates that the removal of Q_B quinone or digestion of Q_B protein increases the accessibility of Q_A^- to protons on the time scale of photosynthetic back reactions.

3. Above pH 7.5 two different interpretations are possible:

(A) If we assume that $E_m(S_2/S_1)$ is pH independent, the peak temperature shift of the Q band formally reflects an increase in $E_m(Q_A/Q_A^-)$ at pH 7.5-8.5, and a similar but smaller increase in $E_m(Q_B/Q_B^-)$. This is consistent with the data of Robinson and Crofts [2] regarding the decrease in the difference between $E_m(Q_A/Q_A^-)$ and $E_m(Q_B/Q_B^-)$ at high pHs but is contradictory to their assumption of the constant "working potential" for Q_A/Q_A^-. Obviously, this interpretation is unlikely since a protolytic effect on Q_A is hardly conceivable at alkaline pHs.

(B) If we assume that $E_m(Q_A/Q_A^-)$ is pH independent and keeps a constant value at high pHs as well as low pHs as proposed in the "working potential" hypothesis, the peak temperature change would be a reflection of decrease in $E_m(S_2/S_1)$. If this is the case, the decrease in $E_m(S_2/S_1)$ should affect also the peak temperature of the B band arising from $S_2Q_B^-$ recombination. In fact, the B-band peak temperature shows a slight upward shift between pH 7.5 and 8.5, although the extent of the shift is far smaller

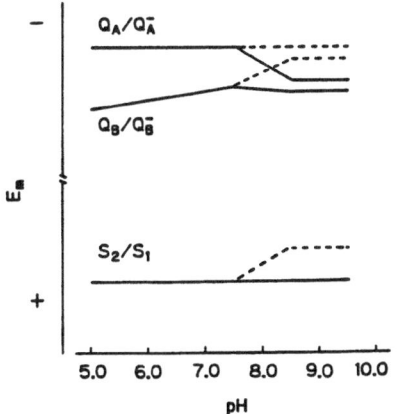

FIGURE 4. Schematic presentation of the alternative interpretation of the unusual reponse of the Q band at alkaline pHs. pH dependences of midpoint potentials (E_m) of S_2/S_1, Q_B/Q_B^- and Q_A/Q_A^- couples were calculated based on two different assumptions; constant $E_m(S_2/S_1)$, solid line and constant $E_m(Q_A/Q_A^-)$ (working potential hypothesis), broken line.

than that of the Q band. The weaker reflection of the change in $E_m(S_2/S_1)$ on $S_2Q_B^-$ pair (B band) could be due to the superposition of the protolytic effect on Q_B^- and the pH dependent changes in $E_m(S_2/S_1)$; the former will effect a downward shift of the peak temperature up to pH 7.9, the reported pK of Q_B^- [2], while the latter will effect an upward shift similar to that observed for the Q band. These relations are schematized in Fig. 4. This interpretation is more likely, even though a direct protolytic effect on S_2 as the cause for the change in $E_m(S_2/S_1)$ is contradictory to the established view on the proton release pattern during water cleavage [5,17-20].

Based on these, we at present consider that the anomalous response may not be due to a direct protolytic effect on Q_A or S_2, but is due to an indirect effect of high pH on the donor side of PSII at pH 7.5 - 8.5 [21]. More detailed examinations on this problem will be published elsewhere.

ACKNOWLEDGEMENT

This work was supported by the research grant on Solar Energy Conversion by Means of Photosynthesis given by the Science and Technology Agency of Japan (STA) to The Institute of Physical and Chemical Research (RIKEN) and partly by Grants-in-Aid (60304093 and 59380029) from the Ministry of Education, Science and Culture (MESC).

REFERENCES

1. Knaff DB (1975) FEBS Lett 60: 331-335
2. Robinson HH and Crofts AR (1984) In: Advances in Photosynthesis Research (Sybesma, C., ed.) Vol. I. pp.477-480, Matinus Nijhoff/Dr. W. Junk Publishers, The Hague
3. Renger G and Tiemann R (1979) Biochim Biophys Acta 545: 316-324
4. Bouges-Bocquet B (1980) Biochim Biophys Acta 594: 85-103
5. Bowes J and Crofts AR (1981) Biochim Biophys Acta 637: 464-472
6. Vermaas WFJ, Renger G and Dohnt G (1984) Biochim Biophys Acta 764: 194-202
7. Rutherford AW, Crofts AR and Inoue Y (1982) Biochim Biophys Acta 682: 457-465
8. Rutherford AW, Renger G, Koike H and Inoue Y (1984) Biochim Biophys Acta 767: 548-556
9. Demeter S and Vass I (1984) Biochim Biophys Acta 764: 24-32
10. Vass I, Horváth G, Herczeg T and Demeter S (1981) Biochim Biophys Acta 634: 140-152
11. Demeter S, Vass I, Hideg E and Sallai A (1985) Biochim Biophys Acta 806: 16-24
12. Arntzen CJ and Ditto CL (1976) Biochim Biophys Acta 449: 259-279
13. Vass I, Koike H and Inoue Y (1985) Biochim Biophys Acta 810: 302-309
14. Droppa M, Horváth G, Vass I and Demeter S (1981) Biochim Biophys Acta 638: 210-216
15. Renger G (1976) FEBS Lett 69: 225-230
16. Renger G and Weiss W (1982) FEBS Lett 137: 217-221
17. Fowler CF (1977) Biochim Biophys Acta 462: 414-427
18. Saphon S and Crofts AR (1977) Z Naturforsch 32c: 617-626
19. Velthuys BR (1980) FEBS Lett 115: 167-170
20. Damoder D and Dismukes GC (1984) FEBS Lett 174: 157-161
21. Völker M, Ono T, Inoue Y and Renger G (1985) Biochim Biophys Acta 806: 25-34

Photosynthesis Research 10: 437–444 (1986)
© Martinus Nijhoff Publishers, Dordrecht

pH DEPENDENT CONFORMATIONAL CHANGES AND ELECTROSTATIC EFFECTS IN PLASTOCYANIN

E.L. GROSS, J.E. DRAHEIM[*], G.P. ANDERSON, D.G. SANDERSON and S.L. KETCHNER
The Department of Biochemistry, The Ohio State University, Columbus, Ohio, 43210 [*]Present address: Dept. of Chemistry, University of Toledo, Toledo, Ohio.

1. ABSTRACT

Reduction of plastocyanin (PC) caused a change in the electric field at the surface of the molecule which resulted in a 0.3 pH unit increase in the pK_a of a nitrated derivative of Tyr 83. This change in electrical potential could alter the affinity for cytochrome f which is known to bind at this site. Conversely, properties of the copper center, including the pH dependence of the reduction potential, are regulated by the charge on the surface of the molecule. Both the reduction potential and conformation (as measured by near-UV circular dichroic spectra) were pH dependent. Thus the conformation and electrostatic behavior of PC are dependent on oxidiation state, pH and surface charge, raising the possibility that its redox activity is controlled by the pH gradient.

2. INTRODUCTION

Plastocyanin[1] (PC) is a 10.5 kD "blue" copper protein which functions in chloroplast electron transport shuttling electrons between cytochrome f (cyt f) and P700 (1). The crystal structures of oxidized (2), reduced (1) and apo- (3) poplar PC have been determined. The copper center (Fig. 1) is ligated to four residues (histidines 87 and 37, a cysteine and a methionine) in a distorted tetrahedral geometry. PC contains two binding sites for redox agents (4-7). Negatively-charged molecules such as ferricyanide bind at the top of the molecule at His 87. Positively-charged molecules, on the other hand, bind on the "east face" in the vicinity of Tyr 83. This is reasonable since Tyr 83 is surrounded by negative charges at residues #42-45 and #59-61. These negatively-charged residues are highly conserved in higher plant PC's (8). There is evidence that cyt f, which contains a positively-charged binding site (9), binds at the east face (10,11).

The activity of PC is regulated by the charge on the molecule, the pH and the salt concentration of the medium. For example, cations are required for the interaction of PC with negatively-charged PSI (13-16). In contrast, salts inhibit the favorable interaction with cyt f (10, 17). The activity of reduced PC with inorganic electron acceptors decreases upon lowering the pH due to the protonation of His 87 (18). This decrease in reaction rate causes a positive shift in the reduction potential since the

[1] Abbreviations: CD, circular dichroism; cyt f, cytochrome f; EDA, ethylenediamine; FPLC, Fast Protein Liquid Chromatography; NT, nitrotyrosine; PC, plastocyanin; UV, ultraviolet.

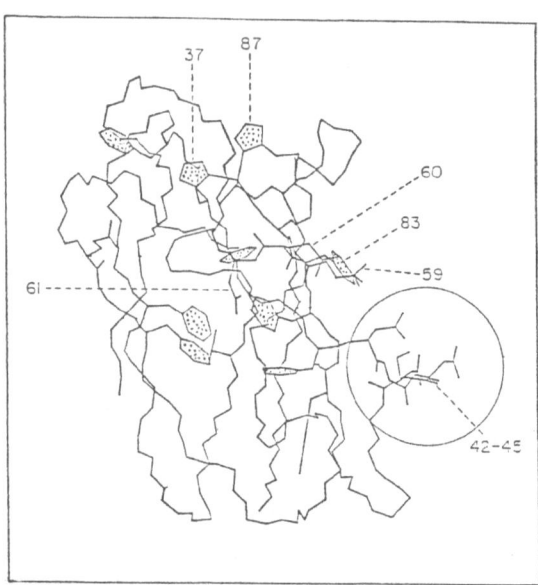

Figure 1.   Computer graphics representation of oxidized poplar (Populus
            nigra) plastocyanin. Crystallographic coordinates were obtained
            from the Brookhaven Protein Data Bank and displayed using a
            commercial graphics package called IDEAS (33). The histidine
            ligands to the copper center; the aromatic residues including
            Tyr 83 and the negatively-charged residues at #42-45 and #59-61
            are shown.

rate of reduction of oxidized PC is pH independent. The decrease in the
reactivity of reduced PC is paradoxical for two reasons. First, the rate
of reaction of reduced PC with its natural reaction partner, P700, is
optimal at low pH (ca. 4.7) (19). Second, PC is located within the
intrathylakoid space (20) and, therefore, must be able to be active at low
pH in illuminated chloroplasts.

Thus, the charge on PC and the protonation state of ionizable residues
are important in regulating its activity. In this paper, we will ask three
questions concerning the effect of charge and protonation state on the
conformation and electrostatic behavior of the PC molecule itself. First,
does pH affect the conformation of PC? Second, does the oxidation state of
the copper affect the electric field and hence the electrical potential
(See Equation 2) at the surface of the molecule in the region of Tyr 83.
This is important since the electrical field will affect the protonation
state of surface residues. Third, does the charge on the "east face" alter
the pH dependence of the reduction potential?

3. METHODS

PC was isolated from spinach and poplar (Populus deltoides) according

to the methods of Davis and San Pietro (21) and Graziani et al. (22)
respectively. Further purification was carried out by FPLC using a Mono-Q
anion exchange column + a linear NaCl gradient. Near-UV circular dichroic
(CD) spectra were measured using a JASCO 500A spectropolarimeter (23).
Tetranitromethane modification was carried out as described by Gross et al.

(24) and Anderson et al. (25). The nitrotyrosine (NT) derivative containing a single modification at Tyr 83 was used for these studies. The pK$_a$'s for nitrotyrosine 83 were determined by measuring the absorption spectrum of the nitrotyrosine derivative as a function of pH using a Cary 118 C spectrophotometer.

PC containing a single molecule of ethylenediamine (EDA) at residues #42-45 was prepared using a modification of the method of Burkey and Gross (26). PC (200-500 M) was incubated at 4°C for one hour in the presence of 20 mM EDA + 5 mM 1-ethyl-3-(3-(dimethylamino)propyl) carbodiimide at pH 6. The products were separated by FPLC using a Mono-Q anion exchange column + a linear NaCl gradient. The location of the label was determined by tryptic hydrolysis + amino acid analysis (11). The derivative used for these studies contains a single molecule of EDA in the region of residues #42-45. Reduction potentials were determined using a thin-layer filar electrode (27).

4.0 RESULTS AND DISCUSSION

4.1 <u>The effect of pH on PC conformation.</u> The pH dependence of PC conformation is important both because PC is located within the intrathylakoid space and must experience large changes in pH during light-dark cycles and also because pH has been shown to affect the activity of reduced PC (18, 28). Near-UV circular dichroism (CD) was used to study the effect of pH on PC conformation (Fig. 2.). The near-UV CD spectrum of PC can be divided into three regions. The region below 270 nm is attributed to phenylalanine residues. That between 270 and 285 nm is attributed to tyrosine residues. The region above 285 nm may be due to charge transfer bands possibly associated with histidine residues (28). Changing the oxidation state of PC primarily affects the phenylalanine and charge tranfer regions (22, 29). Lowering the pH also affects these two regions. The changes are most dramatic for reduced PC which is consistent with the observation that lowering the pH affects the activity of reduced PC to a much greater extent than oxidized PC (18, 28). The changes <u>ca.</u> 290 nm may be directly attributable to the protonation of His 87 whereas those at 260 nm represent protein conformational changes which may occur

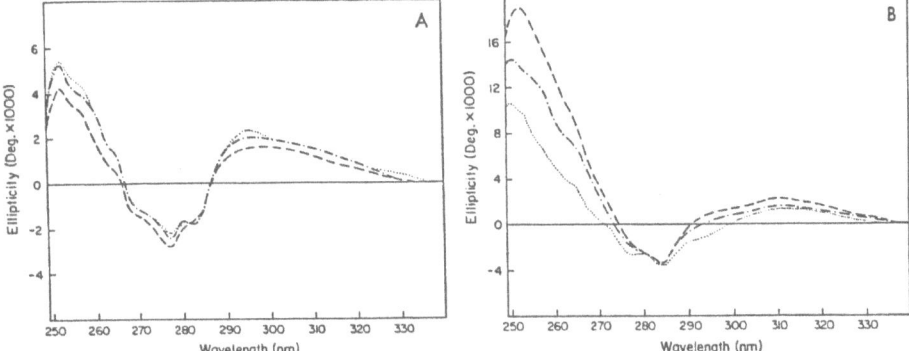

Figure 2. The effect of pH on the near-UV CD spectrum of spinach plastocyanin. The near-UV CD spectrum was measured as a function of pH for oxidized (A) and reduced (B) PC. PC (at 50 uM) was suspended in 10 mM succinate buffer (---) pH 6.0, (- · -) pH 5.5 and (.....) pH 5.0. Other conditions were as described in Methods.

indirectly as a result of the protonation of His 87. These results show
that the conformation of PC is sensitive to both pH and oxidation state.
Both these factors may provide a mechanism for regulating electron
transport at the level of PC.

4.2 <u>The effect of oxidation state on the pK_a of nitrotyrosine
83.</u> Selective nitration was used to study the effect of oxidation state
and species on the electric field at Tyr 83. Tyr 83 is important since it
is located at of the binding site for cyt f (10, 11). Changes in electric
field in this region could change the pK_a's of "east face" residues which
could, in turn, alter the binding constant for cyt f. Thus, the binding
constant of cyt f could be different for oxidized and reduced PC. This
would be a useful property if PC is a mobile electron carrier as has been
suggested (30). To answer this question, PC was reacted with
tetranitromethane according to Eqn. 1.

$$P\text{-}\underset{}{\bigcirc}\text{-OH} + C(NO_2)_4 \longrightarrow P\text{-}\underset{}{\bigcirc}\overset{NO_2}{\text{-OH}} + (NO_2)_3 C^- + 2H^+ \quad (1)$$

When PC was reacted with tetranitromethane, a single nitro group was
incorporated into Tyr 83 (24, 25). This is reasonable since Tyr 83 is the
only exposed tyrosine in spinach PC (Fig. 1). The pK_a of the nitrotyrosine
derivative can be determined since the protonated and non-protonated forms
have different absorption spectra. The absorption maximum for the
protonated form is at 360 nm compared to 428 for the deprotonated form.
Both spinach and poplar PC showed a decrease in the pK_a for nitrotyrosine
83 of 0.3 pH units upon oxidation of the protein (Table I.). This can be
attributed to the increase in positive charge on the copper in the oxidized
state. The change in the electrical potential necessary to produce a 0.3
pH unit change in pK_a is 18 mV. This value for the potential can be used
to calculate the dielectric constant for PC using Eqn 2.

$$\Delta V = \frac{\Delta n\, q}{4\pi e\, e_0\, r} \quad (2)$$

where ε is the apparent dielectric constant, Δn is the unit change in
charge on the copper, q is the charge on an electron, ε is the
permittivity of free space, ΔV is the change in potential and r is the
distance from the copper to Tyr 83. A value of 12 Å was obtained for this
distance using the crystallographic coordinates supplied by the Brookhaven
Protein Data Bank. A value of 70 was obtained for the dielectric constant
which is reasonable for a protein most of whose residues extend into the
aqueous medium (31).

TABLE I.

The Effect of Species and Oxidation State
on the pK_a of Nitrotryosine 83

| Species | pK_a Oxidation State | | ΔpK_a |
	Oxidized	Reduced	Upon Reduction
Spinach[a]	8.3	8.6	0.3
Poplar	7.7	8.0	0.3
Species Difference	0.6	0.6	

The absorption spectrum of nitrotyrosine 83 was determined as a function of pH as described in METHODS after which the pK was calculated. The error is ± 0.05 pH units.

[a] Taken from Ref. 25.

The values of the pK_a's obtained for both spinach and poplar PC are much higher than the pK_as of ca. 7 which are normally observed for nitrotyrosine peptides in water (32). The high pK_a's are undoubtedly due to the effect at Tyr 83 of the negative charges on the "east face" at residues #42-45 and #59-61. Thus, the pK_a of Tyr 83 should be dependent on the number of negative charges on the "east face". In fact, poplar PC should show a lower pK_a for nitrotyrosine 83 than spinach since it contains a serine rather than a glutamate at residue #45 (8). The electrical field at Tyr 83 due to a single charge change at residue #45 can be calculated using Eqn. 2 together with the dielectric constant calculated above. Using a distance of 10Å, a calculated potential of 22 mV is obtained. This would correspond to a ΔpK_a of 0.4 pH units which accounts for most of the 0.6 pH units obtained for both oxidized and reduced PC. Thus, the protonation state at the binding site for cyt f is dependent on both the oxidation state of the copper and the charge on nearby residues. This is likely to affect its interaction with cyt f. In fact, the pH dependence of the interaction of PC with cyt f has been shown to be dependent on the protonation state of residues on the "east face" (10).

4.3 The effect of chemical modification of the "east face" of PC on the pH dependence of the reduction potential. The nitrotyrosine results show that the charge configuration on the "east face" can affect the pK_a of Tyr 83. The next question concerns whether the negative charges on the "east face" can also affect the properties of the copper center. Effects on the copper center can be determined by measuring effects on the reduction potential. Moreover, specific effects on His 87 can be studied by measuring the pH dependence of the reduction potential since the positive shift in reduction potential observed upon lowering the pH has been correlated with protonation of His 87 (18). Previously, Burkey and Gross (25), showed that chemical modification of PC using ethylenediamine + a water-soluble carbodiimide (Eqn. 3) caused a shift in reduction

$$
\underset{\overset{\|}{\text{P - CO}^-}}{\overset{O}{}} + NH_2(CH_2)_2NH_3^+ \longrightarrow \underset{\overset{\|}{\text{P-CONH}_2(CH_2)_2NH_3^+}}{\overset{O}{}} \tag{3}
$$

potential of ca. +40 mV when as few as two molecules of EDA were
incorporated into PC. These studies have been extended using PC in which
only one molecule of EDA is inserted specifically into the
negatively-charged patch at residues #42-45. It can be seen (Table II.) A
change in charge of two units (from -1 to +1) is sufficient to cause a + 35
mV shift in reduction potential. This change in potential occurs without a
decrease in the rate of PC oxidation (33). These results show that changes
in charge configuration on the "east face" does alter the redox behavior of
PC. This suggests the possibility that binding of cyt f, which would also
change the electrostatic charge configuration of the "east face", could
alter the reduction potential of PC and, hence, the equilibrium point of
the cyt f-PC reaction.

TABLE II.

The Effect of Carboxyl Group Modification and pH on
The Formal Potential of Spinach PC

pH	$E^{\circ\,\prime}$ (mV)		
	Control PC	EDA-modified PC	Change Upon Modification
7.0	+372	+407	+35
5.0	+435	+398	-37
Change Upon Lowering the pH	+ 63	- 8	

PC (at 0.6 mM) was incubated in 10 mM phosphate buffer (pH 7.0) or 10 mM
phosphate-citrate buffer (pH 5.0) + 33 mM NaCl. 1-1´-bis(hydoxymethyl)
ferrocene was used as the mediator. The error is $\pm$ 5 mV.

Chemical modification of the "east face" also alters the pH dependence
of the reduction potential. Control PC shows the positive shift in
reduction potential upon lowering the pH associated with inactivation of
reduced PC (18, 28). In contrast, the reduction potential of modified PC
is independent of pH to below pH 5. Direct measurements of the rate of PC
oxidation by ferricyanide showed that EDA-modification prevented
inactivation of the protein at low pH (33). These results indicate that
chemical modification of the east face prevents protonation of His 87. In
conclusion, altering "east face" residues can significantly change the
redox properties of PC.

5.0 CONCLUSIONS

The oxidation state of the copper can affect the protonation of residues on the "east face" in the vicinity of Tyr 83 which is thought to be the binding site for cyt f. Conversely, alteration of the "east face" via chemical modification alters the redox properties of the copper center. Both the conformation of the protein and its electron transfer properties are affected by pH. The results indicate that PC is a flexible molecule which can respond to changes in its environment which might, in turn, provide a site of regulation for electron transport.

Acknowledgments

This research was suppported in part by Grant 1R01 GM30560 from the N.I.H. We wish to thank Dr. Larry Anderson of the Chemistry Dept. at O.S.U. for his help in the development of the filar thin-layer electrode.

References

1. Freeman HC (1981) Coordination Chem 21:29-51
2. Guss JM and Freeman HC (1983) J Mol Biol 169:521-563
3. Garrett TPJ, Clingeleffer DJ, Guss J, Rogers SJ and Freeman HC (1984) J Biol Chem 259:2822-2825
4. Handford, PM, Hill HAO, Lee RW, Henderson RA and Sykes AG (1980) J Inorg Biochem 13:83-88
5. Lappin AG, Segal MG, Weatherburn DC and Sykes AG (1979) J Am Chem Soc 101:2297-2302
6. Chapman, SK, Knox CV, Kathirgamanathan P and Sykes AG (1984) J Chem Soc Dalton Trans 1984:2769-1773
7. Chapman SK, Knox CV and Sykes AG (1984) J Chem Soc Dalton Trans 1984:2775-80
8. Boulter D, Haslett BG, Peacock D, Ramshaw JAM and Scawen MD (1977) Int'l Rev Biochem 13:1-40
9. Beoku-Betts D and Sykes AG (1985) Inorg Chem 24:1142-7
10. Beoku-Betts D, Chapman SK, Knox CV and Sykes AG (1985) Inorg Chem 24:1677-1681
11. Anderson GP, Sanderson DG and Gross EL (In preparation)
12. Lien S and San Pietro A (1979) Arch Biochem Biophys 294:128-137
13. Lockau W (1979) Eur J Biochem 34:365-373
14. Davis DJ, Krogmann DW and San Pietro A (1980) Plant Physiol 65:697-702
15. Haehnel W, Propper A and Krause H (1980) Biochim Biophys Acta 592:536-45
16. Niwa S, Ishakawa H, Nakai S and Takabe T (1980) J Biochem (Tokyo) 88:1177-1183
17. Sinclair-Day JD, Sisley MJ, Sykes AG, King GC and Wright PE (1985) J Chem Soc Cehm Commun 1985:505-7
18. Takabe T, Ishikawa H, Niwa S and Itoh S (1983) J Biochem (Tokyo) 94:1901-11
19. Hauska G, Berzborn RJ, McCarty RE and Racker E (1971) J Biol Chem 246:3524-3531
20. Davis DJ and San Pietro A (1979) Anal Biochem 95:254-259
21. Graziani MT, Agro AF, Rotilio G, Barra D and Moldavi B (1974) Biochem 13:804-809
22. Draheim JE, Anderson GP, Pan RL, Rellick LM, Duane JW and Gross EL (1985) Arch Biochem Biophys 237:110-117

23. Gross EL, Anderson GP, Ketchner SL and Draheim JE (1985) Biochim Biophys Acta 808:437-447
24. Anderson GP, Draheim JE and Gross EL (1985) Biochim Biophys Acta 810:123-131
25. Burkey KO and Gross EL (1982) Biochem 21:5886-5890
26. Sanderson DG and Anderson LB (1985) Anal Chem 57:2388-2393
27. Chapman SK, Davies DM, Watson AD and Sykes AG (1983) ACS Symp Ser 211:177-197
28. Draheim JE, Anderson GP, Duane, JW and Gross EL (1986) Biophys J (In press)
29. Takano M, Takahashi M and Asada K (1982) Arch Biochem Biophys 218:369-375
30. Matthew JB and Richards FM (1982) Biochem 21:4999-5009
31. Means GE and Feeney RE (1971) Chemical Modification of Proteins. Holden Day San Francisco
32. Sanderson DG, Anderson, LB and Gross EL. Submitted to Inorg Chem
33. Duane JW, Rellick LM and Gross EL (1986) Eng Design Graphics J (In press)

Photosynthesis Research 10: 445–450 (1986)
© *Martinus Nijhoff Publishers, Dordrecht*

STIMULATION AND INHIBITION OF PHOTOSYSTEM II ELECTRON TRANSPORT IN CYANOBACTERIA BY IONS INTERACTING WITH THE CYTOPLASMIC FACE OF THYLAKOIDS

G. SOTIROPOULOU and G. C. PAPAGEORGIOU
Greek Atomic Energy Commission, Department of Biology, Aghia Paraskevi-Attiki, 153 10 Greece

Abstract: The mechanism by which suspension medium ions regulate the rate of photoinduced electron transport across photosystem II was investigated with ion permeabilized cells of the cyanobacterium Anacystis nidulans. Electron transport was measured as the reduction of the electroneutral acceptor dichlorophenol indophenol, whose surface concentration is independent of electrostatic membrane potential. Potassium salts stimulate photoinduced electron transport at low concentrations and inhibit it at higher concentrations. No inhibition is observed when an antichaotropic anion is associated with potassium, while the inhibition is more severe the stronger the chaotropic character of the anion. Neutralization of the surface charge by potassium ions ligated to negatively charged membrane sites at the cytoplasmic side is a prerequisite for the expression of the chaotropic inhibition of photosystem II electron transport.

1. INTRODUCTION

Several investigators have reported that metal cations play important regulatory roles in the photosynthetic electron transport of cyanobacteria. Due to the ion impermeability of the bacterial envelope (cell wall plus cell membrane), such studies were performed either with isolated thylakoid fragments (1-7) or with chemically or enzymatically permeabilized cells (8-15). Polypeptide-complexed or sterically trapped ions (Mn^{2+}, Cl^-, Ca^{2+}) are important for the operation and the stability of the O_2-evolving complex of PS II (reviewed in 16-18). We shall exclude, however, these specifically immobilized ions from our discussion restricting it to the effects of diffusible medium ions.

Metal cations stimulate, in general, photoinduced O_2 evolution in the presence of the trivalent anion FeCN. The ranking of cation effectiveness in the order trivalent > divalent > monovalent led several authors (7, 11, 13-15) to invoke electrostatic counterion screening as the principle of the observed stimulation. Several other observations, however, are not entirely in pace with this interpretation. For example, rate stimulation by cations has been observed also with electroneutral (p-benzoquinone, DCIP) and electropositive acceptors (methyl viologen; 3, 6, 8-10, 14, 15). In addition, cation concentration curves of Hill reaction rates are often bell shaped. Counterion screening can only suppress asymptotically the membrane surface potential to zero (conventionally the aqueous bulk phase potential). Therefore, it cannot account for the inhibition at high electrolyte concentration.

The present paper examines the role of medium anions on the inhibition of PS II electron transport in cyanobacteria at elevated electrolyte concentrations. The system studied, ion-permeable Anacystis nidulans cells, is particularly advantageous since thylakoids are virtually intact. Therefore, only the cytoplasmic face of thylakoids is accessible to medium ions. Our results suggest that chaotropic anions modify components of this surface and this is expressed as an inhibition of the photosynthetic electron transport. Neutralization of the membrane surface charge by bound cations appearrs to be a prerequisite for the chaotropic inhibition.

2. PROCEDURE

Anacystis cells were rapidly permeabilized with lysozyme as in (20). The modified cells retained the original shape and were osmoresistant, indicating virtually intact cell walls. To distinguish them from the osmoresponsive spheroplasts, which are produced after complete hydrolysis of the cell wall peptidoglycan by lysozyme, they will be referred to as permeaplasts. They are characterized by a fully functional photosynthetic electron transport chain and photosynthetic control ratios exceeding 2, but their phycobilisome light harvesting system is totally destroyed as a result of the free ion passage through the cell envelope (20). Permeaplasts were resuspended and maintained in 0.5 M sorbitol, 0.05 M Hepes.NaOH, pH 7.5 at a cell density corresponding to 0.45 mg Chl a/ml. Assays were performed after transferring the cells to a low electrolyte medium consisting of 0.5 M sorbitol, 0.005 M Hepes.NaOH, pH 7.5.

DCIP photoreduction was measured by difference absorption spectrophotometry with a Hitachi Model 557 spectrophotometer. Saturating, heat--filtered actinic light (470 w/m^2; $\lambda > 640$ nm) was directed to the sample by a fiber optical guide. The photomultiplier cathode was protected by an interference filter ($\lambda = 580$ nm; $\Delta\lambda = 12$ nm) and a color glass filter (Corning C.S. 4-96). Samples contained permeaplasts equivalent to 5 µg Chl a/ml, 30 µM DCIP, and depending on the assay 0.5 mM DPC and 20 µM DCMU. An absorbance coefficient of 19.8/mM.cm was determined and used to convert absorption readings to DCIP concentrations. All assays were performed at room temperature.

3. RESULTS AND DISCUSSION

Figure 1 shows that gramicidin D more than doubles the rate of

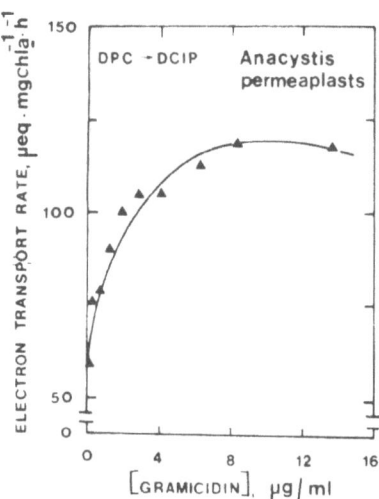

FIGURE 1. Stimulation of photoinduced electron transport in Anacystis permeaplasts by gramicidin D.

photoinduced electron transport at 8 µg/ml. This antibiotic is known to form water-filled transmembrane channels, through which H^+ and other monovalent ions can pass from one side of the membrane to the other (21). Rate stimulation by gramicidin suggests tightly coupled electron transport. This requires a membrane of low passive ion permitivity in order to maintain sufficiently high electrochemical potential difference across it. This observation coroborates similar conclusions based on other experiments (15, 20). We may infer, therefore, that in the absence of ionophores and channel forming compounds, permeaplast thylakoids are impassable to ions. Accordingly, medium electrolytes have access to the outer (cytoplasmic) thylakoid surface only.

Figure 2 illustrates the effects of potassium salts, present in the suspension medium, on photoinduced electron transport terminating in the electroneutral acceptor DCIP. Two cases are examined: Electrons are supplied either to the O_2 evolving complex, located on the inner side of thylakoids, or to a membrane phase intermediate, functioning between H_2O and the PS II reaction center by the lipophilic donor DPC.

Electrolyte effects are more prominent in the first case. The cation concentration curves appear to be strongly influenced by the nature of the associated anion. In the presence of the antichaotropic SO_4^{2-} no inhibition is observed at high K^+ concentrations. On the other hand, pronounced inhibition is observed in the presence of the moderately strong chaotrop NO_3^-. Cl^-, whose chaotropic potency is between SO_4^{2-} and NO_3^- (22) has an intermediate inhibitory effect.

Since DCIP is electroneutral, rate stimulation at low cation concentrations cannot be ascribed to a higher surface area concentration of the acceptor as a result of counterion screening of the negative surface charge. This would have been true for an electronegative acceptor such as FeCN (23). It appears form the data displayed in Figure 2 that the optimum of the

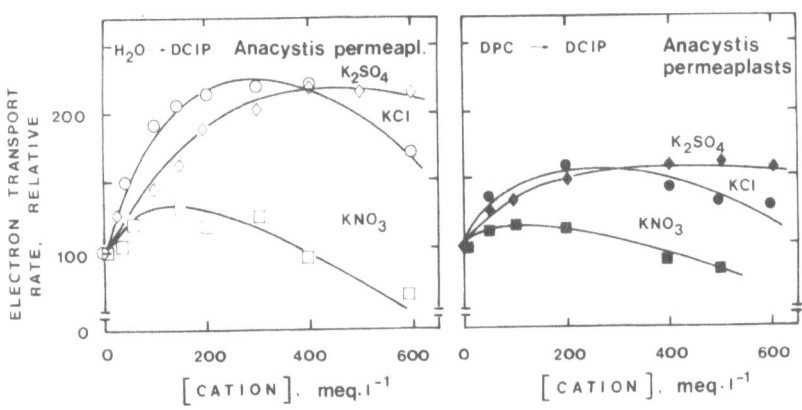

FIGURE 2. Effects of various potassium salts present in the suspension medium on the photoinduced electron transport by <u>Anacystis nidulans</u> permeaplasts from H_2O to DCIP (left) and from DPC to DCIP (right).

TABLE 1. Relative stimulation and inhibition of photoinduced electron transport across photosystem II in <u>Anacystis</u> <u>nidulans</u> permeaplasts by 0.6 M of various potassium salts added to the suspension medium.

K$^+$ salt added	H$_2$O $\longrightarrow$ DCIP Relative rate	DPC $\longrightarrow$ DCIP Relative rate
None	100[1]	100[2]
K$_2$HPO4	195	171
K$_2$SO$_4$	190	144
KCl	149	110
KBr	139	91
KNO$_3$	63	42
KSCN	0	0

[1]Typical rate, 50 electron μequiv/mg Chl.h
[2]Typical rate, 80-100 μequiv/mg Chl.h

cation concentration curves shifts to lower concentrations and maximal rates are lower the more chaotropic the associated anion is.

Table 1 provides further evidence for this regularity. It lists relative rates of photoinduced electron transport to DCIP in the presence of 0.6 M K. Anions can be ranked in the order SCN$^-$ > NO$_3^-$ > Br$^-$ > Cl$^-$ > HPO$_4^{3-}$ = SO$_4^{2-}$ with respect to their inhibitory action, and this order coincides with their ranking as chaotrops (22). Here, as well in Figure 2, the electrolyte effects on the rate of photosynthetic electron transport are more prominent whenever the O$_2$-evolving complex is involved.

A necessary condition for the chaotropic inhibition of the photosynthetic electron transport is a sufficiently high surface concentration of anions. Like all biological membranes, thylakoid membranes are negatively charged on both surfaces at pH > 4 (24), while cyanobacteria thylakoids are even more negatively charged than higer plant thylakoids (25). Taking -0.1 C/m^2 as representative surface charge density for Anacystis thylakoids (25,26) and assuming cation interactions in the Gouy-Chapman sense only (i.e. accumulation in the diffuse double layer of aqueous phase adjacent to the surface, but no permanent contact with the membrane) we may calculate a monovalent anion concentration at the surface of 0.09 M when the aqueous bulk phase concentration of potassium salt is 0.6 M.

This concentration is too low, however, to suffice for local chaotropic effects. On the other hand, results displayed in Figure 2 and Table 1 clearly show that such effects do indeed take place in the presence of suitable anion and high electrolyte content in the suspension medium. We think this is due to the fact that cations interact with membrane surfaces in the Gouy-Chapman-Stern sense (26), i.e. both by screening and by neutralizing the negative surface charge.

Charge neutralization is caused by quasi-permanent ligation of metal cations to exposed anionic sites of the membrane surface. The fraction of such sites occupied by cations at any time depends on the respective dissociation constants. In this manner, the concentration of anions near the surface may reach levels that suffice for chaotropic effects.

The fact that suppression of the negative surface potential alone by means of counterion charge screening does not satisfy the requirements for the

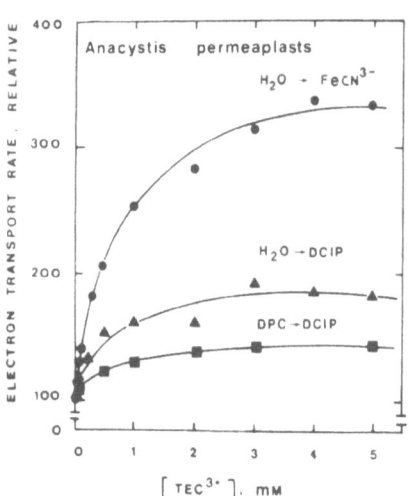

FIGURE 3. Effect of the trivalent tris-ethylene diamine cobalt complex (TEC^{3+}) on photosystem II electron transport in Anacystis permeaplasts.

expression of the anion chaotropic effect is further supported by the experiment shown in Figure 3. Here we plot the effect of the trivalent cobalt complex TEC^{3+} on the rate of electron transport in <u>Anacystis nidulans</u> permeaplasts. In the complex, central cobalt exists already within a stable inner sphere of three ethylene dianine molecules, so that it cannot come into direct contact with thylakoid surface anions.

   Assuming again a typical surface charge density of -0.1 $C.m^{-2}$ we may calculate using the Gouy-Chapman equation (19, 24) that the charge screening effect of 5 mM TEC^{3+} is approximately equivalent to that of 450 mM KCl. Whereas, however, such KCl levels suffice for noticeable inhibition of the Hill reaction rate (see Figure 2) no inhibition appears in the case of TEC^{3+}. It is noteworthy, also, that lowering the negative surface potential to about -48 mV, by adding 5 mM TEC^{3+} to the suspension medium causes a large increase in the photoreduction rate of the trivalent anion FeCN. This result is entirely due to the increase in the surface area concentration of the acceptor (23).

Acknowledgements: We thank Professor J Barber for a gift of TEC^{3+}.

Abbreviations: Chl, chlorophyll; DCIP, 2,6-dichlorophenol indophenol; DCMU, 3-(3',4'-dichlorophenyl)-1,1-dimethyl urea; DPC, 1,5-diphenyl carbazide; FeCN, ferricyanide anion; Hepes, 4-(2-hydroxyethyl)-1-piperazine ethanesulfonic acid; PS, photosystem; TEC^{3+}, tris ethylene diamine cobalt cation.

REFERENCES
1. Fredricks WW and Jagendorf AT (1964) Arch Biochem Biophys 104: 39-49
2. Susor WA and Krogmann DW (1964) Biochim Biophys Acta 88: 11-19
3. Binder A, Tel-Or E and Avron M (1976) Eur J Biochem 67: 187-196
4. Piccioni RG and Mauzerall DC (1978) Biochim Biophys Acta 504: 384-397
5. Piccioni RG and Mauzerall DC (1978) Biochim Biophys Acta 504: 398-405

6. Brand JJ (1979) FEBS Letters 103: 114-117
7. DeRoo CLS and Yocum CF (1981) Biochem Biophys Res Commun 100: 1025-1031
8. Wavare RA and Mohanty P (1982) Photobiochem Photobiophys 3: 327-335
9. Wavare RA and Mohanty P (1983) Photobiochem Photobiophys 6: 189-199
10. Wavare RA, Subbalakshmi R and Mohanty P (1983) Indian J Biochem Biophys 20: 301-303
11. Sotiropoulou G, Lagoyanni T and Papageorgiou GC (1984) In: Advances in Photosynthesis Research (Sybesma C, ed) Vol II, pp 663-666, Martinus Nijhoff/Dr. W. Junk Publishers, the Hague
12. Kalosaka K, Sotiropoulou G and Papageorgiou GC (1984) Biochim Biophys Acta 808: 273-279
13. Papageorgiou GC, Sotiropoulou G, Lagoyanni T and Kalosaka K (1985) In: Creation and Deactivation of Excited States of Biological Molecules (Frackowiak D, ed) pp 11-16, Polish Academy of Sciences, Poznan Polytechnic
14. Papageorgiou GC, Kalosaka K, Lagoyanni T and Sotiropoulou G (1985) In: Recent Advances in Biological Membrane Studies: Structure and Biogenesis, Oxidation and Energetics (Packer L, ed) pp 369-391, Plenum Press, New York
15. Sotiropoulou G and Papageorgiou GC (1985) In: Ion Interactions in Energy Transfer Biomembranes (Papageorgiou GC, Barber J and Papa S, eds) Plenum Press, New York (in press)
16. Renger G and Govindjee (1985) Photosynth Research 6: 33-55
17. Critchley C (1985) Biochim Biophys Acta 811: 33-46
18. Ghanotakis DF and Yocum CF (1985) Photosynth Research 7, 97-114
19. McLaughlin S (1977) In: Current Topics in Membrane and Transport (Bronner F and Kleinzeller A, eds) Vol 9, pp 71-144, Academic Press
20. Papageorgiou GC and Lagoyanni T (1985) Biochim Biophys Acta 807: 230-237
21. McCarty RE (1980) Methods Enzymol 69: 719-728
22. Hatefi Y and Hanstein WG (1969) Proceed Natl Acad Sci US 62: 1129-1139
23. Itoh S (1979) Biochim Biophys Acta 548: 579-595
24. Barber J (1982) Ann Rev Plant Physiol 33: 261-295
25. Kalosaka K and Papageorgiou GC (1984) In: Advances in Photosynthesis Research (Sybesma C, ed) Vol II, pp 707-710, Martinus Nijhoff/Dr. W. Junk Publishers, the Hague.
26. Kalosaka K, Sotiropoulou G and Papageorgiou GC (1985) Biochim Biophys Acta 808: 273-279
27. Eisenberg M, Gresalfi T, Riccio T and McLaughlin S (1979) Biochemistry 18: 5213-5223

V.   Oxygen Evolution; Components; and Mechanisms

Figure 10. Warren L. Buttler, Bill Rutherford, Pierre Joliot, H. Koike and Y. Inoue .... photo taken at Riken, Japan, 1983.

Figure 11. Warren L. Butler and Govindjee preparing to go out for a ride in Warren's boat .... to discuss oxygen evolution from California waters.

Photosynthesis Research 10: 453–471 (1986)
© Martinus Nijhoff Publishers, Dordrecht

REACTION SEQUENCES FROM LIGHT ABSORPTION TO THE CLEAVAGE
OF WATER IN PHOTOSYNTHESIS
- Routes, Rates and Intermediates -

H.T. WITT, E. SCHLODDER, K. BRETTEL and Ö. SAYGIN

ABSTRACT
The reaction sequence between the primary electron acceptor,
the oxidized Chlorophyll-a$_{II}$, and the terminal electron donor,
the water splitting enzyme system S, is being described in the
range from nanoseconds to milliseconds. For the cleavage of
water Chlorophyll-a$_{II}^+$ extracts four electrons in four turnovers from the enzyme system S responsible for the water oxidation. For each extraction the electron is moved step by step
along the chain that connects the Chlorophyll-a$_{II}$ center with
that of S. Beginning with the transfer from the immediate
donor, D_1, to Chl-a$_{II}$, the subsequent transfer from D_2 to D_1^+
ends in the electron transfer from S to D_2^+. This final act
establishes in S the oxidizing equivalent, probably in the
form of oxidized manganese. Coupled with these acts is an intrinsic proton release and a surplus charge formation. After
the generation of the 4th oxidizing equivalent in a concerted
final action the evolution of O_2 from water takes place. Cor-
relations between the events are described quantitatively.

1. The Chlorophyll-Quinone couple as generator for the oxidizing equivalents of water cleavage. The fundamental reactions
for the oxidation of water in photosynthesis take place in
system II of the functional membrane. This process is started
in the reaction center II by the photooxidation of Chloro-
phyll-a$_{II}$ (P-680)(1,13).The ejected electron is trapped by the
first stable acceptor, a plastoquinone molecule, Q_A or X-320
(2). Only this and no other stable acceptor is active in
system II (3). From Q_A via Q_B the electron is transferred to
a pool of plastoquinones, Q (2). In system I Chlorophyll-a$_I$
(P-700)(4) is photooxidized. The ejected electron is trapped
by the terminal acceptor, NADP$^+$. The oxidized Chl-a$_I$ takes up
the electron from the plastoquinone pool. The electron ejec-
tion from Chl-a$_{II}$ and Chl-a$_I$ is vectorial and sets up an electric field across the membrane (5,6). This vectorial charge
separation first observed through electrochromism gave evidence for the asymmetric arrangement of donors and acceptors
within the membrane (6). As a consequence of this charge
separation, one proton per electron is translocated across the
membrane at each system (7). The created pH gradient is used
energetically for ATP formation (8,9). The oxidized Chl-a$_{II}$
is re-reduced lastly by extraction of electrons from water. In
order to use the univalent oxidizing power of Chl-a$_{II}$ for the

cleavage of 2 H_2O into 4 H^+, 4 e^- and the evolution of one O_2, respectively, four Chl-a_{II}^+ oxidations have to cooperate to supply water with the necessary four oxidizing equivalents. First, in three Chl-a_{II} turnovers, three electrons are sequentially extracted step-by-step from a water splitting manganese-containing enzyme system S. The three oxidizing equivalents thereby generated are stabilized and stored. After a fourth photooxidation of Chl-a_{II} and electron extraction, resp., the formation of a further oxidizing equivalent takes place, followed by the evolution of one O_2 (10,11). The four unknown oxidation states of S are termed S-states (see Fig. 1). (The numbers n indicate the number of electrons extracted.) The reaction circle starts with water in S_0. The states S_0,S_1, S_2 and S_3 are stable. S_3 is followed by an unstable S_4 state which reverts to S_0 within 1 ms, together with the O_2 evolution (12). In the dark adapted system the major state is S_1. Therefore, under these conditions, the reaction starts with S_1 and the sequence is $S_1 \rightarrow S_2 \rightarrow S_3 \rightarrow (S_4) \rightarrow S_0 \rightarrow S_1$; i.e., O_2 evolution takes place after the 3rd flash (11).

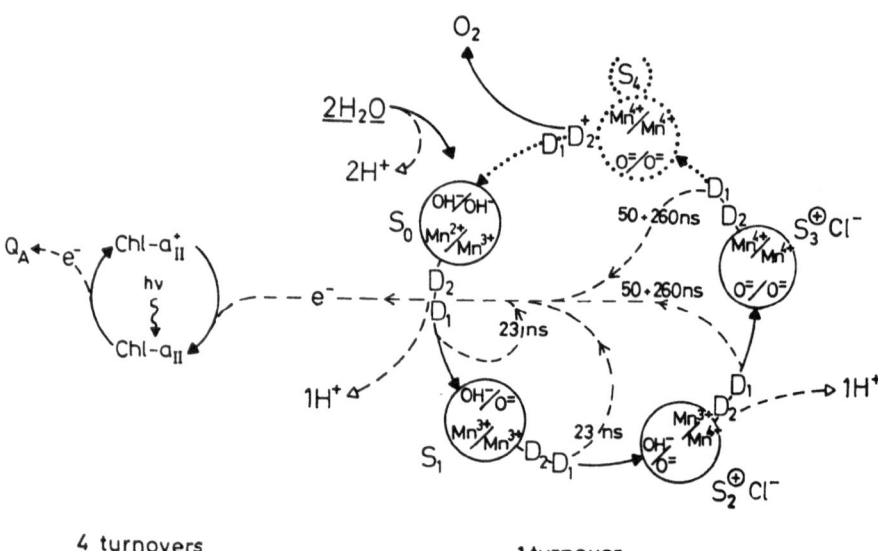

FIGURE 1. "Clock" of water cleavage and its "driver" Chlorophyll-a_{II}. In respect to the state of water one of the two possibilities discussed in § 5 is depicted in the cycle of the S states.

The existence of Chl-a_{II} and its photooxidation was observed through absorption changes which are coupled with the Chl-a_{II} turnovers (1,13). Advances have been made recently concerning (a) time resolution and sensitivity of the flash absorption spectrometer (14-17) and (b) the isolation of a suitable PS II oxygen-evolving complex (18,19). Through refined purification a complex of only 6 protein subunits from the cyanobacterium Synechococcus sp. has been obtained recently which is fully active in water oxidation. Thus precise measurements become possible down to the ns-range. The Chl-a$_{II}$ photooxidation takes place in <1 ns. A short-lived intermediate on the way to Q$_A$ is a pheophytin (20). The oxidized Chl-a$_{II}$ extracts an electron from the electron carrier chain which links Chl-a$_{II}$ with the O$_2$-evolving enzyme. This takes place in the time range of nanoseconds (14-17). Upon repetitive excitation the Chl-a$_{II}$ re-reduction kinetics could be resolved into a multiphasic exponential decay. The ns phases are 20 ns and 250 ns (see Fig. 2B)(15,16). The spectrum is shown in Fig. 2D (16).

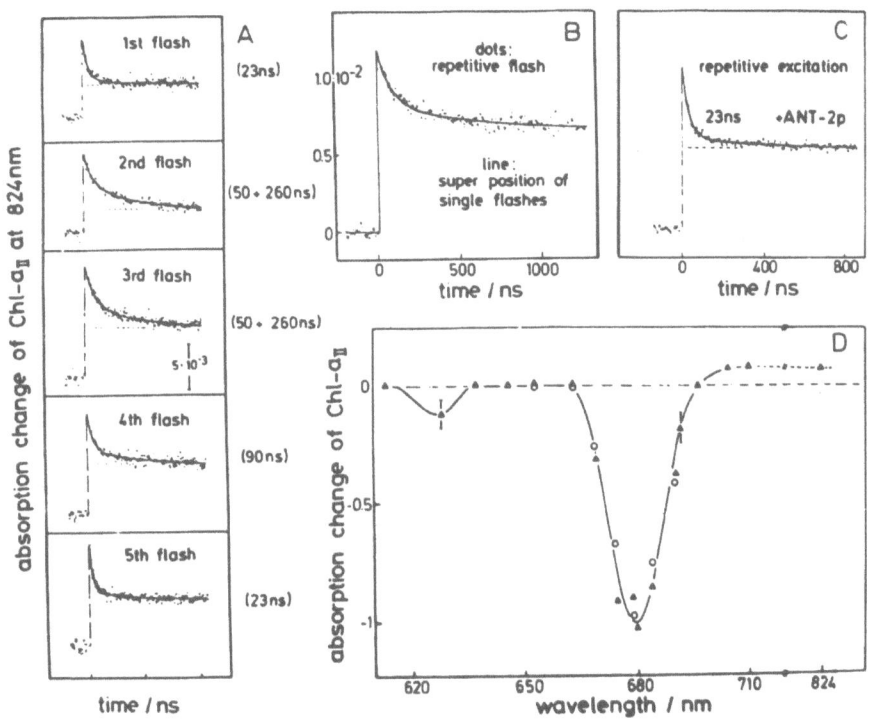

FIGURE 2. Nanosecond kinetics and spectrum of Chlorophyll-a$_{II}$ in system II at different conditions (17). The slow changes of absorption (~50%) are due to the 10 ms re-reduction of Chl-a$_I^+$ (P-700$^+$) in system I (details see § 1 & 2).

2. Quarternary kinetics of the primary electron extractions through Chlorophyll-a$_{II}$. As a consequence of the refined techniques, the Chl-a$_{II}$ reduction can be measured even in one flash with nanosecond time resolution. This offers the possibility of using the reduction kinetics of Chl-a$_{II}^+$ as a probe

for monitoring the reaction pattern on the donor side leading
to the decomposition of water (17). Fig. 2A shows the time
course of $Chl-a_{II}^+$ reduction after each of five single turnover
flashes given to dark adapted samples. In each flash a $Chl-a_{II}$
oxidation takes place. This indicates that with the quanta
uptake 1:1:1:1 the stoichiometry of electron extraction is
1:1:1:1 (see Table I). This is by no means trivial. The
changes after the 1st flash can be fit with an exponential decay of 23 ns, those after the 2nd and 3rd flash biphasically
with values around   50 ns and 260 ns. The change in the
4th flash has an intermediate half-life time. The 5th flash
again induces a decay time similar to the 1st flash. The times show
quarternary oscillations, indicating a coupling with the four
S states. The significant differences in the time are unexpected (but see below).

3. Correlation of the primary electron extraction with the
four S states. The amplitudes from the different phases of the
kinetics in single turnover flashes are compared with the
population of the different S states in dependence on the flash
number (see Fig. 3). The $Chl-a_{II}^+$ re-reduction within $\tau \approx 23$ ns
is correlated with the states S_0 and S_1 (Fig. 3A,B); whereas,
the biphasic reduction within ≈ 50 ns and ~ 260 ns is correlated
with state S_2 as well as S_3 (Fig. 3C,D)(17). As a first consequence of this result, the multiphasic kinetics under repetitive excitation (see § 1)   can be examined. Under these
conditions the four S states are equally populated. Therefore,
these kinetics should be a superposition of the individual
kinetics of the S states, i.e., a superposition of 50% of 23 ns
(due to 25% S_0 and 25% S_1) and 50% of 50 ns + 260 ns (due to
25% S_2 and 25% S_3). This is demonstrated in Fig. 2B (17). A
second consequence can be checked in experiments with addition
of ANT-2p. This substance shifts the states S_2 and S_3 to S_1
so that at appropriate flash frequencies after addition of ANT-
2p practically only the S_1 state is present (21). Therefore,
in the presence of ANT-2p the $Chl-a_{II}^+$ reduction should occur
under repetitive excitation only monophasically with the time
attributed to S_1, that is, within 23 ns. This is documented
in Fig. 2C (17).
The dependence of the primary electron extraction on the S
states is explained by different surplus charges in the different S states (17). These charges may modify the electron
transfers through Coulomb attraction (see next paragraphs). A
consequence of such a dependence is that all subsequent electron transfers along the pathway up to the O_2-evolving enzyme
S should also depend on the S states.
In PS II complexes selected for most efficient O_2 evolution
(one O_2 per 50 Chl molecules in four flashes) $\sim 15\%$ of Chloro-
phyll-a_{II}^+ is reduced under repetitive excitation in the microsecond time range (85% in the ns range, see above) and this in
three exponential phases with 5, 35 and 200 µs (22). It is
only the amplitude of the 35 µs-phase ($\sim 7\%$) that oscillates
with the S states. The 5 µs and 200 µs phases ($\sim 8\%$) may belong
to PS II particles that are inactive in O_2 evolution. Several
former publications on samples with much higher µs fractions

(up to 50%) are therefore obviously not representative of the water splitting part of photosynthesis.

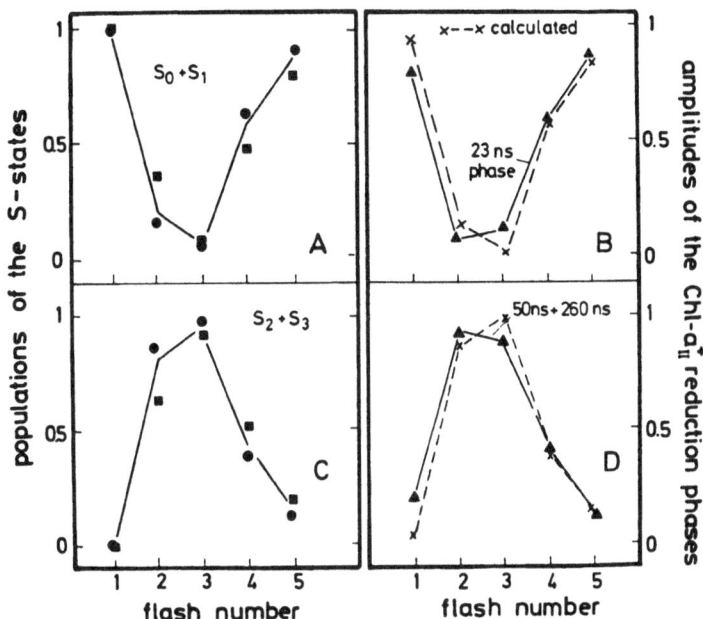

FIGURE 3. Correlation between the population of the S states, S_n, and the different phases of the Chlorophyll-a$_{II}^+$ reduction kinetics as a function of the flash number, N (17) (calculated curves according to eq. (9))

4. Sequence of the surplus charge variation of the S states and the stoichiometry of proton release. - Electrochromic identification. The electron transfer to Chl-a$_{II}^+$ is retarded in states S_2 and S_3 compared to S_0 and S_1 (see § 3). For an explanation of this effect it is assumed that the O_2-evolving complex in states S_2 and S_3 is more positive by one elementary charge than in S_0 and S_1. The sequence of charges should be thus $0:0:+:+$ and the stoichiometry of the variation of these charges $0:+1:0:-1$ for $S_0 \to S_1 \to S_2 \to S_3 \to S_0$ (s. Fig. 1 and Table I). It is expected that the electric field of the surplus charges at S_2 and S_3 should induce a Stark effect and electrochromism, resp.; i.e., a shift of the absorption bands of the PS II complex (6). The shift should be indicated through corresponding quarternary absorption changes. As the S_0 to S_3 states are stable for seconds and minutes (see § 1), such absorption changes should be observable in the time range of seconds. Fig. 4A

shows the measurements performed on a dark-adapted system in the red region, starting with S_1 (see § 1). A quarternary oscillation is observed which corresponds to the expected stoichiometry of 0:+1:0:-1 (see Table I). The changes as a function of wavelength result in a difference spectrum, indicating an electrochromic band shift of antennae Chlorophyll-a (23,24). The results support the conclusion obtained from the analysis of the ns-reduction kinetics of $Chl-a_{II}^+$ in paragraphs 2, 3 and 9.

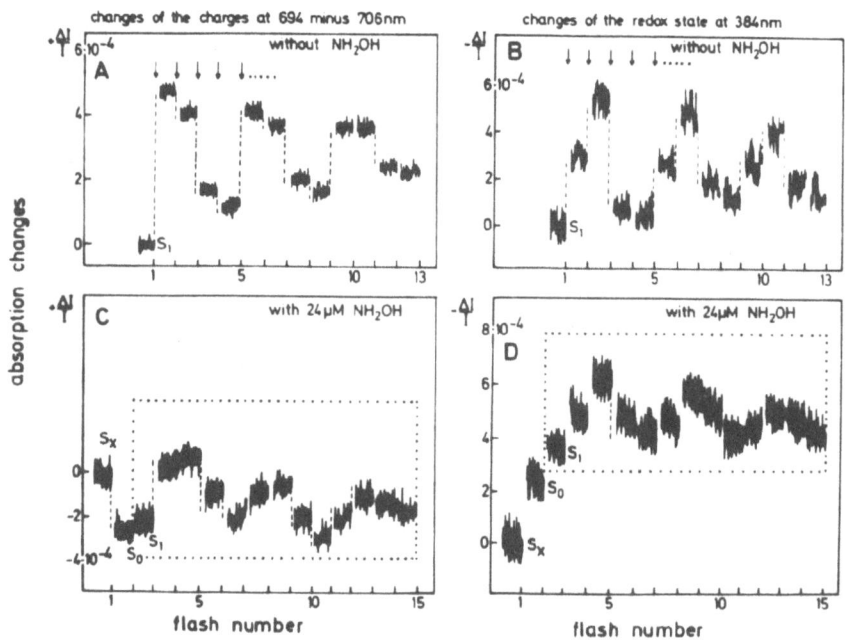

FIGURE 4.
A: Quarternary oscillation of the changes of charges in dependence on the flash number (23,24)
B: Quarternary oscillation of the redox changes of the oxidizing equivalents in dependence on the flash number (32)
C: The same as A but with addition of NH_2OH shifting the S states two steps backwards
D: The same as B but with addition of NH_2OH shifting the S states two steps backwards

The changes of charges, coupled with the elementary acts in the O_2-evolving complex, provide information on the release of those protons immediately connected with this process (so-called intrinsic protons). The sequence of electron extraction, 1:1:1:1 (see § 2), together with the measured

quarternary oscillation of charges 0:+1:0:-1 is only understandable if the stoichiometry of the proton dissociation and
intrinsic release, resp.,is 1:0:1:2 (see Fig. 1 and Table I).
The absence of a proton release during the electron extraction
from S_1 to S_2 sets up the surplus charge. The release of two
protons together with the electron extraction from $S_3 \rightarrow S_0$
would cause the disappearance of this charge. It should be
pointed out that the extrinsic proton release measured in the
outer water phase by pH electrodes and pH indicators, resp.,
must not necessarily correspond to the intrinsic proton release, because the latter one may not be in equilibrium with
the external phase. It is, however, the stoichiometry of the
intrinsic proton formation that is of importance in the examination of the mechanism of water cleavage. Extrinsic proton
release measurements were controversial for years. At present
a pattern of 1:0:1:2 is accepted (25-27); i.e., the extrinsic
sequence corresponds to the intrinsic one.

Table I

Stoichiometry and states of five events in the turnover of the
water splitting cycle. In (S_4) creation of a further oxidizing equivalent (*) takes place with immediate reduction, together with the three oxidizing equivalents accumulated in
S_3 (see text). If instead of two manganeses four are engaged
in water oxidation, see § 6 for the action of the two
additional Mn's.

	$S_0 \longrightarrow$	$S_1 \longrightarrow$	$S_2 \longrightarrow$	$S_3 \xrightarrow{(S_4)}$	S_0
absorbed quanta hν	1	1	1	1	
electron extrac- tion Chl-a$_{II}^+ \leftarrow$ e$^-$.... S_n	1	1	1	1	
changes of positive surplus charges	0	1	0	-1	
intrinsic H$^+$ release	1	0	1	2	
possible states of water	OH$^-$ OH$^-$	OH$^-$ O$^=$	OH$^-$ O$^=$	O$^=$ +O$_2$ O$^=$ -2 H$_2$O	OH$^-$ OH$^-$
possible states of oxidizing equivalents	Mn^{2+} Mn^{3+}	Mn^{3+} Mn^{3+}	Mn^{3+} Mn^{4+}	Mn^{4+} Mn^{4+} (*)	Mn^{2+} Mn^{3+}

5. Sequence of the redox changes of the oxidizing equivalents and the possible states of water. The participation of manganese in the water splitting process is undisputed. Four manganese atoms per reaction center have been analyzed but it may be that only two manganeses are engaged in the water oxidation (see below). In the UV region absorption changes were observed which have been attributed to a valence state change of manganeses (28). Dekker et al. proposed that three such successive Mn^{3+}/Mn^{4+} transitions are accumulated on each of the S_0-S_3 steps and reversed in the $S_3 \rightarrow S_0$ step; i.e., the pattern of the change of the valence states should be $+1:+1:+1:-3$ (29). However, according to former results from Velthuys (30), the pattern should be $0:+1:0:-1$. Lavergne (31) supported the Velthuys results. Renger and Weiss reported on two patterns, $0:2:0:-2$ for Mn and $1:-1:+1:-1$ for an unknown species (47,48). However, the spectra published in (47,48) are superimposed by a contribution from a binary oscillation of plastoquinones located at the acceptor side (private communication by Renger). In all experiments in (29-31,47,48), through unavoidable misses and double hits, the populations of the four S states become progressively mixed with increasing number of flashes, obscuring the true pattern. Conclusions as to the true pattern are therefore only possible by disintegration of the mixed S states through theoretical calculations. The latter are based on the values of double hits, misses, the initial S_0/S_1 ratio in the dark-adapted state, and the available precision. We have shown that one can fit different true patterns with the same values of double hits and misses. Therefore, the results in (29-31,47, 48) which are based on such fitting procedures do not permit an unambiguous discrimination. For clarification, we have therefore additionally introduced an independent method: through a chemical modification with NH_2OH the S states are shifted backwards by two full units. In this way, heavy mixtures of the S states are prevented (see below). Without NH_2OH we measured oscillation patterns in the UV at 384 nm as depicted in Fig. 4B (32)(binary oscillations due to Q_B were eliminated through addition of DCMU + SiMo). The pattern is different from that of the charges (Fig. 4A). As the pattern is also a quarternary one and one of stable states only, it must be a property of the S states. Absorption changes at 384 nm are very unlikely due to oxidation products of H_2O. Therefore, redox changes of the oxidizing equivalents should be responsible for the absorption changes shown in Fig. 4B. Because of two reasons the most critical point in both patterns (Fig. 4A,B) is the transition in the 4th flash, $S_0 \rightarrow S_1$; i.e., the last step of the quarternary period: 1. This transition is most heavily mixed with simultaneous transitions from $S_3 \rightarrow S_0$ (20-45%) and also even from $S_2 \rightarrow S_3$. 2. As can be seen in Fig. 4B, especially $S_3 \rightarrow S_0$ has a sign opposite to all other transitions; in this transition the reduction of all oxidizing equivalents takes place together with the O_2 evolution. This situation can be avoided, however, when (a) it becomes possible to observe the $S_0 \rightarrow S_1$ transition in its beginning, when practically no mixing takes place and when (b) also the other transitions, $S_1 \rightarrow S_2$ and $S_2 \rightarrow S_3$, can be observed before the opposing jump,

$S_3 \rightarrow S_o$, takes place. This was now realized through chemical reduction of the S states with NH_2OH which shifts the S states backwards by two units to a state S_x (33). The backward shift is easily recognized in the pattern of the charges as well as in the pattern of redox changes of the oxidizing equivalents (compare the framed patterns in Figs. 4C and 4D with Figs. 4A and 4B). Therefore, with the 2nd flash the cycle starts with S_o followed by the S_o-S_3 transitions preceding $S_3 \rightarrow S_o$.

Before describing the results, the chemical meaning of S_x, the state at the beginning, before S_o, has to be discussed. (a) According to (33) NH_2OH reduces S_1 in the dark to S_o, whereby instead of water NH_2OH is bound; i.e., $S_x = S_x'(NH_2OH)$. This explanation has been criticized (32). (b) On the other hand, it may be possible that NH_2OH is unbound but has reduced manganese in S_1 via S_o; i.e., $S_x = S_x''(n \, Mn^{2+})$.

Considering now first the pattern of the change of charges: In Fig. 4C after the 2nd flash, i.e., transition from S_o to S_1, a negligible change takes place similar to the result shown in Fig. 4A after the 4th flash. After the 1st flash (Fig. 4C) one observes a negative jump. Its amplitude corresponds to the formation of a negative charge. This indicates that, after the removal of one electron from S_x in the 1st flash and the formation of S_o (water), resp., the state S_o is left with one more negative charge than in S_x. This can be explained, if with the removal of one electron from S_x a release of two protons is coupled with the formation of S_o. The release of two H^+ has been observed also by measurements of pH changes in the presence of NH_2OH (45). Concerning the source of these 2 H^+, in case of $S_x \triangleq S_x'$ (see above), one proton may be released if in the 1st flash bound NH_2OH is oxidized with the transition $S_x' \rightarrow S_o$; then the second proton should be released with the binding of water in S_o (scheme (a)). In case $S_x = S_x''$, one of the n Mn^{2+} is oxidized with the transition $S_x'' \rightarrow S_o$; then both protons should be released with the binding of water in S_o (scheme (b)).

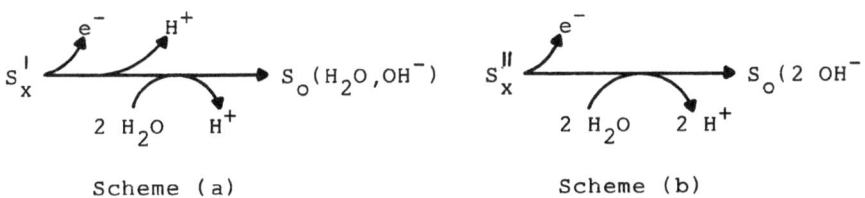

Scheme (a)   Scheme (b)

This means that at least one water must be in dissociated form in state S_o (also see Andreasson et al. (46)). The configuration of water in the further states, S_1-S_3, consequently can be obtained by considering the sequence of the intrinsic proton release, and if one considers that water is not oxidized during the $S_o \rightarrow S_3$ transitions (see below). A clarification as to which of the two states is realized has to wait until the chemical meaning of S_x has been analyzed (in progress). One of the two possibilities is outlined in Fig. 1 and Table I.

With regard to the redox reaction in Fig. 4D after the 2nd
flash a jump is observable from $S_o \rightarrow S_1$. This is not seen in
the $S_o \rightarrow S_1$ transition after the 4th flash shown in Fig. 4B,
which is heavily mixed with other transitions. In flashes 3
and 4 of Fig. 4D we also see a positive absorption changes and
a reverse after the 5th flash. The three-fold absorption
changes in flashes number 2, 3 and 4 indicate that with transi-
tions $S_o \rightarrow S_1 \rightarrow S_2 \rightarrow S_3$ a stepwise formation and stabilization of
three oxidizing equivalents takes place. This points out that
water may be left unoxidized during the $S_o \rightarrow S_1 \rightarrow S_2 \rightarrow S_2$ transition. Water may be oxidized only after the 5th flash, in the
$S_3 \rightarrow (S_4) \rightarrow S_o$ transition, through a concerted 4-equivalent
oxidation with simultaneous O_2 release (29).

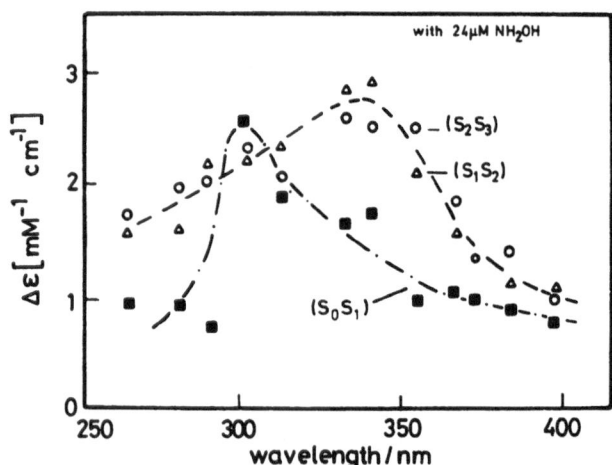

FIGURE 5. Difference spectrum of the oxidizing equivalents at
different S-state transitions in the presence of 24 µM NH_2OH (44)

6. <u>Spectrum of the redox changes of the water oxidizing</u>
<u>equivalents - possible states of manganese*</u> The patterns in
Fig. 4D obtained in the presence of NH_2OH (24 µM) have been
measured as a function of wavelength between 250-400 nm. Under
the NH_2OH-conditions the heavy mixtures of the S states are
prevented (see § 5). The spectrum after the 2nd flash therefore presents nearly the true spectrum of the transitions
$S_o \rightarrow S_1$. It shows a marked deviation from the spectra of the
$S_1 \rightarrow S_2$ and $S_2 \rightarrow S_3$ transitions. The latter two are practically
equal within our measuring capacity. The $S_1 \rightarrow S_2$ and $S_2 \rightarrow S_3$
transitions may be due to a valence state change of Mn^{3+} to
Mn^{4+} as stated by Dekker et al. (29). The $S_o \rightarrow S_1$ transition
which is shifted to a shorter wavelength may be due to a
change from Mn^{2+} to Mn^{3+}. This

*The results of this chapter are unpublished and are presented
here as a first note by Ö. Saygin and H.T. Witt. A detailed
publication is in preparation (44).

suggestion may be supported by the fact that also in vitro the difference spectrum of Mn^{2+}/Mn^{3+} is located at a shorter wavelength than that of Mn^{3+}/Mn^{4+} (49). Furthermore, an Mn^{2+} to Mn^{3+} valence state change for S_o S_1 fits the following argumentation. With addition of the electron donor, NH_2OH (Fig.4D) one may assume that all manganese are reduced down to the state Mn^{2+}; i.e., S_x is realized as S_x (n Mn^{2+})(see § 5). If only two manganeses are engaged in the oxidation of water (35, 36), the S_o-state should be present as S_o (Mn^{2+}/Mn^{3+}), because one electron is extracted with $S_o \rightarrow S_o$. The other states must then be S_1(2 Mn^{3+}), S_2(Mn^{3+}/Mn^{4+}) and S_3(Mn^{4+}/Mn^{4+})(see Fig. 1 and Table I). The oxidizing equivalent created with the last transition from S_3 to the unstable (S_4) might be the electron carrier D_2^+ (see § 7) oxidized in the 4th flash and reduced immediately in unison with the oxidized manganese present in state S_3 (see Fig. 1).
If four manganeses are engaged in the oxidation of water (50) besides the two manganeses mentioned above, a third one might be oxidized in (S_4) but remains "invisible" because of the immediate re-reduction together with the oxidized two Mn's in S_3. The fourth Mn might function only as stabilizer in a tetrameric Mn-cluster.
The principal difference in regard to the spectra of Dekker et al. in (29), which were obtained by the fitting procedure method, is given with our result, showing that the spectrum of $S_o \rightarrow S_1$ is not identical with that of $S_1 \rightarrow S_2$ and $S_2 \rightarrow S_3$. This principal difference has been demonstrated by us also without NH_2OH, i.e., without the backward shifting of the S state, by two pieces of evidence: (1) Using the fitting procedure and disintegration treatment of the other authors (a method which alone is not unequivocal, see above), e.g., at 360 nm where practically no changes of other components besides manganese take place, it was found, also through this procedure, that the change with the transition $S_o \rightarrow S_1$ is different from that of $S_1 \rightarrow S_2$. (2) Furthermore, the analysis of the kinetics of the S-state formations at this wavelength shows also without NH_2OH that the $S_o \rightarrow S_1$ spectrum is different from the others (see (44)).
The proposed states of manganese and of water in the different S states may be arranged in form of dimers (51). Cuban-like structures or adamantane-like complexes (52) have been also discussed (53).

7. Reaction sequence between Chlorophyll-a$_{II}$ and the S states. Two electron carriers D_1 and D_2. We discussed on the one hand the primary electron extraction through Chl-a$_{II}$ and, on the other hand, the events thereby induced within the S states,i.e., (a) variation of charges, (b) intrinsic proton release, (c) states of water and (d) redox reaction due to manganese. What is, however, the bridge between Chl-a$_{II}$ and the S-state components? One electron carrier, D, has been characterized by ESR (37) and optical measurements (38-40). It is suggested that D^+ is a plastoquinol derivative cationic radical (37). The oxidation time of D is not known. The re-reduction of D^+ by the O_2-evolving complex depends on the S states prior to

the flash and the half-life times are $(S_o) < 100$ µs, $(S_1) < 100$ µs $(S_2) \approx 400$ µs, $(S_3) \approx 1$ ms (41). If D were the only donor between Chl-a$_{II}$ and the O_2-evolving complex, the following scheme I should represent the reaction sequence on the donor side of PS II:

$$\text{Chl-a}_{II}^{+}\text{DS}_n \underset{k_{-1}}{\overset{k_1}{\rightleftharpoons}} \text{Chl-a}_{II}\text{D}^{+}\text{S}_n \underset{k_{-2}}{\overset{k_2}{\rightleftharpoons}} \text{Chl-a}_{II}\text{DS}_{n+1}$$

(Scheme I)

($\dot{n} = 0,1,2,3$). The k-values should depend on the S states.

According to scheme I, Chl-a$_{II}^{+}$ would decay biphasically down to an equilibrium level with one phase essentially determined by $k_1 + k_{-1}$ and a second phase corresponding to the reduction of D^{+}, which has been measured in state S_2 and S_3 as 400 µs and 1 ms (see above). In contrast to this prediction the experiments indicate that in state S_2 as well as in state S_3 Chl-a$_{II}^{+}$ is reduced biphasically with half-life times of 50 ns and 260 ns, resp. (see § 2); i.e., more than 10^3 times faster than expected. Consequently, D should not be the immediate donor to Chl-a$_{II}^{+}$. Therefore, we have to assume at least two donors, D_1 and D_2 (17). D_2 would be the species giving rise to EPR signal II$_{vf}$ (37) and D_1 would be an as yet undetected donor to Chl-a$_{II}^{+}$. Restricting ourselves to a linear electron transfer scheme, we get the following four electron transfer routes for the four S states:

$$\text{Chl-a}_{II}^{+}\text{D}_1\text{D}_2\text{S}_0 \underset{k_{-1}}{\overset{k_1}{\rightleftharpoons}} \text{Chl-a}_{II}\text{D}_1^{+}\text{D}_2\text{S}_0 \underset{k_{-2}}{\overset{k_2}{\rightleftharpoons}} \text{Chl-a}_{II}\text{D}_1\text{D}_2^{+}\text{S}_0 \underset{k_{-3}}{\overset{k_3}{\overset{\text{H}^{+}}{\rightleftharpoons}}} \text{Chl-a}_{II}\text{D}_1\text{D}_2\text{S}_1$$

A

$$\text{Chl-a}_{II}^{+}\text{D}_1\text{D}_2\text{S}_1 \underset{k_{-1}}{\overset{k_1}{\rightleftharpoons}} \text{Chl-a}_{II}\text{D}_1^{+}\text{D}_2\text{S}_1 \underset{k_{-2}}{\overset{k_2}{\rightleftharpoons}} \text{Chl-a}_{II}\text{D}_1\text{D}_2^{+}\text{S}_1 \underset{k_{-3}}{\overset{k_3}{\rightleftharpoons}} \text{Chl-a}_{II}\text{D}_1\text{D}_2\text{S}_2^{\oplus}$$

B

$$\text{Chl-a}_{II}^{+}\text{D}_1\text{D}_2\text{S}_2^{\oplus} \underset{k_{-1}}{\overset{k_1}{\rightleftharpoons}} \text{Chl-a}_{II}\text{D}_1^{+}\text{D}_2\text{S}_2^{\oplus} \underset{k_{-2}}{\overset{k_2}{\rightleftharpoons}} \text{Chl-a}_{II}\text{D}_1\text{D}_2^{+}\text{S}_2^{\oplus} \underset{k_{-3}}{\overset{k_3}{\overset{\text{H}^{+}}{\rightleftharpoons}}} \text{Chl-a}_{II}\text{D}_1\text{D}_2\text{S}_3^{\oplus}$$

C

$$\text{Chl-a}_{II}^{+}\text{D}_1\text{D}_2\text{S}_3^{\oplus} \underset{k_{-1}}{\overset{k_1}{\rightleftharpoons}} \text{Chl-a}_{II}\text{D}_1^{+}\text{D}_2\text{S}_3^{\oplus} \underset{k_{-2}}{\overset{k_2}{\rightleftharpoons}} \text{Chl-a}_{II}\text{D}_1\text{D}_2^{+}\text{S}_3^{\oplus} \underset{k_{-3} \; 2\text{H}_2\text{O}}{\overset{k_3 \; (\text{S}_4) \; O_2}{\rightleftharpoons}} \text{Chl-a}_{II}\text{D}_1\text{D}_2\text{S}_0 \; 2\text{H}^{+}$$

D

(Scheme II)

The k-values should depend on the number of the S states (see § 3) and the charges of these states, resp. (see § 4).

In the next paragraph and in Fig. 6 it is shown that D_1, the immediate donor to Chl-a$_{II}^+$, acts in the ns range. Recently, we have shown that with addition of acetate the re-reduction of Chl-a$_{II}^+$ can be reversibly retarded 10^4 times (42). If in this modified system D_1 is still the same, this opens the possibility for analyzing the chemistry of D_1 under extremely simplified conditions.

8. Kinetics of the components of the four electron transfer routes. The time course of Chl-a$_{II}$, D_1 and D_2 can be calculated according to scheme II (17). The solution of the corresponding differential equations is presented in Eqs. (1)-(8) (n = 0,1,2,3); $k_{-2} \ll k_2$. The resulting time courses for the redox reactions of Chl-a$_{II}$, D_1 and D_2 in the four electron transfer routes are illustrated in Fig. 6 (solid lines). Whether these calculations are correct can be checked by comparing the Chl-a$_{II}^+$-reduction kinetics in dependence on the flash number N (Fig. 3B & C, solid lines) with the calculated ones (Fig. 3B & C, broken lines). The latter are given by Eq. (9). The calculation in Fig. 3 fits the experimental values (17).

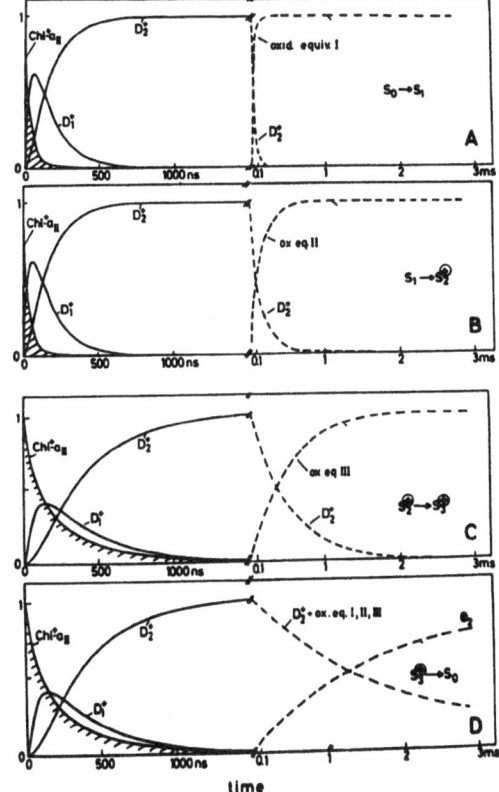

FIGURE 6. Time course of the redox reactions of the components of the four electron transfer routes and S states, resp.

$$\text{Chl-a}_{II}^{+}(t,S_n) = A_o \left\{ \underbrace{\frac{k_1 - r_-}{r_+ - r_-} e^{-r_+ t}}_{a_1} + \underbrace{\frac{r_+ - k_1}{r_+ - r_-} e^{-r_- \cdot t}}_{a_2} \right\} \quad (1)$$

$$D_1^{+}(t,S_n) = A_o \frac{k_1}{r_+ - r_-} \left\{ e^{-r_- t} - e^{-r_+ t} \right\} \quad (2)$$

$$D_2^{+}(t,S_n) = A_o \left\{ 1 - \frac{r_+}{r_+ - r_-} e^{-r_- t} + \frac{r_-}{r_+ - r_-} e^{-r_+ \cdot t} \right\} \quad (3)$$

for $k_3 \ll r_+$, r_-; (k_3 see scheme II); i.e., Eq. (3) describes the time course only up to the maximum D_2 oxidation. The re-reduction of D_2^{+} is outlined in paragraph 10.

$$A_o = \text{Chl-a}_{II} = D_1 = D_2 \text{ at } t = 0 \quad (4)$$

$$r_{\pm} = \frac{1}{2} \left\{ k_1 + k_{-1} + k_2 \pm \sqrt{(k_1 + k_{-1} + k_2)^2 - 4 \, k_1 \cdot k_2} \right\} \quad (5)$$

$$k_1 = a_1 \cdot r_+ + a_2 \, r_- \qquad\qquad k_2 = \frac{r_+ \cdot r_-}{k_1} \quad (6)(7)$$

$$k_{-1} = r_+ + r_- - k_1 - k_2 \quad (8)$$

$$\text{Chl-a}_{II}^{+}(t,N) = \sum_{n=0}^{3} \left[s_n \right]_N \text{Chl-a}_{II}^{+}(t,S_n) \quad (9)$$

it is $\sum_{n=0}^{3} \left[s_n \right]_N = 1$, where $\left[s_n \right]_N$ denotes the population of state S_n prior to the flash N. $\left[s_n \right]_N$ has to be taken from Fig. 3A & 3C. The dependence on S_n (see § 3) is taken into account by the following rate constant values gathered from the measurements in Fig. 3: for S_0 and S_1 it is $r_+ = \ln 2/23$ ns, $r_- = \ln 2/100$ ns, $a_1 = 0.95$ and $a_2 = 0.05$; for S_2 and S_3 it is $r_+ = \ln 2/50$ ns, $r_- = \ln 2/260$ ns, $a_1 = 0.5$ and $a_2 = 0.5$.

Before completing the kinetics by including also the final D_2^{+} reduction and the formation of the oxidizing equivalents, resp. (broken lines in Fig. 6), it will be shown that the retardation of the kinetics of Chl-a$_{II}^{+}$ re-reduction in states S_2 and S_3 can be explained quantitatively by the Coulomb forces of the surplus charges induced through an asymmetric pattern of H^{+} release (see § 4).

9. Coulomb force of surplus charges as mechanism for the rate dependence on the S states. To support the suggestion of a Coulomb mechanism one has to evaluate first the energetic difference due to the different S-dependent k-values. The characteristic difference should be located between the routes without (S_0 and S_1) and with the surplus charge (S_2 and S_3). According to scheme II these are, for example, routes B and C. They are presented in more detail in Fig. 7. It has to be shown that the Coulomb forces of the surplus charges can be made responsible for the energy difference calculated from the k-values.

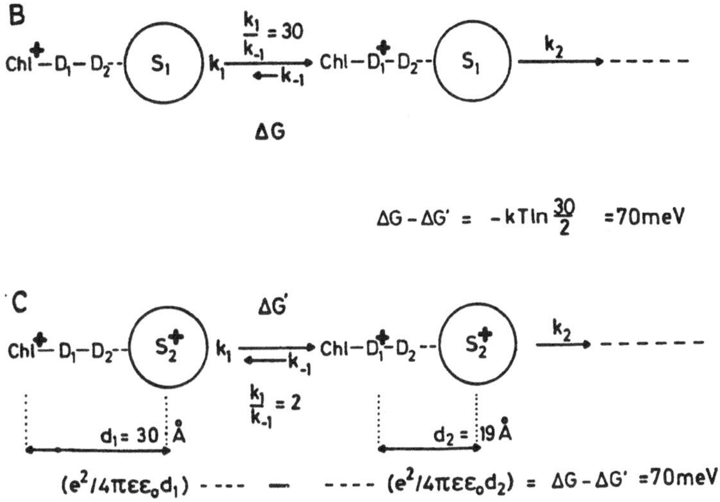

FIGURE 7. Energetic difference between the electron transfer routes without (B) and with (C) a surplus charge on the S states (see scheme II in § 7).

From the k-values in § 8 it follows that there is a change of the equilibrium constant, $k_1/k_{-1} = K$ from $K \approx 30$ in states S_0 and S_1 to $K \approx 2$ in states S_2 and S_3; that is, by a factor of 15. The corresponding difference of standard free-energy change, $G^o = -kT \ln K$ is therefore:

$$\Delta G^o_{S_{2,3}} - \Delta G^o_{S_{0,1}} = -kT \left(\ln \frac{k_1}{k_{-1}} (S_{2,3}) - \ln \frac{k_1}{k_{-1}} (S_{0,1}) \right) = -kT \ln \frac{30}{2} = 70 \text{ meV} \tag{10}$$

If this difference is due to the Coulomb energy of an elementary charge more positive in states S_2 and S_3 as in S_0 and S_1 (§ 4), this value can be estimated as follows (neglecting entropy changes). We have to calculate only the difference of the Coulomb energy due to the surplus charge in the S states before and after the electron transfer from D_1 to $Chl-a_{II}^+$. In states S_0 and S_1 this energy is zero (no surplus charge). Therefore, according to Fig. 7:

$$\Delta G^O_{S_{2,3}} - \Delta G^O_{S_{0,1}} = \frac{e^2}{4\pi\epsilon_0\,\epsilon}\left(\frac{1}{d_{D_1-S}} - \frac{1}{d_{Chl-a_{II}-S}}\right) - 0 = 70 \text{ meV} \quad (11)$$

d_{D_1-S} and $d_{Chl-a_{II}-S}$ are distances between D_1-S and $Chl-a_{II}-S$, respectively, With $\epsilon \approx 4$ and plausible distances, e.g., $d_{D_1-S} = 19$ Å and $d_{Chl-a_{II}-S} = 30$ Å (i.e., $d_{Chl-a_{II}-D_1} = 11$ Å we get 70 meV for the difference in Eq. (11); i.e., the Coulomb model satisfies the free-energy change evaluated from the kinetics and thermodynamics, resp., in Eq. (10).

10. Kinetics of the components of the S states. The cleavage of water is initiated by four electron extractions out of the reaction center of water. For each of these extractions the electron has to be moved step-by-step along the chain that connects the $Chl-a_{II}$ reaction center with that of water. Beginning with the transfer from D_1 to $Chl-a_{II}^+$, the subsequent transfer from D_1 to D_2^+ ends in the final one, the electron transfer from S to D_2^+. This final act should establish the oxidizing equivalent. Fig. 4D shows the threefold step-by-step oxidation of a substance, probably manganese, during the $S_0 \to S_1 \to S_2 \to S_3$ transition. Water may be unoxidized up to state (S_4)(see § 5). The four turnovers result in a "clock" of water cleavage as depicted in Fig. 1.
The time course of each of the four transitions is depicted in Fig. 6. The course of $Chl-a_{II}^+$, D_1^+ and the formation of D_2^+ (solid lines), i.e., the "traffic" of the defect electrons from $Chl-a_{II}^+$ to the S-system, is the result of an analysis of the $Chl-a_{II}$ kinetics in § 8. Within the S-system the final steps before the cleavage of water are the re-reduction of D_2^+, the formation of the oxidizing equivalents, the intrinsic H^+ release, and the variation of charges. The re-reduction times of D_2^+ (dotted lines) are only available for $S_2 \to S_3$ and $S_3 \to S_0$ from measurements by ESR (41)(see Table II). D_2^+ reduction times, however, should correspond to the oxidation times of the oxidizing equivalents. The latter have been measured through the time course of the absorption changes in the UV region in (43,44,47)(see Table II). The times of the intrinsic proton release and charge variation are as yet no available. The times for the extrinsic proton release (45) are slower or at least not faster than the times of the oxidizing equivalent (see Table II). In $S_3 \to S_0$, where the concerted final action takes place, all components react within one and the same time span of ca. 1.2 ms, where oxidation of water takes place with the simultaneous evolution of O_2.

Table II

Half-life times of events coupled with the S-state transitions

Half-Life Times	$S_o \longrightarrow S_1 \longrightarrow S_2 \longrightarrow S_3 \xrightarrow{(S_4)} S_o$				Lit
D_2^+ reduction	?	?	400 μs	1.2 ms	(41)
oxidation and reduction of the oxidizing equivalents	30 μs	110 μs	350 μs	1.2 ms	(43)
	50 μs	40 μs	80–150 μs	1.2 ms	(44)
	50 μs	100 μs	220 μs	1.2 ms	(47)
extrinsic H^+ release	250 μs	–	200 μs	1.2 ms	(45)

ACKNOWLEDGEMENTS
   This work was supported by grants from the Deutsche
Forschungsgemeinschaft (Sonderforschungsbereich 312, Teil-
projekt A3).

The results were presented - inter alia - at the 85. Meeting
of the Deutsche Bunsengesellschaft für Physikalische Chemie at
Heidelberg, May 10, 1986 (Ber. Bunsenges. f. Phys. Chemie, An
International Journal of Physical Chemistry, Vol. 90, No. 10,
October 1986).

REFERENCES
1. Döring, G., Stiehl, H.H. and Witt, H.T. (1967) Z. Natur-
    forsch. 22b, 639–644
2. Stiehl, H.H. and Witt, H.T. (1968) Z. Naturforsch. 23b,
    220–224; Stiehl, H.H. and Witt, H.T. (1969) Z. Naturforsch.
    24b, 1588–1598
3. Brettel, K., Schlodder, E. and Witt, H.T. (1985) Photo-
    biochem. Photobiophys. 9, 205–213
4. Kok, B. (1957) Acta Bot. Neerl. 6, 316–336
5. Junge, W. and Witt, H.T. (1968) Z. Naturforsch. 23b, 244–254
6. Witt, H.T. (1979) Biochim. Biophys. Acta 505, 355–427
7. Schliephake, W., Junge, W. and Witt, H.T. (1968) Z. Natur-
    forsch. 23b, 1571–1578
8. Mitchell, P. (1966) Biol. Rev. 41, 445–502
9. Schlodder, E., Gräber, P. and Witt, H.T. (1982) in: Elec-
    tron Transport and Photophosphorylation (J. Barber, ed.),
    Elsevier Biomedical Press, pp. 107–167
10. Kok, B., Forbush, B. and McGloin, M. (1970) Photochem.
    Photobiol. 11, 457–475
11. Joliot, P., Barbieri, G. and Chabaud, R. (1969) Photochem.
    Photobiol. 10, 309–329
12. Joliot, P., Kok, B. (1975) in: Bioenergetics of Photosyn-
    thesis (Govindjee, ed.), pp. 387–412, Academic Press New
    York

13. Döring, G., Renger, G., Vater, J. and Witt, H.T. (1969)
    Z. Naturforsch. 24b, 1139-1143
14. van Best, J.A. and Mathis, P. (1978) Biochim. Biophys. Acta
    503, 178-188
15. Brettel, K. and Witt, H.T. (1983) Photobiochem. Photobio-
    phys. 6, 253-260
16. Schlodder, E., Brettel, K., Schatz, G.H. and Witt, H.T.
    (1984) Biochim. Biophys. Acta 765, 178-185
17. Brettel, K., Schlodder, E. and Witt, H.T. (1984) Biochim.
    Biophys. Acta 766, 403-415
18. Schatz, G.H. and Witt, H.T. (1984) Photobiochem. Photobio-
    phys. 7, 77-89
19. Schatz, G.H. and Witt, H.T. (1984) Photobiochem. Photobio-
    phys. 7, 1-14
20. Klimov, V.V., Klevanik, A.V., Shuvalov, V.A. and Krasnovsky,
    A.A. (1977) FEBS Lett. 82, 183-186
21. Renger, G., Bouges-Bocquet, B. and Delosme, R. (1973)
    Biochim. Biophys. Acta 292, 796-807
22. Schlodder, E., Brettel, K. and Witt, H.T. (1985) Biochim.
    Biophys. Acta 808, 123-131
23. Saygin, Ö. and Witt, H.T. (1984) FEBS Lett. 176, 83-87
24. Saygin, Ö. and Witt, H.T. (1985) FEBS Lett. 187, 224-226
25. Fowler, C.F. (1977) Biochim. Biophys. Acta 462, 414-421
26. Saphon, S. and Crofts, A.R. (1977) Z. Naturforsch. 32c,
    617-626
27. Förster, V., Hong, Y.-Q. and Junge, W. (1981) Biochim.
    Biophys. Acta 638, 141-152
28. Dekker, J.P., van Gorkom, H.J., Wensink, J. and Ouwehand,L.
    (1984) Biochim. Biophys. Acta 764, 301-309
29. Dekker, J.P., van Gorkom, H.J., Wensink, J. and Ouwehand,L.
    (1984) Biochim. Biophys. Acta 767, 1-9
30. Velthuys, B.R. (1981) in: Photosynthesis II (Akoyunoglou,
    ed.), pp. 75-85, Balaban Intern. Science Serv., Philadel-
    phia
31. Lavergne, J. (1986) Photochem. Photobiol. 43, 311-317
32. Saygin, Ö. and Witt, H.T. (1985) Photochem. Photobiol. 10,
    71-82
33. Radmer, R. (1979) Biochim. Biophys. Acta 546, 418-425
34. Förster, V. and Junge, W. (1985) Photochem. Photobiol. 41,
    191-194
35. Klimov, V.V., Allakhverdiev, S.I., Shuvalov, V.A. and
    Krasnovsky, A.A. (1982) FEBS Lett. 148, 307-312
36. Yamamoto, Y., Nishimura, M. (1983) Biochim. Biophys. Acta
    724, 294-297
37. Blankenship, R.E., Babcock, G.T., Warden, J.T. and Sauer,K.
    (1975) FEBS Lett. 51, 287-293
38. Dekker, J.P., Brok, M. and van Gorkom, H.J. (1984) in:
    Advances in Photosynthesis Research (Sybesma, C., ed.),
    Vol. I, pp. 171-174, Martinus Nijhoff/Dr. W. Junk Publishers
    The Hague
39. Diner, B. and de Vitry, C. (1984) in: Advances in Photo-
    synthesis Research (Sybesma, C., ed.), Vol. I, pp. 407-411,
    Martinus Nijhoff/Dr. W. Junk Publishers, The Hague
40. Weiss, W. and Renger, G. (1984) FEBS Lett. 169, 219-223

41. Babcock, G.T., Blankenship, R.E. and Sauer, K. (1976)
    FEBS Lett. 61, 286-289
42. Saygin, Ö., Gerken, S., Meyer, B. and Witt, H.T. (1986)
    Photosynthesis Research 9, 71-78
43. Dekker, J.P., Plijter, J.J., Ouwehand, L. and van Gorkom,
    H.J. (1984) Biochim. Biophys. Acta 767, 176-179
44. Saygin, Ö. and Witt, H.T. (1986) in: Proceedings of the
    VII. International Congress of Photosynthesis, Martinus
    Nijhoff/Dr. W. Junk Publishers
45. Förster, V. and Junge, W. (1985) Photochem. Photobiol. 41,
    183-190
46. Andréasson, L.E., Hansson, Ö. and Vänngard, T. (1983)
    Chemica Scripta 21, 71-74
47. Renger, G. and Weiss, W. (1986) Biochemical Society
    Transactions, Vol. 14, pp. 17-20
48. Renger, G. and Weiss, W. (1986) Biochim. Biophys. Acta 850,
    184-196
49. Saywer, D.T., Bodini, M.E., Willis, L.A., Riechel, T.L.
    and Magers, K.D. (1977) in: Bioinorganic Chemistry
    (Raymond, K.N., ed.), Vol. 2, pp. 330-349, American
    Chemical Society, Washington, D.C.
50. Murata, N., Miyao, M., Omatoa, T., Matsunami, H. and
    Kuwabara, T. (1984) Biochim. Biophys. Acta 765, 363-369
51. Boucher, L.J. and Coe, C.G. (1975) Inorg. Chem. 14,
    1289-1294
52. Wieghardt, K., Bossek, U. and Gebert, W. (1983) Angew.
    Chem. Int. Ed. Engl. 22, 328-329
53. Brudvig, G.W. and Crabtree, R.H. (1986) Proc. Natl. Acad.
    Sci. USA 83, 4586-4588

Photosynthesis Research 10: 473–481 (1986)
© *Martinus Nijhoff Publishers, Dordrecht*

MODIFICATION OF OXYGEN EVOLVING CENTER BY TRIS-WASHING

T. YAMASHITA

Institute of Biological Sciences, Tsukuba University
Sakura-mura, Ibaraki 305, Japan

ABSTRACT. Tris-washing inhibits the O_2-evolving center of chloroplasts and their particles specifically and reversibly, and it was applied to many investigations on O_2-evolving center and PS II reaction center. In this review are introduced the various photosynthetic investigations in which Tris-washing was applied and are also discussed briefly on the site and the mechanism of Tris-inactivation, properties of P680 and Z, characteristic change in fluorescence and delayed light emission, and reactivation of O_2-evolving center by DCPIP·H_2-treatment and photo-reactivation of Tris-washed chloroplasts and their particles.

I. INTRODUCTION
Tris-washing inactivates the O_2-evolving center specifically and reversibly [1-17]. The inactivated chloroplasts and particles could be reactivated by Mn^{2+}-addition [20-21], DCPIP·H_2-treatment [12, 17, 22-31] and photo-reactivation [26-27, 29, 31-36]. 0.8 M Tris-buffer at pH 8.0 was usually used for the inactivation, however, 0.2 M buffers at pH 8.0 [38] and at pH 10.0 [59] could also be used. Chloroplasts were uncoupled by Tris-washing at pH 7.2 [104]. Scince the discovery of Tris-inactivating effect on chloroplasts [1], Tris-washed chloroplasts and their particles were the good subject to study (A). light-induced electron transport in the presence of artificial electron donors such as TCPIP·H_2 [1], DCPIP·H_2 [3-4, 123], phenylenediamine [4, 9, 37, 59, 64-65, 81, 83, 86, 91, 113-114], hydroquinone [8, 10, 22, 37-38, 51-52, 81, 83, 86, 115-116], benzidine [8, 64, 86, 113-114], semicarbazide [8, 113], diphenylcarbazide [38-44, 54], ascorbate [9, 37, 79, 86-87], Mn^{2+} [10, 50, 52, 54, 64, 83-84], H_2O_2 [46-48], hydrazobenzene [44] and tetraphenylboron [45, 54], (B). variable fluorescence [4-5, 8, 10, 12, 22, 33, 46, 50-61, 123] and delayed light emission (delayed fluorescence, luminescence) [27, 52, 60-65], (C). modification of Mn-binding site, Mn-extraction and reconstruction [5, 9-12, 15-36], (D). EPR-multisignal of polynuclear manganese in S_2-state [15, 66], (E). effect of chloroplast Mn on the proton spin-lattice relaxation rate [67-69], (F). signals of light-induced oxidation and reduction of P680 [70-77], D_1 [78], Z [61, 79-96], C550 [28, 111-112], cytochrome b_{559} [28, 31, 111-114], X320 [115], X591 [28], A475 [46], A515 [46, 72, 107] and A520 [116], (G). peptides of O_2-evolving center [6, 17, 97-103], (H). uncoupling of photophosphorylation [104-108], (I), modification of cytochrome b_{559} [31, 110] and thylakoid membranes [109, 117-118], (J). effect of Mg^{2+} on the energy transfer from PS II to PS I [123]. These results, accumulated in many years , must be understood by referring to the recent improved informations.

II. THE SITE AND THE MECHANISM OF TRIS-INACTIVATION

The study of Nakamoto et al [1] in FMN and O_2-dependent chloroplast photophosphorylation was the first torch to have thrown a light upon the profile
of PS II specific Tris-inactivation. Addition of (TCPIP + NADH) system at
aerobic condition or addition of PMS at anaerobic condition restored the
photosynthetic ATP formation of Tris-inactivated or DCMU-poisoned chloroplasts. The site of Tris-inactivation was estimated to be on the oxidizing
side of PS II and distinguished from the site of inhibition by DCMU or o-
phenanthlorine [3]. This was known from the fact that the addition of
these inhibitors decreased the activity of non-cyclic photophosphorylation
and the $P/2e^-$ ratio of Tris-washed and $DCPIP \cdot H_2$ donated grana to about a
half. However, it was the study of variable chlorophyll fluorescence under
the aid and the suggestion of W. L. Butler that provided the definite evidence to restrict the inactivation site on the oxidizing side of PS II [4,
8]. The variable fluorescence, an indicator of the photoreduction of the
primary electron acceptor of PS II [118], was decreased by Tris-washing of
chloroplasts and recovered by adding DCMU or PS II-specific electron donors.
Later investigations and data identified the Tris-inactivation site at the
Mn-binding site of PS II [5, 9-12, 26, 32-33] in the S_2-state [12-13] or on
the O_2-evolving side over Z (D_1) [78, 82, 85]. By a careful study of the
flash induced oxidation of cytochrome b_{559} and X591 [28], Tris-washing was
found to have another effect to inhibit the electron transport between
PS II_b and PS II_a and to accelerate the electron transport on the reducing
side of PS II_a. Tris-washing at pH 7.2 was not inhibitory for the O_2-evolu-
tion but uncoupled the photophosphorylation [104-108]. This uncoupling was
removed by using 0.8 M Tris-HPO_4 buffer instead of 0.8 M Tris-Cl buffer in
the Tris-washing [106] or by adding 0.1 mM ATP or ADP in the Tris-washing
[108].

Many aspects had been pointed out in explaining the mechanism of Tris-
inactivation. Aged Tris-buffer yielded formaldehyde and inhibited chloroplasts [2]. Tris has a primary amino group which become non-ionic form
($R-NH_2$) at higher pHs and inhibit chloroplasts [7-8, 11]. Inactivation of
chloroplasts by Tris-washing was stimulated by illumination (1000 lux),
especially at pH lower than 8.0 [7]. Single and double flashes were more
effective than triple flashes in stimulating the inactivation. Red light
flashes (650 nm) were 4.2 fold effective than far red light flashes (720 nm
) [12-13]. These results suggested that Tris-NH_2 inhibited the S_2-state of
O_2-evolving center specifically by binding to it. This view was based on
the report that unprotonated base NH_3 binded rapidly to S_2-state by stabilizing the positive charge of S_2 [62]. Tris-inactivated chloroplasts could
be reactivated by $DCPIP \cdot H_2$-treatment (incubating chloroplasts with DCPIP +
ascorbate) [12, 17, 22-31], or incubating the inhibited chloroplasts with
ADRY reagents (FCCP, ANT-2p) or tetraphenylboron [12]. These reagents
would reduce the S_2-Tris complex to give rise S_1 + Tris and then the S_{1-4}
cycle of O_2-evolving center recovered [12]. Tris-washing often decreased
chloroplast Mn content, suggesting that Tris attacked the Mn-site of O_2-
evolving center. In the early studies [5, 9], the Mn content of Tris-washed chloroplasts was reported to be about one third of the Mn of untreated
chloroplasts. Later studies found that they kept more than a half and some
times over 85% of Mn of untreated chloroplasts [11. 16, 18, 23, 25-26, 29-
31]. Considering that Tris-washed chloroplasts could be reactivated by
$DCPIP \cdot H_2$-treatment alone [12, 17, 22-31], the O_2-evolving center should
still have kept necessary Mn in it even after the Tris-washing. This
variety in Mn content would be reflecting the toughness of thylakoid membrane of each investigators' preparations against the damaging in Tris-
washing. Chloroplast Mn content was further decreased by damaging thyla-

koid membrane through Tris-washing and sonication [11], Tris-acetone-wash-
ing [32] and Tris-washing at pH 8.8 [26, 33]. The Mn content variation
could be understood by detecting the increase of EPR hexaquo-Mn^{2+}-signal in
Tris-washed chloroplasts. The 6-line signal of Mn^{2+} appeared by releasing
Mn from O$_2$-evolving center by Tris-washing and disappeared by irradiating
chloroplasts with PS II light [10]. This hexaquo-Mn^{2+} accumulated in the
interior space of thylakoid by Tris-washing and diffused out slowly in 2.5
hr a half) [11]. Reactivation of Tris-inactivated chloroplasts extinguish-
ed the EPR-signal, suggesting the rebinding of Mn^{2+} to the O$_2$-evolving
center [25]. Untreated chloroplasts contained a large amount of unfunctional Mn from which EPR-Mn^{2+}-signal arise when chloroplasts were Tris-
washed. After removal of this unfunctional Mn^{2+} by salt/EDTA-washing,
Tris-washing of chloroplasts released only one hexaquo Mn^{2+} from a unit of
O$_2$-evolving center [16]. S$_2$-state specific EPR-signal was found by cooling
chloroplasts and PS II membrane to the liquid helium temperature [15]. This
signal appeared as the hyperfine multilines and was similar to the signal
of Mn(III)-Mn(IV) bipyridine compound. Tris-washing of chloroplasts extin-
guished this signal [15, 66] and g=4.1 signal [66].
    A peptide like substance, named Oxygen Evolution Factor was extracted by
sonicating grana in 0.8 M Tris-buffer (pH 8.0) [6]. This factor increased
the oxygen evolving activity of 0.05% Triton X-100 treated and sonicated
grana by three fold, but this could not reactivate the Tris-washed grana.
This pioneering work of Huzisige et al suggested that Tris-washing released
some active peptide from O$_2$-evolving center. Recent studies identified the
release of three active peptides (16, 24, 33 KDa) by Tris-washing of PS II
particles [17, 97-103]. However, no peptides were released by Tris-washing
of chloroplasts [14]. Tris-washing released Cu^{2+} from chloroplasts and
formed Tris-Cu^{2+} complex which oxidized thiol group of thylakoid peptides.
This thiol oxidation was stopped by adding the reductant DCPIP·H$_2$ [14].

III. PHOTOOXIDATION AND DARK-REDUCTION OF P680 AND Z

    Tris-inactivation of chloroplasts reduced the decay speed of flash-induc-
ed signals of P680$^+$ [70-77], D$_1^+$ [78] and Z$^+$ [61, 79-96] from nsec to 2-200
usec range. Slowing of these signal decays was useful to detect and analyze the primary photochemical reactions of PS II. The P680$^+$ signal of
Tris-washed chloroplasts was first observed by measuring the absorption
change at 680 nm [70], but later it was detected by observing the DCMU-sen-
sitive absorption change at 820 nm in the presence of ferricyanide and a
far-red background light to keep P700$^+$ in the oxidized state [71-72]. In
Tris-washed chloroplasts, the life time of single flash-induced P680$^+$ was
3.5 μsec at pH 8.0. The life time became longer by lowering pH of reaction
mixture (44 μsec at pH 4.0) suggesting that the electron donor D$_1$ (Z) was a
pH sensitive proton carrier and reduced P680$^+$ faster at higher pH [73-74].
When successive double flashes were given to Tris-washed chloroplasts,
electrons in the oxidizing side of PS II were all exhausted and P680$^+$ was
reduced mostly by the PS II primary acceptor (PQ$_1$) in a speed about 200 μsec
[73-74]. The difference between single and double flashed decay kinetics
would indicate one electron capacity in D$_1$ or Z. By exciting Tris-washed
and inside-out thylakoids with repetitive flashes in the presence of pH
indicator bromocresol purple, D$_1$ was found to bind with H$^+$ in the reduced
state and to become deprotonized in the oxidized state because photooxidized amount of D$_1$ was proportional to the amount of H$^+$ released [75]. The
amplitude of the fast transient component of P680$^+$ (t$_{\frac{1}{2}}$=42 μsec) in Tris-
extracted PS II particles decreased by lowering the redox potential of the
medium (Em'$\approx$ 240 mV), although the transient time was not affected by the
redox potential change [77]. These results suggested the presence of elec-

tron donor Do other than D_1, possibly cytochrome b_{559}, which could reduce $P680^+$ in less than 2 μsec [77]. By comparing the effect of pH changes on the decay time of $P680^+$ (EPR-signal at g=2.003) and the rise time of EPR-signal II_f, Z was identified as D_1 [76]. The absorbance difference spectra between Z^+ and Z (peaks at 260, 300 and 390-450 nm) was measured by subtracting both the spectra of primary acceptor Q^- and ferrocyanide from the flash-induced spectra (50 msec fraction) of Tris-washed PS II particles [61] . A similar result was also obtained by measuring the spectrae change of Tris-washed chloroplasts 3.5 msec after the flash (peaks at 265-270, 300-305 and 350 nm) [78].

The red light-induced EPR signal II of untreated chloroplasts consists of two fractions; a very slow decaying signal II_s ($t_{\frac{1}{2}} \approx 1$ hr) and a fast decaying signal II_{vf} ($t_{\frac{1}{2}}$ =700 μsec) [84]. In Tris-washed chloroplasts, a fast decaying signal II_f ($t_{\frac{1}{2}}$ =490 msec) substituted the signal II_{vf} [81, 84]. Lozier and Butler [79] explained the fast decaying signal II as arising from the photooxidation of PS II species by red light because the decay was further accelerated by adding ascorbate. EPR signal II_f was found to arise from Z^+, an oxidized hypothetical donor to $P680^+$, as signal II_f had a midpoint potential more than 480 mV and addition of DCMU accelerated the decay of the signal by stimulating the back reaction of reduced primary acceptor Q^- with Z^+ [82], similar to delayed light emission of Tris-washed chloroplasts [62]. Hydroquinone and phenylenediamine, the membrane permeable lipophilic PS II electron donor, reduced Z^+ on the inner surface of the thylakoids and shortened the decay of signal II_f. Hydrophilic Mn^{2+} reduced $P680^+$ on the outer surface of thylakoid and decreased the signal II_f magnitude [83]. The effect of other electron donors on the signal was also reported [86]. Addition of phenylenediamine also reduced the decay time ($t_{\frac{1}{2}}$) of signal II_s or D^+ (the oxidized form of electron donor to Z^+) in Tris-washed chloroplasts at pH 8.5 from 600 to 30 sec [90]. It was after the dosage of successive five flashes that the signal II_s intensity attained to the maximal and the signal II_s rising after each flash decreased to zero and that the amount of a flash-induced phenylenediamine oxidation increased to the maximal level. Based on these results Boussac and Etienne proposed a mechanism that D competed with phenylenediamine in reducing Z^+, in an equilibrium $D^+-Z \rightleftarrows D-Z^+$, and that phenylenediamine also reduced D^+ in the dark [91]. The standard redox potential of D^+/D was determined to +760 mV by titrating the signal II_s with K_2IrCl_6. As the equilibrium constant of $D^+-Z/D-Z^+$ was $K=10^4$, the standard redox potential Z^+/Z of Tris-washed chloroplasts was estimated to be 1 V [95]. The standard redox potential Z^+/Z of Tris-washed chloroplasts was found to be shifted to the lower region by at least 120 mV [94]. This shift was estimated by comparing the equilibrium constant K_1 of $P680^+-Z \rightleftarrows P680-Z^+$ reaction of Tris-washed chloroplasts with the constant K_1 of O_2-evolving chloroplasts. This constant K_1 was obtained from the decay curve of signal II, the rate of the back reactions $Z^+-P680-Q_a^- \rightleftarrows Z-P680^+-Q_a^- \rightarrow Z-P680-Q_a$. The power of microwave required for detecting Z^+ of Tris-washed chloroplasts (signal II_f and II_s) saturated at about 25 mW [80, 85, 88, 95]. However it did not saturate even at 200 mW for the signal II_{vf} of O_2-evolving chloroplasts [85] and the signal II_f of NH_3-inhibited thylakoid membranes [88]. These results suggested that the release of the bound Mn through Tris-washing eliminated the paramagnetic interactions between Z^+ and Mn, and resulted in a much easier saturable transient radical [85, 88]. In O_2-evolving PS II particles, flash-induced EPR-signal II_f could be observed with Tris-washed or 33 KD_a peptide-removed particles. Addition of 33 KD_a peptide to 1 M $MgCl_2$-washed particles decreased the magnitude of signal II_f although the recovery of O_2-evolution was small (4%) [96].

IV. VARIABLE CHLOROPHYLL FLUORESCENCE AND DELAYED LIGHT EMISSION

The intensity of variable chlorophyll fluorescence and delayed light emission (luminescence, delayed fluorescence) of Tris-washed chloroplasts was changed to a large extent by adding reductants, oxidants and inhibitors. It has been used as a good indicator in studying the redox reactions of PS II reaction center. Regarding to the variable fluorescence of Tris-washed chloroplasts, Butler [56] stated, that under steady irradiation the fluorescence yield of Tris-washed chloroplasts increased very little because there were not sufficient endogenous electron donors to reduce the pools of redox compounds (PQ) which could oxidize C550 (Q), and that in the presence of DCMU, which blocked electron transport immediately after C550 (Q), the fluorescence yield did increase to its maximal level during irradiation showing that endogenous donor was able to reduce C550 (Q). The other quencher "P680$^+$", whose effect was noticed by Okayama and Butler [120], Butler [121] and Butler et al [122], also decreased the variable fluorescence yield of Tris-washed chloroplasts excited by repetitive flashes in the rate of $\geq$ 1 HZ [64]. In this condition, Z$^+$-P680$^+$-Q$^-$ was rapidly built up and the fluorescence was quenched even in the presence of Q$^-$. Addition of Mn^{2+}, a P680$^+$ specific electron donor, increased the fluorescence [64]. In the presence of phenylenediamine, the dithionite-induced increase in fluorescence of Tris-washed and dark-adapted chloroplasts was larger after irradiating the chloroplasts by an odd number flashes (Q-R + $h\nu$ → Q-R$^-$, Q-R$^-$ + 2e$^-$(dithionite) → Q$^-$-R^{2-}) than the increase after the even number flashes (Q-R + 2$h\nu$ → Q-R^{2-}, Q-R^{2-} + PQ → Q-R + PQ^{2-}, Q-R + 2e$^-$(dithionite) → Q-R^{2-}) [59]. However, in the absence of phenylenediamine, the dithionite-induced fluorescence increased only once after the first flash to the low level of fluorescence which was observed after even number flashes in the presence of phenylenediamine, and no more influenced by the following flashes [59]. These results suggested that in the absence of electron donors, Tris-washed and dark-adapted chloroplasts had been keeping one electron in the oxidizing side of PS II which was movable to Q by the flash excitation and that even after multiple flashes, only one positive charge could be stored at the oxidizing side of PS II.

Tris-washing and addition of DCMU to chloroplasts favored in accumulating P680$^+$ and Q$^-$, and stimulated the initial rapid rise component of delayed light emission. Addition of PS II electron donors to the system depressed the emission by reducing P680$^+$ [52, 62-63]. Addition of 50 mM NH$_4$Cl decreased the delayed light emission rate of untreated and DCMU-poisoned chloroplasts, but it was much less effective in decreasing the emission of Tris-washed and DCMU-poisoned chloroplasts. Velthuys suggested that NH$_3$ bound to untreated chloroplasts and stabilized the charge separation of S$_2$-state but that NH$_3$ was unable to bind to Tris-washed chloroplasts [62]. Later on, binding of Tris to S$_2$-state was demonstrated by stimulating the Tris-inactivation with flash light [12].

V. PHOTO-REACTIVATION (LIGHT-REACTIVATION) OF TRIS-WASHED CHLOROPLASTS

Washing of chloroplasts with 0.8 M Tris-20% acetone buffer (pH 8.4) or 0.8 M Tris-buffer (pH 8.8) inactivated O$_2$-evolving center and released most of their Mn [26]. These Mn-depleted chloroplasts could be reactivated by DCPIP·H$_2$-treatment and subsequent Mn-incorporating process called "photo-reactivation (light-reactivation)". This was obtained by incubating the chloroplasts with Mn^{2+}, Ca^{2+}, Cl$^-$ and DTT under weak light (about 300 lux) for 20 min [26-27, 29, 32-36]. Addition of uncouplers [26, 32-33, 35] inhibited the photo-reactivation. Action spectra of photo-reactivation identified the light-absorbing pigments to be chlorophylls and carotenoids [33].

Flash-analysis of photo-reactivation revealed that more than two sequential light-reactions with dark-reaction in between was necessary to recover the activity [34]. Chloroplasts [33] and grana [36] increased Mn-content after the photo-reactivation. The high potential form of cytochrome b_{559} (cytochrome b_{559HP}) of untreated chloroplasts was changed to its low potential form (cytochrome b_{559LP}) by Tris-washing and returned back to the original high potential form (cytochrome b_{559HP}) by incubating the Tris-washed chloroplasts in DCPIP·H_2 and then exposing them to room light [31].

ACKNOWLEDGEMENT
The author wishes to thank Prof. Govindjee of Univ. Illinois, Urbana-Champaign for his generosity in admitting the modification of the title of this review, Prof. J. H. C. Goedheer of Rijks Univ., and Prof. K. Yanagisawa and Miss T. Sakuta of Tsukuba Univ. for reading this manuscript.

REFERENCES
1. Nakamoto T, Krogmann DW and Vennesland B (1959) J Biol Chem 234: 2783-2788
2. Jacobi G (1961) Naturwiss 48: 577
3. Yamashita T and Horio T (1968) Plant & Cell Physiol 9: 267-287
4. Yamashita T and Butler WL (1968) Plant Physiol 43:1978-1986
5. Homann PH (1968) Biochem Biophys Res Commun 33: 229-234
6. Huzisige H, Isimoto M and Inoue H (1968) In: Comparative Biochemistry and Biophysics of Photosynthesis (Shibata K, Takamiya A, Jagendorf AT and Fuller RC eds.) pp. 170-178, Univ. Park Press, Pennsylvania/Univ Tokyo Press, Tokyo
7. Ikehara N and Sugahara K (1969) Bot Mag Tokyo 82: 271-277
8. Yamashita T and Butler WL (1969) Plant Physiol 44: 435-438
9. Cheniae GM and Martin IF (1970) Biochim Biophys Acta 197: 219-239
10. Lozier R, Baginsky M and Butler WL (1971) Photochem Photobiol 14: 323-328
11. Blankenship RE and Sauer K (1974) Biochim Biophys Acta 357: 262-266
12. Cheniae GM and Martin IF (1978) Biochim Biophys Acta 502: 321-344
13. Frasch WD and Cheniae GM (1980) Plant Physiol 65: 735-745
14. Takahashi M, Takano M and Asada K (1981) J Biochem 90: 87-94
15. Dismukes GC and Siderer Y (1981) Proc Natl Acad Sci US 78: 274-278
16. Yocum CF, Yerkes CT, Blankenship RE, Sharp RR and Babcock GT (1981) Proc Natl Acad Sci US 78: 7507-7511
17. Henry LA, Møller BL, Andersson B and Åkerlund H-E (1982) Carlsberg Res Commun 47: 187-198
18. Selman BR, Bannister TT and Dilley RA (1973) Biochim Biophys Acta 292: 566-581
19. Yamamoto Y and Nishimura M (1983) Biochim Biophys Acta 724: 294-297
20. Yamashita K, Itoh M and Shibata K (1969) Biochim Biophys Acta 189: 133-135
21. Klimov VV, Allakhverdiev SI, Shuvalov VA and Krasnovsky AA (1982) FEBS Lett 148: 307-321
22. Yamashita T, Tsuji J and Tomita G (1971) Plant & Cell Physiol 12: 117-126
23. Yamashita T, Tsuji-Kaneko J, Yamada Y and Tomita G (1972) Plant & Cell Physiol 13: 353-364
24. Huzisige H and Yamamoto Y (1972) Plant & Cell Physiol 13: 477-491
25. Blankenship RE, Babcock GT and Sauer K (1975) Biochim Biophys Acta 387: 165-175
26. Yamashita T and Tomita G (1975) Plant & Cell Physiol 16: 283-296
27. Inoue Y, Yamashita T, Kobayashi Y and shibata K (1977) FEBS Lett 82:

303-306
28. Huzisige H, Doi M and Natuga T (1979) Plant & Cell Physiol 20: 935-946
29. Barber J, Nakatani HY and Mansfield R (1981) Israel J Chem 21: 243-249
30. Mansfield R and Barber J (1982) FEBS Lett 140: 165-168
31. Butler WL and Matsuda H (1983) In: The Oxygen Evolving System of photosynthesis (Inoue Y, Crofts AR, Govindjee, Murata N, Renger G and Satoh K eds.) pp. 113-122, Academic Press, Tokyo
32. Yamashita T and Tomita G (1974) Plant & Cell Physiol 15: 69-82
33. Yamashita T and Tomita G (1976) Plant & Cell Physiol 17: 571-582
34. Yamashita T, Inoue Y, Kobayashi Y and Shibata K (1978) Plant & Cell Physiol 19: 895-900
35. Yamashita T (1982) Plant & Cell Physiol 23: 833-841
36. Yamashita T and Ashizawa A (1985) Arch Biochem Biophys 283: 549-557
37. Ben-Hayyim G and Avron M (1970) Eur J Biochem 15: 155-160
38. Vernon LP and Shaw ER (1969) Plant Physiol 44: 1645-1649
39. Brandon PC and Elgersma O (1973) Biochim Biophys Acta 292: 753-762
40. Reitz G and Ohad I (1976) J Biol Chem 251: 247-252
41. Tischer W and Strotmann H (1979) Z Naturforsch 34c: 992-995
42. Packham NK and Barber J (1983) Biochim Biophys Acta 723: 247-255
43. Barr R, Troxel KS and Crane FL (1983) Plant Physiol 73: 301-315
44. Haveman J, Duysens LNM, Van der Geest TCM and Van Gorkom HJ (1972) Biochim Biophys Acta 283: 316-327
45. Erixon K and Renger G (1974) Biochim Biophys Acta 333: 95-106
46. Inoué H, Wakamatsu K and Nishimura M (1971) Plant & Cell Physiol 12: 457-460
47. Inoué H and Nishimura M (1971) Plant & Cell Physiol 12: 739-747
48. Verthuys B and Kok B (1978) Biochim Biophys Acta 502: 211-221
49. Wydrzynski T, Huggins BJ and Jursinic PA (1985) Biochim Biophys Acta 809: 125-136
50. Itoh M, Yamashita K, Nishi T, Konishi K and Shibata K (1969) Biochim Biophys Acta 180: 509-519
51. Yamashita T and Butler WL (1969) Plant Physiol 44: 1344-1346
52. Itoh S, Katoh S and Takamiya A (1971) Biochim Biophys Acta 245: 121-128
53. Homann PH (1971) Biochim Biophys Acta 245: 129-143
54. Homann PH (1972) Biochim Biophys Acta 256: 336-344
55. Rosenberg JL, Sahu S and Bigat TK (1972) Biophys J 12: 830-850
56. Butler WL (1972) Biophys J 12: 851-857
57. Mohanty P, Braun BZ and Govindjee (1972) FEBS Lett 20: 273-276
58. Huzisige H and Yamamoto Y (1973) Plant & Cell Physiol 14: 953-963
59. Verthuys BR and Amesz J (1974) Biochim Biophys Acta 333: 85-94
60. Bowes JM, Crofts AR and Itoh S (1979) Biochim Biophys Acta 547: 320-335
61. Dekker JP, Van Gorkom HJ, Brok M and Ouwehand L (1984) Biochim Biophys Acta 746: 301-309
62. Velthuys BR (1975) Biochim Biophys Acta 396: 392-401
63. Haveman J and Lavorel J (1975) Biochim Biophys Acta 408: 269-283
64. Jursinic P and Govindjee (1977) Biochim Biophys Acta 461: 253-267
65. Itoh S (1980) Plant & Cell Physiol 21: 885-895
66. Zimmermann JL and Rutherford AW (1984) Biochim Biophys Acta 767: 160-167
67. Wydrzynski T, Zumbulyadis N, Schmidt PG and Govindjee (1975) Biochim Biophys Acta 408: 349-354
68. Wydrzynski TJ, Marks SB, Schmidt PG, Govindjee and Yocum CF (1978) Biochem 17: 2155-2162
69. Sharp RR and Yocum CF (1980) Biochim Biophys Acta 592: 185-195

70. Govindjee, Döring G and Govindjee R (1970) Biochim Biophys Acta 205: 303-306
71. Haveman J and Mathis P (1976) Biochim Biophys Acta 440: 346-355
72. Conjeaud H, Mathis P and Paillotin G (1979) Biochim Biophys Acta 546: 280-291
73. Conjeaud H and Mathis P (1980) Biochim Biophys Acta 590: 353-359
74. Reinman S, Mathis P, Conjeaud H and Stewart A (1981) Biochim Biophys Acta 635: 429-433
75. Renger G and Voelker M (1982) FEBS Lett 149: 230-207
76. Boska M, Sauer K, Buttner W and Babcock GT (1983) Biochim Biophys Acta 722: 327-330
77. Golbeck JH and Warden JT (1985) Biochim Biophys Acta 806: 116-123
78. Weiss W and Renger G (1984) FEBS Lett 169: 219-223
79. Lozier RH and Butler WL (1973) Photochem Photobiol 17: 133-137
80. Warden JT and Bolton JR (1974) Photochem Photobiol 20: 245-250
81. Babcock GT and Sauer K (1975) Biochim Biophys Acta 376: 315-328
82. Babcock GT and Sauer K (1975) Biochim Biophys Acta 376: 329-344
83. Babcock GT and Sauer K (1975) Biochim Biophys Acta 396: 48-62
84. Blankenship RE, Babcock GT, Warden JT and Sauer K (1975) FEBS Lett 51: 287-293
85. Warden JT, Blankenship RE and Sauer K (1976) Biochim Biophys Acta 423: 462-478
86. Yerkes CT and Babcock GT (1980) Biochim Biophys Acta 590: 360-372
87. Yerkes CT and Babcock GT (1981) Biochim Biophys Acta 634: 19-29
88. Yocum CF and Babcock GT (1981) FEBS Lett 130: 99-102
89. Berthold DA, Babcock GT and Yocum CF (1981) FEBS Lett 134: 231-234
90. Boussac A and Etienne AL (1982) FEBS Lett 148: 113-116
91. Boussac A and Etienne AL (1982) Biochem Biophys Res Commun 109: 1200-1205
92. Boska M, Sauer K, Buttner W and Babcock GT (1983) Biochim Biophys Acta 722: 327-330
93. Chanotakis DF and Babcock GT (1983) FEBS Lett 153: 231-234
94. Yerkes CT, Babcock GT and Crofts AR (1983) FEBS Lett 158: 359-363
95. Boussac A and Etienne AL (1984) Biochim Biophys Acta 766: 576-581
96. Franzén L-G, Hansson Ö and Andréasson L-E (1985) Biochim Biophys Acta 808: 171-179
97. Åkerlund H-E and Jansson C (1981) FEBS Lett 124: 229-232
98. Yamamoto Y, Doi M, Tamura N and Nishimura M (1981) FEBS Lett 133: 265-268
99. Kuwabara T and Murata N (1982) Plant & Cell Physiol 23: 663-667
100. Yamamoto Y, Shimada S and Nishimura M (1983) FEBS Lett 151: 49-53
101. Ono T-A and Inoue Y (1983) FEBS Lett 164: 255-260
102. Murata N, Miyao M, Omata T, Matsunami H and Kuwabara T (1984) Biochim Biophys Acta 765: 363-369
103. England RR and Evans EH (1985) Biochim Biophys Acta 808: 323-327
104. Ikehara N and Uribe EG (1970) FEBS Lett 9: 321-323
105. Ikehara N and Uribe EG (1971) Arch Biochem Biophys 147: 717-727
106. Ikehara N and Nishimura M (1973) Plant & Cell Physiol 14: 61-75
107. Ikehara N and Nishimura M (1973) Plant & Cell Physiol 14: 77-90
108. Ikehara N (1974) Plant & Cell Physiol 15: 943-947
109. Takahashi M and Asada K (1985) Plant & Cell Physiol 26: 1093-1100
110. Erixon K, Lozier R and Butler WL (1972) Biochim Biophys Acta 267: 375-382
111. Knaff DB and Arnon DI (1969) Proc Natl Acad Sci US 63: 956-962
112. Knaff DB and Arnon DL (1969) Proc Natl Acad Sci US 64: 715-722

113. Knaff DB and Arnon DI (1970) Biochim Biophys Acta 223: 201-204
114. Arnon DI, Knaff DB, McSwain DB, Chain RK and Tsujimoto HY (1971) Photochem Photobiol 14: 397-425
115. Renger G and Wolff CH (1976) Biochim Biophys Acta 423: 610-614
116. Renger G (1979) Biochim Biophys Acta 547: 103-116
117. Kobayashi Y, Inoue Y and Shibata K (1976) Biochim Biophys Acta 440: 600-608
118. Stemler A (1977) Biochim Biophys Acta 460: 511-522
119. Duysens LNM and Sweers HE (1963) In: Studies on Microalgae and Photosynthetic Bacteria (Japanese Society of Plant Physiologists eds.) Special Issure of Plant & Cell Physiol, pp. 353-372, Univ of Tokyo Press, Tokyo
120. Okayama S and Butler WL (1972) Biochim Biophys Acta 267: 523-529
121. Butler WL (1972) Proc Natl Acad Sci US 69: 3420-3422
122. Butler WL, Visser JWM and Simons HL (1973) Biochim Biophys Acta 292: 140-151
123. Satoh K, Strasser R and Butler WL (1976) Biochim Biophys Acta 440: 337-345

Photosynthesis Research 10: 483–488 (1986)
© *Martinus Nijhoff Publishers, Dordrecht*

CHARACTERIZATION OF A PHOTOSYSTEM II REACTION CENTER COMPLEX
ISOLATED BY EXPOSURE OF PSII MEMBRANES TO A NON-IONIC
DETERGENT AND HIGH CONCENTRATIONS OF NaCl[1]

DEMETRIOS F. GHANOTAKIS and CHARLES F. YOCUM

Division of Biological Sciences, The University of Michigan,
Ann Arbor, MI 48109-1048 USA

ABSTRACT

A highly resolved PSII reaction center complex has been
prepared by exposure of PSII membranes to the detergent
octylglucopyranoside at elevated ionic strengths; oxygen
evolution activity is about 1,000 μmoles O_2/hr/mg Chl in the
presence of $CaCl_2$. A Mn quantitation and a kinetic study of
Z, the donor to P_{680}, reveals that on a Chl basis this new
preparation shows an almost four-fold enrichment in Mn and
the electron transport components of PSII.

1. INTRODUCTION

Research in several laboratories (3, 9, 10, 13) has
produced highly refined preparations of a PSII "core" complex
which, although photochemically active, does not possess the
capacity to evolve oxygen. More recent work (6, 11, 12) has
produced isolated oxygen evolving reaction center complexes
from PSII membranes by dissociating these membranes with non-
ionic detergents such as digitonin or octylglucopyranoside.
The product of these procedures is a purified PSII reaction
center complex depleted of the water soluble 17 and 23 kDa
polypeptides, and these preparations therefore require the
presence of non-physiological concentrations of Ca^{2+} and Cl^-
for optimum oxygen evolution activity (5, 6, 11, 12). In
this communication we report results from a new method (5)
for isolation of a PSII reaction center complex. A manganese
quantitation and a spectroscopic study of Z, the primary
donor to P_{680}, reveals a four-fold enrichment, on a Chl
basis, in the Mn-content as well as the electron transport
components of PSII.

[1] Dedicated to the memory of Warren Butler, whose research
provided new insights and ideas about the structure and
function of PSII.

Abbreviations: BZ, benzidine; Chl, chlorophyll; DCBQ, 2,5-
dichloro-p-benzoquinone; EPR, electron paramagnetic
resonance; LHC, light harvesting complex; OGP, 1-O-n-octyl-b-
D-glucopyranoside; PSII, Photosystem II; R.C.C, reaction
center complex; Tris, 2-amino-2-(hydroxymethyl)-1-3-
propanediol.

2.   MATERIALS AND METHODS

PSII membranes, prepared as in (4), were treated as in
(5) to produce the highly active oxygen evolving PSII
reaction center complex. Tris treated systems were prepared
by exposure of PSII preparations to 0.8M Tris plus 1mM EDTA.
Oxygen evolution activity was measured with a Clark-type
oxygen electrode. Gel electrophoresis was carried out as in
(2) with the modifications described in the figure captions.
EPR spectroscopy was carried out on a Bruker ER-200D
spectrometer operated at X-band and interfaced to a Nicolet
1180 computer. Instrument modifications, as well as the
flash lamp circuitry and the protocol for signal averaged,
flashing-light kinetic experiments, are described in ref. 15.

3.   RESULTS

Fig. 1, lane 2 shows that the new PSII Reaction Center
Complex (R.C.C.) consists of the hydrophobic polypeptides
observed in "core" complex preparations from PSII, along with
two other hydrophobic polypeptides(with molecular weights
which we estimate to be about 20 and 28 kDa), and the hydrophilic 33 kDa species; the complex has been depleted of the
water-soluble 17 and 23 kDa polypeptides and thus the presence of non-physiological concentrations of both Ca^{2+} and
Cl^- are required for oxygen evolution activity (see ref. 5).
In contrast to the preparation described in (6), the PSII
reaction center complex isolated by our procedure is very
active when DCBQ is present as an electron acceptor; $Fe(CN)_6^{3-}$
appears to be a more effective acceptor in this new preparation when compared to control PSII membranes, but is not as
effective as DCBQ (data not shown).

As shown in Table I, an analysis of the manganese content of this new preparation reveals an almost four-fold enrichment of manganese on a Chl basis. Since a series of experiments carried out by Matsuda and Butler (8), has clearly
demonstrated that high potential Cyt b_{559} reveals the structural integrity of the photosynthetic membrane and that dis-
ruption of that integrity causes Cyt_{559} to be modified to
lower potential forms, we studied the state of Cyt_{559} in this
new preparation by use of EPR spectroscopy. We have found
that a significant amount of high potential Cyt_{559} has shifted to lower potential form(s) in the Complex; even though addition of $CaCl_2$ reconstituted oxygen evolution activity, it
did not restore Cyt_{559} to its high potential form(s) (data
not shown).

When a higher ionic strength (1M NaCl) was used during
exposure of PSII membranes to OGP, the PSII reaction center
complex which resulted was also depleted of the water soluble
33 kDa polypeptide (Fig. 2) as well as most of the manganese
(Table I). This preparation shows very little oxygen evolution activity (data not shown). This modified isolation
procedure (35 mM OGP + 1M NaCl) was also applied in Tris-
treated PSII membranes; the polypeptide content of the reaction center complex which results from such a treatment

is shown in Fig. 2. The kinetic behavior of Z^+ in this Tris-PSII R.C.C. was studied by EPR and was compared to the kinetic behavior of Z^+ observed in Tris-PSII membranes. As shown

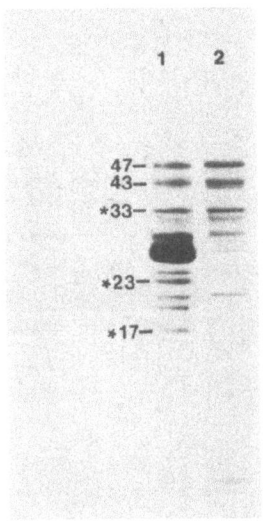

Fig. 1. Gel electrophoresis patterns of (1) untreated PSII membranes; (2) O_2 evolving reaction center complex prepared by exposure of PSII membranes to 35 mM OGP + 0.5 M NaCl. A 15% acrylamide resolving gel was used and 6M urea was present in the gel. (*) denotes water soluble extrinsic polypeptides.

Table II: Mn Content of Various PSII Preparations

Preparation	Mn-Content (atoms/250 Chl)
PSII Membranes	4[a]
PSII Reaction Center Complex (35:1.0)	6.0
Tris-PS II Membranes	0.6
Tris-PSII Reaction Center Complex (35:1.0)[b]	0
PS II Reaction Center Complex (35:0.5)[b]	14.8

a.   See refs. 1 and 4
b.   (x:y); x = concentration of OGP (mM), y = concentration of NaCl (M) used for the isolation of the R.C.C.

in Fig. 3B, Z^+ decays relatively slowly in Tris-treated PSII R.C.C., but its decay is dramatically accelerated upon addition of an exogenous donor such as benzidine. A calculation of the second order rate constant for benzidine from the data of Fig. 3B gives $k = 6.0 \times 10^6$ $M^{-1}.s^{-1}$; this rate constant is higher compared to that observed in Tris PSII membranes ($k = 1.0 \times 10^6$ $M^{-1}.s^{-1}$). A comparison of the amplitude of the kinetic traces of Z^+ in the Tris-PSII reaction center complex (Fig. 3B) with that observed in Tris-PSII membranes (Fig. 3A) reveals that, on a Chl basis, an enrichment of Z^+ approaching four-fold has occurred in the Tris-PSII R.C.C.

2A.   2B.

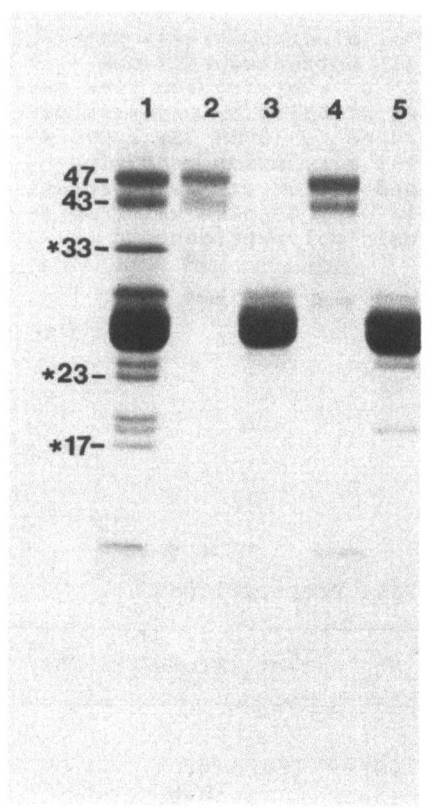
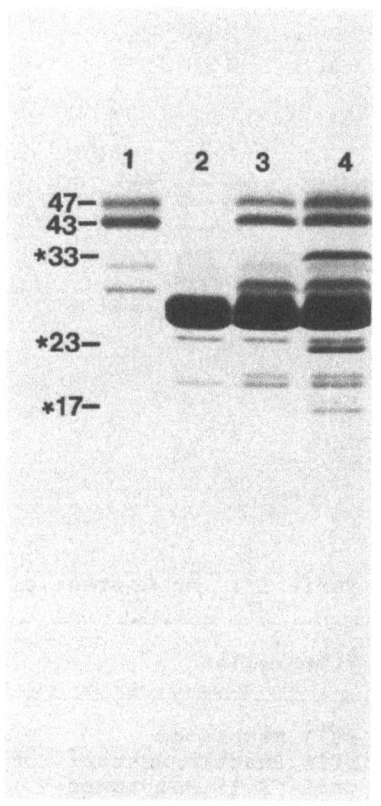

2A. Gel electrophoresis patterns of (1) untreated PSII membranes; (2) the reaction center complex prepared by
exposure of PSII membranes to 35mM OGP plus 1M NaCl
(Fraction B); (3) Fraction A obtained from treatment
in 2; (4) the reaction center complex prepared by
exposure of Tris-treated membranes to 35mM OGP plus
1M NaCl (Fraction A); (5) Fraction A obtained from
treatment in 4. A combination of a 12% (upper part)
and 18% (lower part) acrylamide resolving gel was used
and 6M urea was present in the gel.
2B. Gel electrophoresis patterns of (1) the reaction center
complex prepared by exposure of Tris-PSII (2) Fraction
A obtained from treatment in 1; (3) Tris-treated PSII
membranes; (4) untreated PSII membranes. Conditions
as in Fig. 1.

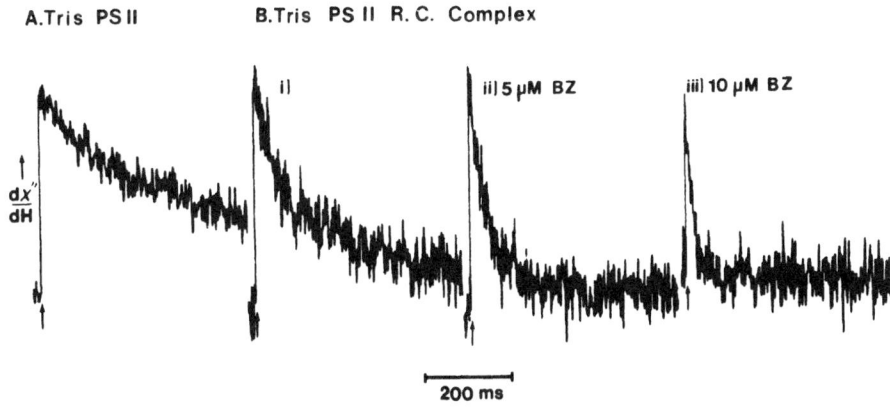

3.   Kinetic transients of Z^+ at room temperature in A) Tris-
      treated PSII membranes (4.16 mg Chl/ml) and B) the PSII
      reaction center complex prepared from Tris treated PSII
      membranes (0.56 mg Chl/ml); i) 0 μM Bz, ii) 5 μM Bz and
      iii) 10 μM BZ. Instrument conditions: time constant
      1 ms, flash rate 0.25 Hz, modulation 4 Gpp, gain 10 x
      10^5 and number of scans averaged 100 (for A) and 200
      (for B). A mixture of 3 mM $Fe(CN)_6^{3-}$ and 3 mM $Fe(CN)_6^{4-}$
      served as an artificial acceptor system.

4.   DISCUSSION

      Although the isolation of Photosystem II membranes (1,
7, 14) has advanced knowledge of the polypeptide composition
of PSII, further resolution of the PSII system is necessary
for effective spectroscopic studies of both the primary
reactions and the water cleavage process. Recently, a series
of techniques have been reported for isolation of oxygen
evolving reaction center complexes (5, 6, 11, 12). As shown
in Table I, the active PSII reaction center complex is
enriched in Mn by a factor of four, and it is therefore a
very attractive system for a spectroscopic characterization
of the Mn-complex. Low temperature ESR and EXAFS experiments, which require very concentrated samples, will benefit
by use of this new system. The kinetic study of Z^+ shown in
Fig. 3 also demonstrates a 3.7-fold enrichment in a component
of the PSII electron transport system. In addition to its
use in spectroscopic studies, the reduced number of polypeptides present in the PSII reaction center complex will
facilitate further elucidation of the structural role of the
various polypeptides as well as their relationship to sites
of catalytic activity.

Acknowledgment: This research was supported by grants from the National Science Foundation (PCM82-14240) and the Competitive Research Office of USDA (G-82-1127).

REFERENCES
1. Berthold DA, Babcock GT and Yocum CF (1981) FEBS Lett. 134, 231-234
2. Chua NH (1980) In (San Pietro, A. ed) Methods in Enzymology vol. 69, pp 434-446, Academic Press, NY
3. Diner BA and Wollman FA (1980) Eur. J. Biochem 110, 521-527
4. Ghanotakis DF, Babcock GT and Yocum CF (1984) Biochim Biophys Acta 765, 388-398
5. Ghanotakis DF and Yocum CF FEBS Lett. in press
6. Ikeuchi M, Yuasa M and Inoue Y (1985) FEBS Lett. 185, 316-322
7. Kuwabara T and Murata N (1982) Plant Cell Physiol 23, 533-539
8. Matsuda H and Butler WL (1983) Biochim Biophys Acta 725, 320-324
9. Satoh K and Butler WL (1978) Plant Physiol 61, 373-379
10. Satoh K, Nakatani, Steinback KE, Watson J and Arntzen, CJ (1983) Biochim Biophys Acta 723, 142-150
11. Satoh K, Ohno T and Katoh S (1985) FEBS Lett. 180, 320-330
12. Tang, X.-S. and Satoh K (1985) FEBS Lett. 179, 60-64
13. Vernon LP and Shaw ER (1971) In (San Pietro, A. ed) Methods in Enzymology, Academic Press, NY Vol. 23, pp 277-289
14. Yamamoto Y, Doi M, Tamura N and Nishimura M (1981) FEBS Lett. 133, 265-268
15. Yocum CF, Yerkes CT, Blankenship RE, Sharp RR and Babcock GT (1981) Proc Natl Acad Sci USA 78, 7507-11

Photosynthesis Research 10: 489–496 (1986)
© *Martinus Nijhoff Publishers, Dordrecht*

LIGHT-DEPENDENT INACTIVATION OF PHOTOSYNTHETIC OXYGEN EVOLUTION DURING NaCl
TREATMENT OF PHOTOSYSTEM II PARTICLES: THE ROLE OF THE 24-kDa PROTEIN

M. MIYAO and N. MURATA
National Institute for Basic Biology, Myodaiji, Okazaki 444, Japan

Abstract. Photosystem (PS) II particles prepared from spinach thylakoids
with Triton X-100 were treated with 1.5 M NaCl either in the light or dark.
Under both conditions, the 24-kDa and 18-kDa proteins were released from
the particles, but rebound to them when the NaCl concentration was reduced
to 34 mM by dilution. Oxygen evolution measured after the dilution was inactivated following NaCl treatment in the light, but not following
treatment in the dark. The inactivation in the light was suppressed when
5 mM $CaCl_2$ was added during or after the NaCl treatment. Based on these
observations, a scheme is proposed for the mechanism of light-dependent
inactivation of oxygen evolution during NaCl treatment of PS II particles
and for the function of the 24-kDa protein in regulating the conformation
of a supposed Ca^{2+}-binding intrinsic protein.

1. INTRODUCTION

Studies on PS II preparations from higher plants have suggested that
three extrinsic proteins of 33 kDa, 24 kDa and 18 kDa participate in photosynthetic oxygen evolution [1]. The 33-kDa protein functions as a Mn-binder
(or Mn-stabilizer) [2,3], and the 18-kDa protein as a Cl^--concentrator
[4,5]. The 24-kDa protein can be substituted for by 5 mM Ca^{2+} [6,7].
Ghanotakis et al. [8] inferred that the 24-kDa protein (and/or 18-kDa protein) provides the oxygen-evolving complex with a high affinity binding
site for Ca^{2+}. Boussac et al. [9] inferred that the 24-kDa protein is a
concentrator of Ca^{2+} at the site of action of the ion in the oxygen-
evolving complex.

In the present study, we observed that illumination is necessary for
inactivating oxygen evolution during the treatment of PS II particles with
NaCl. We propose that illumination of PS II particles depleted of the
24-kDa protein causes Ca^{2+} to be lost from the oxygen-evolving complex,
resulting in the inactivation of oxygen evolution, and that 24-kDa protein
regulates the conformation of a supposed Ca^{2+}-binding protein.

2. MATERIALS AND METHODS

PS II particles were prepared from spinach thylakoids with Triton X-100
[10]. The particles were kept frozen in the presence of 30% (v/v) ethylene
glycol in liquid nitrogen until use. For experiments, they were thawed,
washed twice with 1.0 mM EGTA/10 mM NaCl/300 mM sucrose/25 mM Mes-NaOH (pH
6.5) and then once with 10 mM NaCl/300 mM sucrose/25 mM Mes-NaOH (pH 6.5)
(low-salt medium) by suspension and centrifugation, and were finally suspended in the low-salt medium and kept in the dark for 2 h before treatment.

Abbreviations: Chl, chlorophyll; EGTA, ethyleneglycol-bis-(β-aminoethyl
ether)-N,N,N',N'-tetraacetic acid; Mes, 4-morpholineethanesulphonic acid;
PS, photosystem; SDS, sodium dodecylsulphate.

For the NaCl treatment, the particles were suspended in 1.5 M NaCl/300 mM sucrose/25 mM Mes-NaOH (pH 6.5) at 0.5 mg Chl/ml. After standing for designated periods at 0°C either in the light (white light at an intensity of 3.0 W/m² obtained from a tungsten lamp through optical filters, Toshiba Y-42 and two pieces of Hoya HA-50) or in the dark (dim green light), the suspension was diluted 63-fold with the low-salt medium to give 34 mM NaCl. The diluted suspension was supplemented with 0.3 mM phenyl-p-benzoquinone and 0.05% bovine serum albumin, and the oxygen-evolution activity was measured [10].

For analysis of proteins dissociated from, or reassociated with, PS II particles during the NaCl treatment and the following dilution, the suspension of NaCl-treated particles was centrifuged at 35,000 x g for 5 min either directly or after the dilution. The pelleted particles were subjected to SDS polyacrylamide gel electrophoresis [11]. Before use, all the suspending and dilution media were passed through a column of Chelex X-100 (Bio-Rad Lab.) to remove all traces of divalent cations.

The 24-kDa protein was prepared as described previously [12], and dialyzed for 6 h against 5 mM EGTA and 10 mM Mes/NaOH (pH 6.5) and then further for 6 h against 10 mM Mes/NaOH (pH 6.5).

3. RESULTS

The effect of light during NaCl treatment of PS II particles on the oxygen-evolution activity is shown in Fig 1. When the particles were treated with 1.5 M NaCl in the dark, the activity, measured after reduction of the NaCl concentration to 34 mM by dilution, did not decrease appreciably. When the treatment was done in the light, the activity gradually diminished and, after a 90-min illumination, reached 10% of the original level. This inactivation was a light-dependent, but not light-triggered process, since it did not proceed further when the light was turned off after a 30-min illumination (Fig. 1).

To determine the involvement of the 24-kDa and 18-kDa proteins in the light-dependent inactivation of oxygen evolution during the NaCl treatment, we investigated their dissociation and reassociation during treatment and

FIGURE 1. Effect of light during NaCl treatment of PS II particles on oxygen-evolution activity measured after reduction of the NaCl concentration. PS II particles were suspended in 1.5 M NaCl/300 mM sucrose/25 mM Mes-NaOH (pH 6.5) in the presence or absence of 10 mM EGTA. After designated periods of incubation, the suspension was diluted with the low-salt medium to give 34 mM NaCl, and then the activity was measured. Conditions during NaCl treatment: ●——●, dark and no EGTA; ○——○, light and no EGTA; ■——■, dark and 10 mM EGTA; □– – –□, light and 10 mM EGTA; ▲— – –▲, light for 30 min, then dark, and no EGTA.

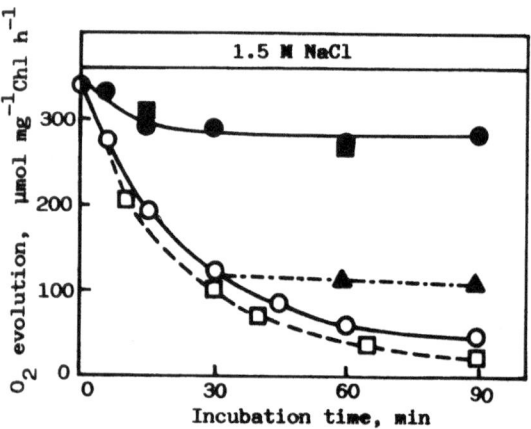

after subsequent reduction of the NaCl concentration. Table 1 shows that when PS II particles were separated from the medium by centrifugation after NaCl treatment for 30 min either in the light or dark but before dilution, only about 5% of each protein was left bound to the particles. This suggests that, in 1.5 M NaCl, these proteins were dissociated from the particles regardless of the light conditions. Oxygen-evolution activity of the particles collected by centrifugation and devoid of the 24-kDa and 18-kDa proteins was about 45% of that before the NaCl treatment when the treatment was done in the dark, and about 25% when the treatment was done in the light. However, the activity of NaCl-treated particles depleted of the 24-kDa protein was unstable: the longer the reaction mixture was exposed to room light before an assay for oxygen evolution, the lower was the activity (data not shown). The 63-fold dilution of the suspension, to give 34 mM NaCl, resulted in the reassociation of almost all the 24-kDa protein to the particles, although illumination during the NaCl treatment slightly reduced the rebinding (Table 1). Even though the rebinding of the 24-kDa protein was almost complete following the dilution after NaCl treatment in the light, oxygen-evolution activity was greatly diminished. These observations suggest that the light-dependent inactivation during the NaCl treatment was independent of the protein binding. Separate experiments revealed that incubation with 1.5 M NaCl for 5 min was sufficient to release more than 90% of each of the 24-kDa and 18-kDa proteins from PS II particles and that rebinding of the released proteins was completed within 5 min after dilution (data not shown).

Table 2 shows that the presence of 5 mM $CaCl_2$ during the NaCl treatment protected the oxygen-evolving complex against the light-dependent inactivation. A protective effect of $CaCl_2$ was observed also when the salt was added after the NaCl treatment in the light. $MgCl_2$ could not substitute for $CaCl_2$ in either case. These observations seem to suggest that the light-dependent inactivation of oxygen evolution in the concentrated NaCl resulted from release of Ca^{2+} from the oxygen-evolving complex. Table 2

TABLE 1.   Dissociation and reassociation of the 24-kDa and 18-kDa proteins in the oxygen-evolving complex during NaCl treatment of PS II particles and after subsequent reduction of the NaCl concentration. For NaCl treatment, PS II particles were treated with 1.5 M NaCl for 30 min either in the light or dark. They were then collected by centrifugation either directly (before dilution) or after reduction of the NaCl concentration to 34 mM by dilution (after dilution). The collected particles were suspended in the low-salt medium, and subjected to determinations of protein contents and oxygen-evolution activity.

Centrifugation	Conditions for NaCl treatment	Protein bound (%)		O_2 evolution ($\mu mol/mg$ Chl/h)
		24-kDa	18-kDa	
Before treatment		100	100	370 (100%)
Before dilution	Dark	6	3	170 (46%)
	Light	4	4	100 (27%)
After dilution	Dark	90	60	290 (78%)
	Light	82	60	120 (32%)

also reveals that the effect of Ca^{2+} in restoring oxygen-evolution activity was observed even when the ion was added after dilution following the NaCl treatment in the light. This is in contrast to the results of Ghanotakis et al. [8] who reported that Ca^{2+} in the outer aqueous medium slowly gains access to its functional site in PS II particles containing the 24-kDa protein but lacking Ca^{2+}. However, in our result, 5 mM $CaCl_2$ could restore the oxygen-evolution activity within 5 min.

A potent chelator of Ca^{2+}, EGTA, had no effect on the light-dependent inactivation of oxygen evolution, nor on the stability of activity in the dark during the NaCl treatment (Fig. 1). 20 mM EGTA eliminated the protective effect of 5 mM $CaCl_2$ against the light-dependent inactivation (data not shown). When the untreated particles containing the 24-kDa protein were incubated in 20 mM EGTA, the chelator also had no effect on the activity either in the light or dark (data not shown).

Fig. 2 shows the effect of $CaCl_2$ on oxygen-evolution activity of PS II particles which were depleted of the 24-kDa protein and Ca^{2+} after NaCl treatment in the light. Upon addition of $CaCl_2$, the activity increased rapidly within the initial 5 min, then more slowly to the maximum level during the next 30 min. In contrast to the ineffectiveness of EGTA added during the NaCl treatment (Fig. 1), the chelator suppressed the the

TABLE 2. Effect of $CaCl_2$ and $MgCl_2$ added before or after NaCl treatment, or after subsequent dilution on light-dependent inactivation of oxygen evolution. To study the effect of salts present during the treatment, PS II particles were treated with 1.5 M NaCl in the light for 60 min in the presence of $CaCl_2$ or $MgCl_2$. Then activity was measured after dilution. To study the effect of salts added after the treatment, the particles were treated with 1.5 M NaCl in the absence of $CaCl_2$ and $MgCl_2$ for 60 min in the light. After addition of $CaCl_2$ or $MgCl_2$ the suspension was incubated for a further 5 min in the dark and then diluted. After dilution, the concentrations of Ca^{2+} and Mg^{2+} were 80 μM and no longer had any effect on oxygen-evolution activity [7]. To study the effect of salts after subsequent dilution, the particles were treated with 1.5 M NaCl for 60 min in the light. The suspension was diluted and incubated for 5 min in the dark. After 5 mM $CaCl_2$ or 5 mM $MgCl_2$ was added, the activity was measured.

Conditions for treatment	Salt addition (5 mM)			O_2 evolution (μmol/mg Chl/h)
	before treatment	after treatment	after dilution	
Before treatment				320 (100%)
Dark	—	—	—	270 (84%)
Light	—	—	—	70 (22%)
Light	$CaCl_2$	—	—	300 (94%)
Light	$MgCl_2$	—	—	80 (25%)
Light	—	$CaCl_2$	—	250 (78%)
Light	—	$MgCl_2$	—	80 (25%)
Light	—	—	$CaCl_2$	260 (81%)
Light	—	—	$MgCl_2$	80 (25%)

oxygen-evolution activity restored by $CaCl_2$ (Fig. 2). The effectiveness of EGTA diminished with the incubation time over a period of hours (Fig. 2).

Fig. 3 shows the effect of EGTA on oxygen-evolution activity of PS II particles which had previously been depleted of the 24-kDa protein and Ca^{2+} upon NaCl treatment in the light, and then reactivated by Ca^{2+} and rebinding of the 24-kDa protein by dilution. In this experiment, the 24-kDa protein was externally added upon the dilution in order to ensure the complete rebinding of the protein. EGTA reduced the activity only slightly, and its effectiveness diminished with the incubation time over a period of minutes. These observations indicate that rebinding of both Ca^{2+} and the 24-kDa protein converted the oxygen-evolving complex to a form which binds Ca^{2+} tightly (insensitive to EGTA).

To determine whether the mode of Ca^{2+} binding to the oxygen-evolving complex of PS II particles was restored by the reconstitution with Ca^{2+} and the 24-kDa protein, the reconstituted PS II particles, as used in the experiment shown in Fig. 3, were collected by centrifugation and subjected to treatment with 1.5 M NaCl in the dark in the presence of 20 mM EGTA. This second NaCl treatment, however, did not appreciably inactivate oxygen-evolution activity measured after dilution (data not shown), suggesting that the tight binding of Ca^{2+} was restored in the reconstituted PS II particles.

4. DISCUSSION

The following characteristics concerning the interaction between Ca^{2+} and the 24-kDa protein in the oxygen-evolving complex of PS II particles have been derived from the present and previous studies: (i) The 24-kDa protein dissociates from the complex in concentrated NaCl and reassociates in a low-salt medium [Table 1, 12,13]; these processes occur within minutes. The reassociation does not require Ca^{2+} [Table 1, 8]. (ii) The binding of Ca^{2+} to the functional site is necessary for oxygen evolution [6,7]. (iii) The Ca^{2+}-binding site is supposed to be on one of the intrinsic proteins of the complex, but not on the 24-kDa or 33-kDa protein,

FIGURE 2. Effect of $CaCl_2$ and EGTA added after NaCl treatment in the light on oxygen-evolution activity measured after reduction of NaCl concentration. PS II particles were treated with 1.5 M NaCl for 60 min in the light. After 5 mM $CaCl_2$ was added, the suspension was incubated in the dark. At a designated time a portion of the suspension was withdrawn and the activity was measured after dilution with the low-salt medium. Addition to the suspension: 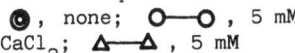, none; O—O, 5 mM $CaCl_2$; △—△, 5 mM $CaCl_2$, and then 20 mM EGTA at the times indicated by arrows.

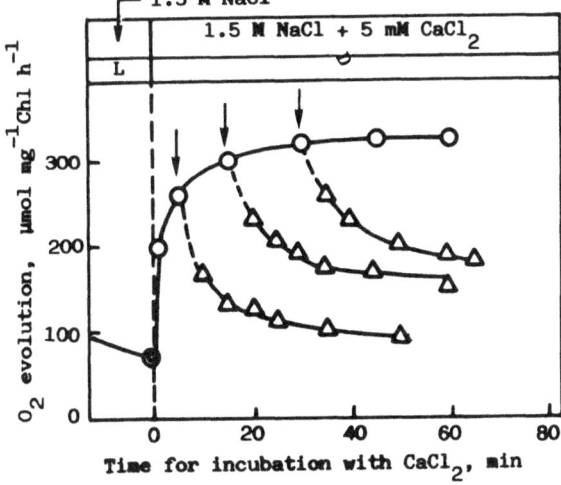

since the effect of Ca^{2+} on oxygen evolution occurs even in PS II particles depleted of all three extrinsic proteins [2]. A protein of 13-15 kDa isolated from lettuce PS II particles by Sparrow and England [14] may correspond to the Ca^{2+}-binding protein. (iv) When the 24-kDa protein is bound to the complex, Ca^{2+} is not released from the complex either in the light or dark. (v) In the complex depleted of the 24-kDa protein, Ca^{2+} is released under illumination, but is tightly bound in the dark, i.e., insensitive to EGTA [Fig. 1, 15]. (vi) In the complex depleted of the 24-kDa protein and Ca^{2+}, the rebinding of Ca^{2+} to the functional site is not tight, since EGTA eliminates the oxygen-evolution activity within minutes [Fig. 2]. However, the EGTA-insensitive part of activity (corresponding to the tight binding) appears gradually [Fig. 2]. (vii) When both Ca^{2+} and the 24-kDa protein rebind to the complex, the Ca^{2+} becomes tightly bound, since oxygen-evolution activity becomes insensitive to EGTA [Fig. 3]. (viii) Moreover, when the trapping of Ca^{2+} once becomes tight in the presence of the 24-kDa protein in the reconstituted system, the tight trapping of Ca^{2+} in the dark does not require the 24-kDa protein [the present study].

Although the total amount of Ca^{2+} bound to PS II particles was determined by Ghanotakis et al. [6], the Ca^{2+} bound to the functional site has not been analyzed quantitatively. It is assumed, therefore, that oxygen-evolution activity is an indirect measurement of Ca^{2+} binding to the functional site. The points listed above and the following arguments are based on this assumption.

In order to explain the light-dependent inactivation of oxygen evolution during the NaCl treatment in relation to the binding of Ca^{2+} and the 24-kDa protein to the oxygen-evolving complex, we propose a scheme presented in Fig. 4. The 24-kDa protein binds directly to the supposed Ca^{2+}

FIGURE 3. Effect of EGTA on oxygen-evolution activity of PS II particles which had previously been inactivated with NaCl treatment in the light, and then reactivated with $CaCl_2$ and rebinding of the 24-kDa protein. PS II particles were treated with 1.5 M NaCl for 60 min in the light and after addition of 5 mM $CaCl_2$, incubated further for 5 min in the dark. The suspension was then diluted 63-fold with the low-salt medium containing the 24-kDa protein (0.3 mg protein/mg Chl) and incubated in the dark. 20 mM EGTA was add-

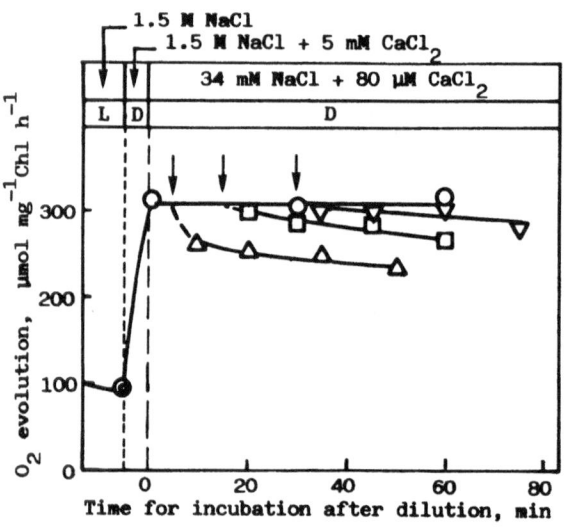

ed at times indicated by arrows, and the suspension was incubated further in the dark. At the designated time, a portion of the suspension was withdrawn, and the activity was measured without further dilution. Addition to the suspension: O—O , none; △—△ , □—□ , ▽—▽ , 20 mM EGTA.

binding intrinsic protein. Upon treatment with 1.5 M NaCl in the dark, the oxygen-evolving complex, which contains the 24-kDa protein and Ca^{2+} at their functional binding sites (1), releases the 24-kDa protein within minutes but retains bound Ca^{2+}, resulting in (2). Upon reduction of the NaCl concentration to 34 mM by dilution, the released 24-kDa protein rebinds to the complex to fully restore the oxygen-evolution activity; i.e., re-conversion from (2) to (1). Ca^{2+} in (2) is tightly bound to the complex, since the activity is insensitive to EGTA added during the NaCl treatment. When the complex in (2) is illuminated, Ca^{2+} is released from the functional site, as previously suggested [15], to form (4) via (3), and oxygen evolution becomes inactivated. The light-dependent release of Ca^{2+} is expressed schematically as a conformational change in the Ca^{2+}-binding intrinsic protein, i.e., the opening of a Ca^{2+}-trapping cavity to the outer medium. Reduction of the NaCl concentration under these conditions results in a rebinding of the 24-kDa protein to form (5), which is inactive in oxygen evolution.

The protective effect of 5 mM Ca^{2+} added during or after the NaCl treatment in the light (Table 2) can be explained by the re-conversion from (4) to (3). Since 5 mM Ca^{2+} can restore oxygen-evolution activity in minutes (Fig. 2), the functional binding site of Ca^{2+} is assumed to be open to the outer aqueous medium as shown in (4), and is accessible to the exogenous Ca^{2+}. When the NaCl concentration is reduced to 34 mM in (3), the protein rebinds to the complex to form (6). When the complex in (3) is kept in darkness, the Ca^{2+}-trapping cavity slowly closes to become insensitive to EGTA (Fig. 2), resulting in re-conversion to (2). In the complex (5), the functional binding site is still accessible to Ca^{2+} (Table 2); conversion to (6) by Ca^{2+}. Ca^{2+} in (6) becomes insensitive to EGTA within

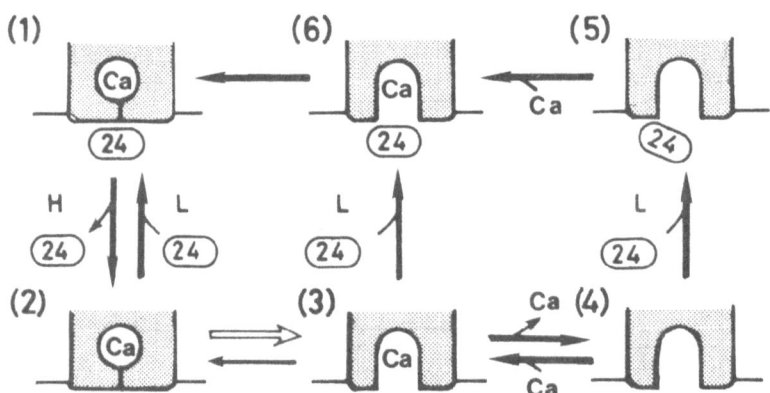

FIGURE 4. A scheme proposed for the mechanism of the effect of NaCl and light on the dissociation and reassociation of Ca^{2+} and 24-kDa protein with the oxygen-evolving complex. In this scheme, the 18-kDa and 33-kDa proteins and the intrinsic proteins other than the supposed Ca^{2+}-binding protein, are all neglected. The white arrow represents the light-dependent process. The bald black arrows represent fast processes which are complete within minutes in the dark, and the thin arrow represents a slow process requiring a period of hours in the dark. H and L stand for high salt (1.5 M NaCl) and low salt (34 mM NaCl) conditions, respectively.

minutes. In addition, the cavity for the Ca^{2+} binding in (6) seems to close within minutes to form (1), since the complex (6) became unsusceptible to NaCl treatment in the dark.

Two research groups have suggested that the 24-kDa protein provides the oxygen-evolving complex with a high affinity binding site for Ca^{2+} [8,9]. However, the present study shows that Ca^{2+} remains tightly bound to the complex depleted of the 24-kDa protein in the dark and is released under illumination. In the presence of the 24-kDa protein, though, Ca^{2+} is not released from the complex either in the light or dark. Therefore, it is concluded that a major function of the 24-kDa protein is to retain Ca^{2+} in its functional site in the light, probably by retaining optimum conformation of the Ca^{2+}-binding protein. This is a prerequisite for the oxygen-evolving complex, since oxygen is evolved only in the light.

There is a discrepancy between the results of several research groups with respect to the effect of the 24-kDa protein on oxygen-evolution activity. After the release of all the 24-kDa protein from PS II preparations, Ghanotakis et al. [8] and we [12] observed about 40 % of the original activity left in PS II particles, whereas in inside-out thylakoid vesicles Akerlund et al. [13] detected practically no activity. This difference can be explained by the scheme presented in Fig. 4, if it is noted that the NaCl treatment was performed at a Chl concentration of 0.5 mg/ml in our experiment [12] and 20 µg/ml in that of Akerlund et al. [13]. The effectiveness of light in releasing Ca^{2+} is supposed to be greater in the diluted suspension of PS II preparation. Therefore, the treatment by Akerlund et al. [13] should have been more effective in releasing Ca^{2+}, and thus inactivating oxygen evolution.

ACKNOWLEDGEMENTS

This work was supported by Grants-in-Aid for Special Research (60040057) and Cooperative Research (60304093) from the Ministry of Education, Science and Culture, Japan, to N.M.

REFERENCES

1. Murata N and Miyao M (1985) Tren Biochem Sci 10: 122-124
2. Miyao M and Murata N (1984) FEBS Lett 170: 350-354
3. Kuwabara T, Miyao M, Murata T and Murata N (1985) Biochim Biophys Acta 806: 283-289
4. Akabori K, Imaoka A and Toyoshima Y (1984) FEBS Lett 173: 36-40
5. Miyao M and Murata N (1985) FEBS Lett 180: 303-308
6. Ghanotakis DF, Babcock GT and Yocum CF (1984) FEBS Lett 167: 127-130
7. Miyao M and Murata N (1985) FEBS Lett 168: 118-120
8. Ghanotakis DF, Topper JN, Babcock GT and Yocum CF (1984) FEBS Lett 170: 169-173
9. Boussac A, Maison-Peteri B, Etienne A-L and Vernotte C (1985) Biochim Biophys Acta 808: 231-234
10. Kuwabara T and Murata N (1982) Plant Cell Physiol 23: 533-539
11. Kuwabara T and Murata N (1983) Plant Cell Physiol 24: 741-747
12. Miyao M and Murata N (1983) Biochim Biophys Acta 725: 87-93
13. Akerlund H-E, Jansson C and Andersson B (1982) Biochim Biophys Acta 681: 1-10
14. Sparrow RW and England RR (1984) FEBS Lett 177: 95-98
15. Dekker JP, Ghanotakis DF, Plijter JJ, van Gorkom HJ and Babcock GT (1984) Biochim Biophys Acta 767: 515-523

Photosynthesis Research 10: 497–503 (1986)
© Martinus Nijhoff Publishers, Dordrecht

THE RELATION BETWEEN THE CHLORIDE STATUS OF THE PHOTOSYNTHETIC WATER
SPLITTING COMPLEX AND THE INHIBITORY EFFECTIVENESS OF AMINES

PETER H. HOMANN

Institute of Molecular Biophysics and Department of Biological Science
Florida State University, Tallahassee, Florida 32306-3015, USA

KEY WORDS: Chloride; amine inhibition; water splitting; polypeptides of
photosystem II; photosystem II

ABSTRACT

The protective role of chloride ions (Cl^-) against inhibition of the
photosynthetic water splitting complex by amines was investigated with
purified photosystem II membrane particles from tobacco chloroplasts.
Seemingly competitive interactions occurred between Cl^- (except at low
concentrations) and Tris, but not between Cl^- and NH_3. The rate of Cl^-
release was not increased by the amines but, instead, may have been limited
by a labilization under the experimental conditions of the extrinsic 23 kDa
polypeptide. An additional detachment of the 18 kDa polypeptide was seen
when SO_4^{2-} ions were present. Tris induced changes of the thermoluminescence patterns of flash illuminated photosystem II particles were
found to be different from those caused by either Cl^- deficiency or high
pH. It is concluded that the protective functions of Cl^- are brought about
not because it is bound to the target site of the inhibitory actions of
Lewis bases like amines and hydroxyl ions. Instead, this effect of Cl^- may
be due to its influence on the tertiary and quaternary structures of the
water oxidizing protein complex.

1. INTRODUCTION

Among the admirably broad expertise and interests of the late Warren
L. Butler, research on the mechanism of photosynthetic O_2-evolution has
occupied a central position. From it, my research has received much
thought-provoking and stimulating impetus. It is with gratitude,
therefore, that I dedicate my article on the interactions of Cl^- with the
water oxidase to the memory of this eminent scientist and esteemed
colleague.

Even though Cl^- ions are known to be essential participants in the
process of photosynthetic water oxidation [1; see review in 2], it is not
known in what way they are associated with the water oxidase, and what
their functions are. Sandusky and Yocum [3] have suggested that Cl^- serves
as an electron-conducting inner sphere bridging ligand in the Mn-cluster of
the water oxidase. We, on the other hand, favor a role of Cl^- in the
protonation-deprotonation events associated with the process of water
splitting [4; see also 6].

In the bridging ligand hypothesis it is implied that, as a Lewis acid, Mn has a high affinity for Lewis bases such as OH^-, amines and Cl^-. Hence, the labilization of functional Cl^- in moderately alkaline media is ascribed to its competitive displacement by OH^- [2]. Similarly, kinetic analyses seem to suggest that the protection afforded by Cl^- against inhibition by amines is due to a competition for the same binding site [3].

In the course of our study on the merits of the two viewpoints about the interactions of Cl^- with the water oxidizing site, a pivotal role has emerged for Cl^- in the binding of the extrinsic 23 kDa polypeptide. The anion, therefore, accomplishes more than serving as a bound cofactor at one specific site.

2. MATERIALS AND METHODS

In order to be able to work with a water oxidizing system that is directly exposed to the bulk phase of the suspension medium, the present studies were done with purified photosystem II (PSII) membranes prepared from tobacco (Nicotiana tabacum) leaves by a somewhat modified version of the method of Berthold et al. [7].

Stock solutions of the amines were prepared by using the amine bases to adjust the pH of Mops (3-(N-morpholino)-propanesulfonic acid) or Hepes (4-(2-hydroxyethyl)-1-piperazineethane sulfonic acid) buffers which were mixed with Mes (4-morpholineethane sulfonic acid) (total buffer concentration 1 M) to provide larger amounts of neutralizable acid. Reaction media were acidified by injections of adequate amounts of 1 M Mes (free acid). All incubations were in dim light.

Oxygen evolution was measured with a Clark-type oxygen electrode in 2000 watt/m^2 red light using p-phenyl-benzoquinone as electron acceptor. Fluorescence measurements were carried out as in previously published work [8], and thermoluminescence recordings were made as reported elsewhere [5]. Polyacrylamide gel electrophoresis of Li-dodecylsulfate solubilized membranes was performed according to published procedures [9].

3. RESULTS

If amines and hydroxyl ions act as competing Lewis bases at a Cl^- binding site as Sandusky and Yocum [3] have suggested, an addition of amines to PSII particles might accelerate Cl^- loss, just as an increase of the pH does. In the experiment of Fig. 1, PSII particles were incubated in media of various pH, and the activity of O_2 evolution was measured after acidification to pH 6, either in the presence or in the absence of Cl^-. The pH shift caused an almost complete protonation of the added amines and rendered them noninhibitory. Thus, we could assess the Cl^- release that had occurred during the incubation by determining at the low pH the extent of Cl^- reversible activity loss. As Fig. 1 shows, such an approach revealed that the Cl^- release was quite independent of the presence of the amine and, instead, mainly a function of the pH.

In another test, the time course of Cl^- loss was determined at a fixed pH but in this case, Na_2SO_4 was omitted from the reaction medium in order to avoid its accelerating influence on Cl^- release. From Fig. 2 it can be seen that an incubation in the presence of the amines NH_3 and Tris caused a considerable degree of irreversible inhibition of O_2 evolving activity

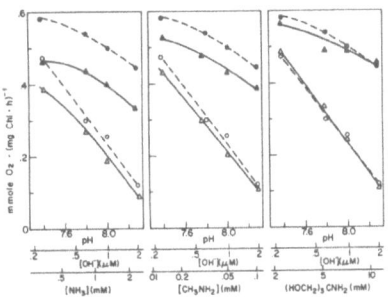
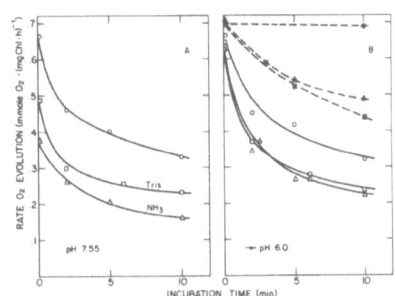

FIGURE 1. pH dependence of loss of O₂-evolving activity in the presence of
amines. Suspension medium contained 0.4 M sucrose, 10 mM Na₂SO₄, Hepes-Mes
buffer adjusted with NaOH (●,○) or the respective amine (▲,△), and PSII
particles corresponding to 10 μg Chl/ml. Incubation for 10 s, and activity
measurement after pH change to pH 6 either in the absence (open symbols) or
in the presence of 20 mM Cl⁻ (closed symbols).

FIGURE 2. Time course of inhibitory action of Tris and NH₃ on O₂-
evolution. Suspension medium contained 0.4 M sucrose, Mops-Mes buffer
adjusted to pH 7.55 with either NaOH (●,○) or Tris (■,□) or NH₃
(▲,△), and PSII particles equivalent to 8 μg Chl/ml. [Tris-base] = 7 mM;
[NH₃] = 0.6 mM. Dashed curves represent samples to which 50 mM Cl⁻ had
been added after incubation. Assay either at pH of incubation (A) or after
a pH change to 6 (B).

which prevented an unequivocal estimate of amine-induced Cl⁻ loss.
However, when comparing for the shortest incubation the rates measured at
the incubation pH, and after the pH shift, it is clear that the appearance
of an inhibition by the amines preceded the loss of Cl⁻. This conclusion
cannot be dismissed by the argument that amines act on the S₂ state [10]
and, consequently, do not inhibit until the light is turned on: we found
that two exposures to a saturating light flash in the course of a one min
incubation did not in any way increase the activity loss that could be
reversed by Cl⁻. On the other hand, when a 5 sec period of continuous
light preceded the acidification to pH 6, most of the lost activity could
no longer be restored by Cl⁻ addition, presumably because of photo-
inhibitory events during the illumination at pH 7.5. These observed
responses to light flashes or to a continuous illumination agreed with
results reported previously by Frasch and Cheniae in their study on the
mechanism of the inhibitory actions of Tris and other amines [11].

    In order to circumvent the problem of irreversible inactivations, we
used the fluorescence induction kinetics of the particles as a sensitive
qualitative assay of the Cl⁻ status after a brief illumination with weak
and, therefore, less damaging light. The traces of Fig. 3 confirm that the
inhibitory effect of the amines exceeded that attributable to Cl⁻ loss.

    All these data, and others, suggested that the displacement of Cl⁻
from its functional site by amines, if it occurred, was not the rate
limiting step of Cl⁻ release into the bulk medium. Since a crucial role in
the diffusion of Cl⁻ to and from its site of binding may be played by 23
and 18 kDa extrinsic polypeptides [2,5], we extended our analyses to PSII

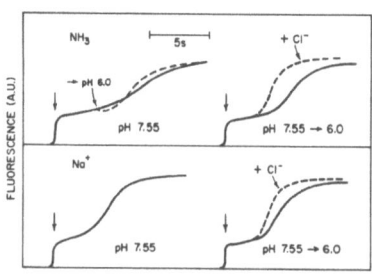 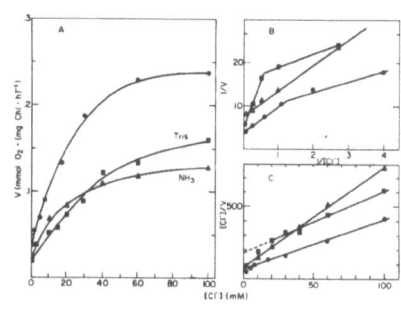

FIGURE 3. Effect of NH3 on the chlorophyll a fluorescence induction curves of intact PSII particles. Experimental conditions as in Fig. 2, 15 s incubation. [Cl⁻], when added, 60 mM. Illumination .05 watt/m² green light. Downward arrows indicate "light on". (A.U. = arbitrary units.)

FIGURE 4. O₂-evolution rate (V) in the presence of NH3 or Tris as a function of [Cl⁻]. Suspension medium as for Fig. 2. PSII particles were freed of their 23 and 18 kDa polypeptides by treatment with 1.5 M NaCl [9], and were incubated 1 min in Na-Mops buffered reaction medium to release Cl⁻, then 1.5 min with the added Cl⁻ and another min with the amine. [Tris base] = 7 mM; [NH3] = 0.6 mM.

particles freed of these two polypeptides and tested the alleged competitive protection rendered by Cl⁻ against inhibition by amines [3]. Fig. 4 presents three modes of display of the measured O₂-evolving activity as a function of Cl⁻ in the absence and presence of inhibitory amines. Two of the presentations (Figs. 4B and 4C) are based on kinetic analyses in which Cl⁻ is treated as an essential activator of the water oxidase [4]. The nonlinearity of the plots of the control data obtained in the absence of an amine can possibly be attributed to the uncertainty of the experimental points at very low Cl⁻ concentrations. In the case of the Tris inhibited sample, the biphasic character of the plots was even more pronounced, presumably due to the irreversible inactivations under conditions of Cl⁻ deficiency [11] (see also Fig. 2). In spite of such shortcomings, an almost identical V_{max} (intercept with the abscissa in Fig. 4B, and same slope $1/V_{max}$ in Fig. 4C) was evident for the control and the Tris inhibited sample. Thus, at moderate Cl⁻ concentrations, our results seem to reflect a competition between Cl⁻ and Tris for a common binding site as Sandusky and Yocum [3] had described for intact thylakoids. In contrast to Tris, NH3 appeared to act as uncompetitive or mixed type inhibitor.

Since these results were not confined to experiments with particles depleted of the 18 and 23 kDa polypeptides but seen also with PSII preparations from which the two polypeptides had not been removed (not shown), we wondered whether perhaps they became detached under our assay conditions as appears to be the case after Cl⁻ removal at high pH [12]. Indeed, as Fig. 5 shows, particles incubated at pH 7.25 had almost completely lost the 23 kDa polypeptide when Cl⁻ was withheld from the medium. In the presence of Na₂SO₄, the 18 kDa species disappeared as well (not shown). The polypeptide loss appeared to be enhanced by the presence of Tris. The seeming ineffectiveness of NH3 might be attributable to its lowered concentration at pH 7.2. At pH 7.5 the 23 kDa polypeptide were lost regardless of the base used in the buffer.

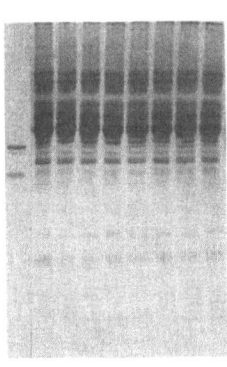

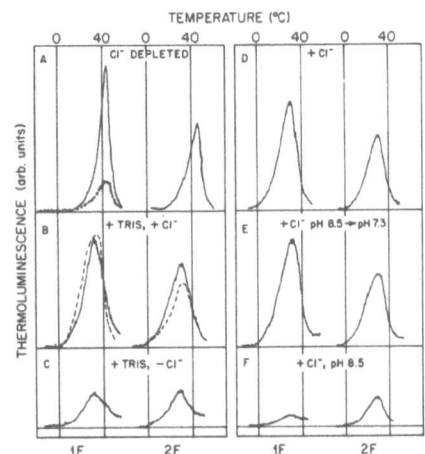

FIGURE 5. Polypeptide pattern obtained by gel electrophoresis after 3 min incubations of PSII particles in the presence or absence of added Cl⁻ under various conditions. 40 mM Cl⁻ added after the incubation of samples in lanes 2,4,6 and 8. [Mops + Mes] = 33 mM. Lane 1: purified 18 and 23 kDa polypeptides from tobacco; lanes 2 and 3, pH 7.5, Na-buffer; lanes 4 and 5, pH 7.25 Tris-buffer ([Tris-base] = 4 mM); lanes 6 and 7, pH 7.25, NH₃ buffer ([NH₃] = 0.3 mM); lanes 8 and 9, pH 7.25, Na-buffer.

FIGURE 6. Thermoluminescence emission from spinach PSII particles after one and two saturating flashes under various conditions. Suspension medium contained 0.4 M sucrose, 25 mM buffer and, when indicated, 40 mM NaCl. When present as buffer neutralizing base, [Tris-base] = 3 mM. A: Cl⁻ depletion by incubation with 75 mM Na₂SO₄ in medium buffered at pH 7.2 with Na-Hepes; dotted line: depletion by high pH shock [4] and assay in Na-Hepes buffered medium at pH 7.9. B: Cl⁻ added to medium with Tris-neutralized Mops (pH 7.2) before (⎯⎯) or after (----) flash illumination; C: like B, no Cl⁻ added. D: Cl⁻ containing medium buffered with Na-Hepes at pH 7.3. E: Na-Tricine buffered medium shifted from pH 8.5 to pH 7.3 with Mes (free acid) after flash illumination. F: like E, no pH shift.

The possible targeting of the Cl⁻ binding site by OH⁻ and Tris was investigated further by measuring the pattern of thermoluminescence after illumination with single light flashes. Recent studies have revealed that regardless of assay pH and the mode of Cl⁻ depletion Cl⁻ deficiency is reflected by an upward shift of the emission temperature [5]. At high pH, the luminescence intensities are very low even in the presence of Cl⁻, and lowest after preillumination with one flash [14]. Fig. 6 reveals that Tris neither shifted the temperature of the emission peak, nor caused the luminescence intensity after one flash to be lower than that after two flashes. However, the presence of Tris resulted in an overall suppression of the emission intensity which could be readily reversed by adding 40 mM Cl⁻ before or after the flash illumination.

4. DISCUSSION

Our data show that, in spite of some common traits, the response of PSII to OH⁻, Tris, and NH$_3$ cannot be explained with a common mechanism of action. With regard to the protecting role of Cl⁻ it appears that, where it occurs, it is an expression of a rather general effect of this anion on the organization of the water oxidizing site. Indeed, Cl⁻ mitigates not only detrimental effects of amines but also actions of high pH, heat and excessive light intensities [2]. We show that one aspect of this protective function can be correlated with the stabilization of the binding of the 23 and perhaps the 18 kDa polypeptides. From experiments with PSII particles lacking these polypeptides, however, we know [2] that the action of Cl⁻ is also directed at entities closer to the active site of the water oxidase. Yet, the presently available data, including the ones presented in this article, are insufficient evidence for the conclusion that Cl⁻ and inhibitory amines share a common binding site, e.g. as inner sphere ligands of the functional Mn [3]. This is true even if one acknowledges that there are multiple actions of any one amine [3]. All are nucleophiles but NH$_3$, for example, may also act as a water analog, and Tris is an amino compound endowed with oxy-groups capable of H-bonding. The latter property could account for the influence of Tris on the conformational integrity of the S$_2$ state as postulated by Frasch and Cheniae [11].

The interplay of Cl⁻ and inhibiting amines may well be indirect. We envisage a crucial role of Cl⁻ in the maintenance of the tertiary and quaternary organization of the proteins in the water oxidizing complex which not only is critical for the operation of the active site, but also modifies its susceptibility to an inhibitory attack by amines. As previously proposed [15], the negatively charged Cl⁻ ion may act by stabilizing protonated groups in the water oxidase. However, the correlation of the chaotropic nature of other anions with their effectiveness as Cl⁻ substitutes [1] suggests that the hydration energy of is an important functional asset as well. An organizing role of Cl⁻ would be quite analogous to its activator function in -amylase which Levitzki and Steer [16] have attributed to local conformational changes of the enzyme.

ACKNOWLEDGMENTS

These studies were aided by NSF grant DMB 8304416. The thermo-luminescence measurements were made possible during a research visit at the Institute of Physical and Chemical Research in Wako, Saitama, Japan (aided by NSF grant INT 8407399) through the generosity of the Solar Energy Research group and its Director, Dr. Yorinao Inoue. The excellent technical assistance by Ms. Michelle Currier is gratefully acknowledged.

REFERENCES

1. Gorham PR and Clendenning KA (1972) Arch Biochem Biophys 37:199-223
2. Critchley C (1985) Biochim Biophys Acta 811: 33-46
3. Sandusky PO and Yocum CF (1984) Biochim Biophys Acta 766:603-611
4. Homann PH (1985) Biochim Biophys Acta 809:311-319
5. Homann PH and Inoue Y (1986) in: Ion Interactions in Energy Transfer Systems (Papageorgiou GC, Barber J, and Papa S, eds.) pp. 279-290, Plenum Publ. Co., New York

6. Govindjee, Kambara T and Coleman W (1985) Photochem Photobiol 42:187–210

7. Berthold DA, Babcock GT and Yocum CF (1981) FEBS Lett 134:231–236

8. Richter ML and Homann PH (1983) Arch Biochem Biophys 222:67–77

9. Miyao M and Murata N (1984) Biochim Biophys Acta 765:253–257

10. Delrieu MJ (1976) Biochim Biophys Acta 440:176–188

11. Frasch WD, and Cheniae, GM (1980) Plant Physiol. 65:735–745

12. Izawa S, Muallem A and Ramaswamy NK (1983) in: The Oxygen Evolving System of Photosynthesis (Inoue Y, et al., eds) pp. 293–302, Academic Press, Tokyo

13. Cammarata K and Cheniae G (1985) Plant Physiol 77 (Suppl), 16

14. Vass I, Koike H and Inoue Y (1986) Biochim Biophys Acta, in the press

15. Johnson JD, Pfister VR and Homann PH (1983) Biochim Biophys Acta 723:256–265

16. Levitzki A and Steer ML (1974) Eur J Biochem 41:171–180

VI.   Photosynthetic Bacteria; Metabolism

Figure 12. David Knaff, Warren L. Butler, Shnuel Malkin, Dick Malkin, Jeanette Brown, Paul Mathis, Bacon Ke, Harry Frank, and George Hoch. No one could go in the Jacuzzi due to age limitation but all problems dealing with bacterial metabolism were solved.

Photosynthesis Research 10: 507–514 (1986)
© *Martinus Nijhoff Publishers, Dordrecht*

507

ACTIVE TRANSPORT IN PHOTOTROPHIC BACTERIA

DAVID B. KNAFF

Department of Chemistry and Biochemistry,
Texas Tech University, Lubbock, Texas 79409-4260 (U.S.A.)

ABSTRACT
    Phototrophic bacteria utilize light-driven, cyclic electron flow to pump protons out of their cytoplasm, creating an electrochemical proton gradient, $\Delta \bar{\mu}_H+$, outside acid and positive. These bacteria exchange external protons for internal cations (Na^+, K^+ and Ca^{+2}), allowing the cells to maintain a nearly constant internal pH while maintaining the electrical component of $\Delta \bar{\mu}_H+$. Na^+/H^+ exchange also establishes an electrochemical Na^+ gradient. Phototrophic bacteria are able to utilize these electrochemical gradients as energy sources for the uptake of a wide variety of metabolites (e.g., sugars, organic acids and amino acids) via metabolite/cation symports.

1. INTRODUCTION
    The study of active transport in bacteria occupies a central position in current research in bioenergetics. While a wealth of information is available concerning transport processes in prokaryotes that derive the energy for active transport from respiration or fermentation, until recently little data had been collected on transport phenomena in phototrophic bacteria. However, more recently, a significant body of information on light-dependent transport processes has accumulated. This overview will concentrate on transport by phototrophic purple sulfur and purple non-sulfur bacteria. Some data obtained from studies with halobacteria will be included because, although halobacteria are not truly phototrophic prokaryotes, there are many similarities between light-dependent transport in halobacteria and in true phototrophic bacteria. The majority of transport systems that will be discussed meet criteria usually set forth for active transport in that substrate transport requires an energy source and proceeds without chemical modification of the transported substrate. A brief discussion of sugar phosphorylation with concomitant phosphorylation (via the PTS system) in a purple non-sulfur bacterium will also be included. A considerable body of evidence suggests the transport of ions and of non-PTS substrates (e.g., amino acids) in these light-utilizing prokaryotes occurs via "chemiosmotic" mechanisms [41,73]. The evidence indicates that light does not serve as the direct source of energy for active transport but rather the role of light is confined to the generation of an electrochemical gradient of protons ($\Delta \bar{\mu}_H+$). The electrochemical proton gradient (or in some cases an electrochemical gradient of Na^+ ions) then serves as the actual energy source for secondary active transport.

2. THE ELECTROCHEMICAL PROTON GRADIENT IN PHOTOTROPHIC BACTERIA
    There are generally considered to be five distinct families of

phototrophic bacteria [69]. As little is known about active transport in the two families of green bacteria (the Chlorobiaceae and Chloroflexaceae), this overview will discuss transport in only the three families of purple bacteria (Rhodospirillacea, Chromatiacea and Ectothiorhodospiracea). In these bacteria, light-driven cyclic electron flow results in proton efflux from the cells, producing a $\Delta\bar{\mu}_H+$ with both ΔpH (outside acidic) and membrane potential (outside positive) components [32,62]. Membrane potential ($\Delta\Psi$) and ΔpH measurements [7,14,21,57,58,60,61] have been made using intact cells of representative purple sulfur (Chromatium vinosum) and purple non-sulfur (Rhodopseudomonas sphaeroides and Rhodopseudomonas capsulata) bacteria. The most detailed measurements of ΔpH in intact cells have been made with C. vinosum and Rps. sphaeroides. In both bacteria, the ΔpH maintained by cells decreased with increasing external pH. For example, in Rps. sphaeroides pH decreased from 2 units at an external pH = 5 to 0 at pH 8.0 [61]. Both bacteria maintain a significantly smaller ΔpH when suspended in Na^+ containing medium, compared to K^+ containing medium. Furthermore, in C. vinosum it was demonstrated that the largest ΔpH values were maintained in media free of alkali or alkaline earth metal cations and that addition of Na^+, K^+ or Ca^{2} significantly decreased ΔpH [14]. These effects of metal cations have been attributed to the action of Ca^{2}/H^+ and K^+/H^+ exchange systems (antiports) in C. vinosum [12,13] and to the possible presence of Na^+/H^+ antiports in C. vinosum [14] and Rps. sphaeroides [26]. (See below.) Light-induced $\Delta\Psi$ has been measured in cells of Rps. sphaeroides, Rps. capsulata and C. vinosum [57,58,79,14,21,39] and shown to increase with increasing external pH. For example, in C. vinosum $\Delta\Psi$ (inside minus outside) increased from -60 mV at pH 5.5 to -190 mV at pH 8.0. The $\Delta\Psi$ maintained by these cells was relatively unaffected by the ionic composition of the medium. In C. vinosum, reciprocal changes in ΔpH and $\Delta\Psi$ with increasing external pH produce a net $\Delta\bar{\mu}_H+$ that varies little as a function of external pH. In potassium phosphate buffer, for example, $\Delta\bar{\mu}_H+$ values never exceed -220 mV nor fall below -180 mV as the external pH was varied from 5.5 to 8.0 [14]. Cells of both Rps. sphaeroides and C. vinosum [21,39] also can maintain a $\Delta\Psi$, (but apparently not a ΔpH) in the dark under anaerobic conditions. This dark $\Delta\Psi$ appears to result from proton pumping coupled to ATP hydrolysis by the F_oF_1 ATPase [39].

Membrane vesicles with opposite membrane sidedness compared to whole cells (chromatophores) can be prepared from a number of phototrophic purple bacteria [29,36,55,58]. Light-driven cyclic electron flow in chromatophores generates a $\Delta\bar{\mu}_H+$, inside acidic and negative, the opposite polarity to that produced by intact cells. The magnitudes of $\Delta\bar{\mu}_H+$ and of its two components, $\Delta\Psi$ and ΔpH, have been measured in chromatophores prepared from three species of photosynthetic purple non-sulfur bacteria Rps. sphaeroides [22,57,58], Rps. capsulata [3,6,8] and Rhodospirillum rubrum [5,34,47]. Results obtained from the three different species agree reasonably well qualitatively, but there are significant quantitative differences in the values reported for $\Delta\Psi$ and ΔpH. It is likely that many of these differences can be attributed to the different techniques used to measure ΔpH and $\Delta\Psi$. Measurement of ΔpH using the fluorescent amine, 9-aminoacridine, gave larger values for ΔpH than did techniques using the distribution of radioactively labeled, permeant weak bases [5,58]. Estimations of $\Delta\Psi$ based on the amplitude of band shifts of intrinsic membrane carotenoids or on the response of an extrinsic fluorescent probe [5-8] were higher than values estimated from the distribution of radioactively labeled, permeant anions [22,58]. Recently attempts have been made to quantitate possible sources of error in the ΔpH and $\Delta\Psi$ measurements and to reconcile these differences in $\Delta\Psi$ and ΔpH measurements [1,4,16,48]. One general feature observed in these studies of $\Delta\bar{\mu}_{H^+}$ in chromatophores was that the values of ΔpH and $\Delta\Psi$ maintained by chromatophores were dependent on the anion composition of the suspension medium [47,57,58], with the presence of a permeant anion resulting in a decrease in $\Delta\Psi$ and a concomitant increase in ΔpH.

Halobacterium halobium in contrast to phototrophic bacteria, contains a membrane bound, bacteriorhodopsin-containing protein that couples the absorption of light to the formation of an electrochemical proton gradient by direct proton pumping without intermediate electron transfer reactions [18,28,70]. The light-induced ΔpH and $\Delta\Psi$ generated by intact cells of H. halobium have been quantitated as a function of external pH [2,56]. The magnitude of $\Delta\Psi$ and ΔpH and the effects of medium constituents on these parameters described above for C. vinosum and Rps. sphaeroides are similar to those observed in H. halobium.

3. CATION TRANSPORT IN PHOTOTROPHIC BACTERIA

Light-induced Ca^{2+} movements have been observed with Rps. capsulata chromatophores [31] and with cells and chromatophores of C. vinosum and R. rubrum [12]. While the mechanism of the Ca^{2+} flux seen with Rps. capsulata is not known, Ca^{2+} efflux in C. vinosum and R. rubrum has been shown to occur via an electrogenic Ca^{2+}/H^+ antiport ($H^+:Ca^{2+} > 2.0$). These antiports do not respond to Mg^{2+}, Zn^{2+}, or monovalent cations. However Sr^{2+}, Mn^{2+} and, to a lesser extent, Ba^{2+} can substitute for Ca^{2+}. Rps. capsulata also possess two distinct systems for the energy-dependent uptake of Mg^{2+} and Mn^{2+} [31]. Ca^{2+} efflux has been seen in H. halobium vesicles and shown to be catalyzed by a Ca^{2+}/Na^+ antiport. The energy for Ca^{2+}/Na exchange is provided by a Na^+ gradient ($[Na^+]_{out} > [Na^+]_{in}$) maintained by illuminated vesicles (see below). The H. halobium Na^+/Ca^{2+} antiport involves K^+ as a regulatory agent [42].

Light-induced K^+ movements have been observed in three phototrophic bacteria. K^+ flux in cells and chromatophores of C. vinosum was shown to occur via an electroneutral K^+/H^+ antiport [13]. This antiport probably results in the extrusion of K^+ from the cell under physiological conditions. The rate of the antiport shows considerable cooperativity with respect to $[K^+]$. This cooperativity may reflect the cell's requirement to maintain relatively high internal $[K^+]$. C. vinosum also contains an ATP-driven K^+ uptake system [15] which appears responsible for maintaining appropriate intracellular K^+ levels and allowing the cell to adjust its internal osmotic strength. Energy-independent K^+ accumulation has also been observed in Rps. capsulata [30]. In Rps. sphaeroides, K^+ uptake is energized by $\Delta\Psi$ via an electrogenic transport system that utilized neither ΔpH nor ATP as an energy source [26]. Light-induced K^+ uptake by H. halobium was shown to occur via a $\Delta\Psi$-driven K^+ uniport [23,71].

Na$^+$ transport is of particular interest, because in many bacteria active transport of solutes can be either directly driven or stimulated by the presence of a Na^+ gradient [42]. Na^+/H^+ exchange has also been shown to play a key role in the regulation of internal pH by bacteria [63]. Light-induced, uncoupler-sensitive Na^+ extrusion has been observed from intact cells of C. vinosum and R. rubrum and, as expected from the reversed membrane sidedness of chromatophores, Na^+ uptake is seen with chromatophores from C. vinosum, R. rubrum and Rps. sphaeroides [26,37]. The mechanisms by which these Na$^+$ movements occur is not yet fully understood but much of the available data is consistent with the operation of a Na$^+$/H$^+$ antiport [14,26]. Light-induced Na$^+$ efflux from H. halobium also occurs via a Na$^+$/H$^+$ antiport [17,44]. Na$^+$ flux via the antiport is a gated process requiring a threshold $\Delta\mu_H$+ of -130 to -155 mV for rapid Na$^+$ extrusion [45]. Data from a series of mutants established that the Na$^+$/H$^+$ antiport is responsible for the large majority of the light-induced Na$^+$ efflux and for maintaining the Na$^+$ gradient in H. halobium [52].

4. METABOLITE UPTAKE COUPLED TO ION CO-TRANSPORT IN PHOTOTROPHIC BACTERIA

Light-induced uptake of glucose and its non-metabolizable analog, α-methylglucoside has been observed in C. vinosum [38] and considerable evidence exists to support the proposal that uptake occurs via a glucose/H$^+$ symport [38,64]. C. vinosum appears to lack a phosphoenolpyruvate dependent phosphotransferase transport system (PTS) for glucose [38].

Reports of light-dependent, uncoupler-sensitive uptake of amino acids by cells of the purple non-sulfur bacteria Rps. sphaeroides [27] and R. rubrum [74] and the purple sulfur bacterium C. vinosum [9,11,35,64] suggested that both families of bacteria possess $\Delta\mu_H$+-driven transport systems for amino acids. For C. vinosum, direct evidence has been obtained for H$^+$ co-transport with uptake of the non-polar amino acids L-phenylalanine, L-leucine, L-isoleucine and L-valine [11] and of L-aspartate [9]. There appear to be three separate C. vinosum amino acid/H$^+$ co-transport systems (symports) involved in the uptake of the three non-polar amino acids — one for phenylalanine, one for leucine and a third that can transport either valine or isoleucine. The uptake of D-alanine and of L-alanine (and its non-metabolizable analog, α-aminoisobutyrate) in C. vinosum occurs via separate Na$^+$/amino acid symports rather than via H$^+$ symports [11,37,64]. There also appears to be a second aspartate transport system in C. vinosum, of lower affinity for aspartate than the aspartate/H$^+$ symport, that involves Na$^+$ as the co-transported cation [9]. Of considerable interest was the finding, that while isoleucine/valine transport in C. vinosum exhibits the attributes of an amino acid/H$^+$ symport at pH values below 7.5, the transport system appears to be able to utilize either H$^+$ or Na$^+$ as the co-transported ion at pH > 7.5 [11]. For several of these transport systems, Na$^+$ gradients also produce allosteric effects on the apparent affinities for the transported amino acids [9,11,64]. Separate co-transport and allosteric sites for Na$^+$ are apparently present. The most surprising result obtained from this series of amino acid transport studies in C. vinosum came from experiments suggesting that glycine uptake was accompanied by K$^+$ co-transport [10]. Should further research support this interpretation, it would represent the first documentation of a substrate/K$^+$ symport in a prokaryote. The C. vinosum amino acid transport systems appear to be constitutive and, although the amino acids are incorporated into a number of cellular constituents after uptake, the amino acids cannot serve as sole carbon sources for growth [25,72]. Relatively little is known about transport in the Ectothiorhodospiracea, but light-dependent proline and glutamate have been observed with Ectothiorhodospira halophila. Transport of these amino acids in E. halophila exhibits Na$^+$ stimulation but it is unclear whether a Na$^+$ symport exists for these amino acids [66].

Light-induced accumulation of amino acids has also been extensively studied in H. halobium, with at least partial characterization having been accomplished for the transport of 19 amino acids [41,42,53,54]. All 19 amino acids studied are transported via Na$^+$ symports, with no direct involvement of ATP. With the exception of glutamate, amino acid uptake is electrogenic and can be energized either by $\Delta\Psi$ or ΔpNa$^+$. The $\Delta\Psi$ (outside positive) is initially generated by the bacteriorhodopsin-mediated, light-driven H$^+$ efflux. The ΔpNa$^+$ results from the subsequent exchange of internal Na$^+$ for external H$^+$ via the Na$^+$/H$^+$ antiport. The glutamate/Na$^+$ symport is electroneutral and thus can only be energized by ΔpNa$^+$ [44,46]. A gating effect has been observed for the aspartate/Na$^+$ and serine/Na$^+$ symports of H. halobium, with the stoichiometries of these systems changing at $\Delta\Psi$ = -35 mV [45].

Light-induced accumulation of a number of other metabolites has been observed in phototrophic bacteria. Separate systems for the light-dependent uptake of C$_4$ dicarboxylic acids and of pyruvate have been reported for Rps. sphaeroides [24]. Ectothiorhodospira shaposhnikovii cells take up succinate, fumarate and malate in light and Na$^+$-dependent reactions. Evidence has been presented for a Na$^+$/succinate symport in E. shaposhnikovii [33]. Light-dependent, uncoupler-sensitive uptake of malate, succinate, acetate and propionate has been observed in C. vinosum [35] but these transport systems were not characterized further. Malate transport has also been observed in R. rubrum and the quantum yield for both malate and alanine uptake measured in this purple non-sulfur bacterium [74]. Identical values of 1 molecule accumulated per 60 to 70 absorbed quanta were obtained for both alanine and malate. These very high quantum requirements suggest that a relatively small portion of R. rubrum's available energy is devoted to the uptake and accumulation of organic substrates in comparison to other energy-requiring processes (e.g., ATP synthesis and motility).

While there has been general agreement that $\Delta\Psi$ or ΔpH and ΔpNa provide the driving force for secondary active transport in phototrophic bacteria, recent quantitative studies by Konings and co-workers suggest a lack of correlation between the magnitude of $\Delta\bar{\mu}_H^+$ and the rate of solute uptake under certain conditions. These studies (carried out with intact Rps. sphaeroides cells) have raised the possibility of a direct involvement in the control of solute uptake by the electron transport system. In conventional chemiosmotic theories of transport, one would predict that electron flow and active transport are not directly connected but rather are indirectly coupled through $\Delta\bar{\mu}_H^+$ or $\Delta\bar{\mu}_{Na}^+$ [42,73]. The Rps. sphaeroides studies indicated that cyclic electron flow above a threshold level was necessary before alanine uptake occured, independent of the magnitude of $\Delta\bar{\mu}_H^+$ [19-21]. Above this threshold, the rate of alanine uptake was proportional to the light intensity (the rate of electron flow was assumed to be proportional to light intensity) at a given $\Delta\Psi$. (The experiments were conducted under conditions where ΔpH = 0, so $\Delta\bar{\mu}_H^+$ = $\Delta\Psi$). Under certain conditions, the rate of light-dependent alanine uptake actually decreased with increasing $\Delta\Psi$ [19-21]. These results, while not ruling out a central role for $\Delta\bar{\mu}_H^+$ as the energy source for secondary active transport in phototrophic bacteria, point to a possible direct interaction between electron transfer and active transport, perhaps by changing the redox state of dithiol/disulfide groups in the transport proteins [40].

5. PHOSPHOTRANSFERASE (PTS) SYSTEMS IN PHOTOTROPHIC BACTERIA

PTS systems have been known to be present in phototrophic bacteria for a considerable time, with the earliest report describing PTS systems for fructose, but not for glucose, in R. rubrum and Rps. sphaeroides grown under aerobic, non-photosynthetic conditions [40]. Recently Robillard and co-workers have performed a series of interesting mechanistic studies on the fructose-specific PTS system in phototrophically-grown Rps. sphaeroides [50,51]. The Rps. sphaeroides PTS system [50,51] is simpler than those of E. coli or S. typhimurium in that it consists of only two protein components (one membrane-bound and one soluble) compared to the three or four proteins found in E. coli and S. typhimurium PTS systems for different sugars [65]. In particular, the Rps. sphaeroides PTS system lacks the small HPr protein that serves as the intermediate phosphoryl carrier in E. coli and S. typhimurium [65]. It appears that the phosphoryl group of phosphoenolpyruvate is transferred to the Rps. sphaeroides soluble protein (which complexes to the membrane-bound protein) and then to the membrane-bound protein component. Phosphorylation of the membrane-bound component induces the formation of a high-affinity fructose binding site by a "ping-pong" mechanism [51]. The mechanistic similarities between the Rps. sphaeroides and E. coli/S. typhimurium PTS systems, despite the simpler composition of the former, have been interpreted in terms of the Rps. sphaeroides PTS representing an early stage in the evolution of PTS systems [51,67].

6. ACKNOWLEDGEMENTS

Work carried out in the author's laboratory has been supported, in part, by grants from the National Science Foundation (PCM-8109635 and PCM-8408564). The author would like to thank Prof. W.N. Konings, B.A. Melandri, J.B. Jackson and T. A. Krulwich for their helpful and stimulating discussions and Prof. Konings and Prof. G.T. Robillard for access to manuscripts prior to publication. The author would especially like to acknowledge a great personal debt to the late Warren Butler for valuable advice and considerable inspiration during the early part of his scientific career.

7. REFERENCES

1. Baccarini-Melandri A, Casadio R and Melandri BA (1981) In Sanadi DR, ed. Current Topics in Bioenergetics, Vol. 12, pp. 197-258. New York: Academic Press.
2. Bakker EP, Rottenberg H and Caplan SR (1976) Biochim Biophys Acta 440: 557-572.
3. Casadio R, Baccarini-Melandri A and Melandri BA (1974) Eur J Biochem 47: 121-128.
4. Casadio R and Melandri BA (1985) Arch Biochem Biophys 238:219-228.
5. Cirillo VP and Gromet-Elhanan Z (1981) Biochim Biophys Acta 636:244-253.
6. Clark AJ, Cotton NPJ and Jackson JB (1983) Biochim Biophys Acta 723: 440-453.
7. Clark AJ, Cotton NPJ and Jackson JB (1983) Eur J Biochem 130:575-580.
8. Clark AJ and Jackson JB (1981) Biochem J 200:389-397.
9. Cobb AD and Knaff DB (1983) Arch Biochem Biophys 225:86-94.
10. Cobb AD and Knaff DB (1984) Biochim Biophys Acta 777:117-122.
11. Cobb AD and Knaff DB (1985) Arch Biochem Biophys 238:97-110.

12. Davidson VL and Knaff DB (1981) Biochim Biophys Acta 636:53-60.
13. Davidson VL and Knaff DB (1982) Photobiochem Photobiophys 3:167-174.
14. Davidson VL and Knaff DB (1982) Photochem Photobiol 36:551-558.
15. Davidson VL and Knaff DB (1982) Arch Biochem Biophys 213:358-362.
16. Demura M, Kamo N and Kobatake Y (1985) Biochim Biophys Acta
    812:377-386.
17. Eisenbach M and Caplan SR (1979) Curr Top Membrane Trans 12:166-248.
18. Eisenbach M, Cooper S, Garty RM, Johnstone RM, Rottenberg H and Caplan
    SR (1977) Biochim Biophys Acta 465:599-613.
19. Elferink MGL, Friedberg I, Hellingwerf KJ and Konings WN (1983) Eur J
    Biochem·129:583-587.
20. Elferink MGL, Hellingwerf KJ, van Belkum MJ, Poolman B and Konings WN
    (1984) FEMS Microbiol Lett 21:293-298.
21. Elferink MGL, Hellingwerf KJ and Konings WN (1986) Biochim Biophys
    Acta 848:58-68.
22. Ferguson SJ, Jones OTG, Kell DB and Sorgato MC (1979) Biochem J
    180:75-85.
23. Garty H and Caplan SR (1977) Biochim Biophys Acta 459:532-545.
24. Gibson J (1975) J Bact 123:471-480.
25. Gibson J (1984) Ann Rev Microbiol 38:135-159.
26. Hellingwerf KJ, Friedberg I, Lolkema JS, Michels PAM and Konings WM
    (1982) J Bact 150:1183-1191.
27. Hellingwerf KJ, Michels PAM, Dorpema JW and Konings WN (1975) Eur J
    Biochem 55:397-406.
28. Henderson R (1977) Ann Rev Biophys Bioeng 5:87-109.
29. Hochman A, Bittan V and Carmeli C (1978) FEBS Lett 89:21-25.
30. Jasper P (1978) J Bact 113:1314-1322.
31. Jasper P and Silver S (1978) J Bact 133:1323-1328.
32. Junge W and Jackson JB (1982) In Govindjee, ed. Photosynthesis, Vol.
    I, pp. 589-646. New York: Academic Press.
33. Karzanov VV and Ivanovsky RN (1980) Biochim Biophys Acta 598:91-99.
34. Kell DB, Ferguson SJ and John P (1978) Biochim Biophys Acta
    502:111-126.
35. Knaff DB (1978) Arch Biochem Biophys 189:225-230.
36. Knaff DB and Carr JW (1979) Arch Biochem Biophys 193:379-384.
37. Knaff DB, Davidson VL and Pettitt CA (1981) Arch Biochem Biophys 211:
    234-239.
38. Knaff DB and Whetstone R (1980) Arch Biochem Biophys 193:379-384.
39. Knaff DB, Whetstone R and Carr JW (1979) FEBS Lett 99:283-286.
40. Konings WN and Robillard GT (1982) Proc Natl Acad Sci USA
    79:5480-5484.
41. Lanyi JK (1978) Microbiol Rev 42:682-706.
42. Lanyi JK (1979) Biochim Biophys Acta 559:377-397.
43. Lanyi JK, Helgerson SL and Silverman MP (1979) Arch Biochem Biophys
    193: 329-339.
44. Lanyi JK, Rentahl R and MacDonald RC (1976) Biochem 15:1603-1610.
45. Lanyi JK and Silverman MP (1979) J Biol Chem 254:4750-4755.
46. Lanyi JK, Yearwood-Drayton V and MacDonald (1976) Biochem
    15:1595-1603.
47. Leiser M and Gromet-Elhanan Z (1977) Arch Biochem Biophys 178:79-88.
48. Lolkema JS, Abbing A, Hellingwerf KJ and Konings WN (1983) Eur J
    Biochem 130:287-292.

[368]

49. Lolkema JS, Hellingwerf KJ and Konings WN (1982) Biochim Biophys Acta 681: 85-94.
50. Lolkema JS and Robillard GT (1985) Eur J Biochem 147:69-75.
51. Lolkema JS, Ten Hoeve-Duurkens RH and Robillard GT (1985) Eur J Biochem 149:625-631.
52. Luisi BF, Lanyi JK and Weber HJ (1980) FEBS Lett 117:354-358.
53. MacDonald RE, Greene RV and Lanyi JK (1977) Biochem 16:3227-3235.
54. MacDonald RE and Lanyi JK (1975) Biochem 14:2882-2889.
55. Matsurba K and Nishimura M (1977) Biochim Biophys Acta 459:483-491.
56. Michel H and Oesterhett D (1976) FEBS Lett 65:175-178.
57. Michels PAM, Hellingwerf KJ, Lolkema JS, Friedberg I and Konings WN (1981) Arch Microbiol 130:357-361.
58. Michels PAM and Konings W (1978) Eur J Biochem 85:147-155.
59. Nicolay K, Hellingwerf KJ, van Gemerden H, Kaptein R and Konings WN (1982) FEBS Lett 138:249-254.
60. Nicolay K, Kaptein R, Hellingwerf KJ and Konings WN (1981) Eur J Biochem 116:191-197.
61. Nicolay K, Lolkema J, Hellingwerf KJ, Kaptein R and Konings WR (1981) FEBS Lett 123:319-323.
62. Ort DR and Melandri BA (1982) In Govindjee, ed. Photosynthesis, Vol. I, pp. 537-587. New York: Academic Press.
63. Padan E, Zilberstein D and Schuldiner S (1981) Biochim Biophys Acta 650: 151-166.
64. Pettitt CA, Davidson VL, Cobb A and Knaff DB (1982) Arch Biochem Biophys 216:306-313.
65. Postma PW and Lengeler JW (1985) Microbiol Rev 49:232-250.
66. Rinehart CA and Hubbard JI (1976) J Bact 127:1255-1264.
67. Saier MH (1977) Bact Rev 41:856-871.
68. Saier MH, Feucht BU and Roseman S (1971) J Biol Chem 246:7819-7821.
69. Stackebrandt E and Woese CR (1981) Soc Gen Microbiol Symp 32:1-31.
70. Stoeckenius W, Lozier RH and Bogomolni RA (1979) Biochim Biophys Acta 505: 215-278.
71. Wagner G, Hartmann R and Osterhett D (1978) Eur J Biochem 89:169-179.
72. Wagner BJ, Miovic ML and Gibson J (1973) Arch Mikrobiol 91:255-272.
73. West IC (1980) Biochim Biophys Acta 804:91-126.
74. Zebrower M and Loach PA (1982) J Bact 150:1322-1388.

Photosynthesis Research 10: 551–518 (1986)
© Martinus Nijhoff Publishers, Dordrecht

ACCUMULATION OF SILVER BY Chromatium vinosum FROM SOLUTIONS CONTAINING
SILVER THIOSULFATE

MASAO KITAJIMA

Research Laboratories, Asaka, Fuji Photo Film Co., Ltd., Asaka-shi,
Saitama-ken 351, Japan

Key words:Chromatium vinosum, photosynthetic sulfur bacteria, sulfur
bacteria, silver resistant bacteria, silver accumulation by bacteria.

1. ABSTRACT

The photosynthetic sulfur bacterium, Chromatium vinosum, was cultured
in inorganic photographic processing solutions containing silver thiosulfate complex salt $(AgNa_3(S_2O_3)_2)$ under light. It was found that
Chromatium was resistant to Ag and accumulated granular silver in the
membrane during growth. The amount of Ag accumulated in the cells depended
on the initial concentrations of the Ag salt in the culture solution. When
the concentration of Ag was 300 mg/l, the bacteria accumulated Ag as high
as 30% of the dry cell weight. The size of the granules was 0.1 to 0.3 μm.
Results from X-ray microanalysis indicated that these granules consisted
mostly of Ag^0 with small fractions of Ag_2S and $AgCl$.

2. INTRODUCTION

Ag is one of the most toxic heavy metals to bacteria. Its inhibitory
action is stronger than that of Cd, Zn and Hg. Antimicrobial agents such
as silver sulfadiazine were widely used as drugs for treatment of infected
burns and wounds. Several bacteria which are resistant to silver sulfadiazine were isolated from burns and their properties were studied [1].
The metabolism of Ag in Pseudomonas stutzeri isolated from a silver mine
was studied. The Ag accumulation was closely correlated with the presence
of a plasmid [2].

We earlier reported that Thiobacillus novellus and Thiobacillus
thioparus, as well as Chromatium vinosum, were grown in used photographic
processing solutions containing silver thiosulfate complex salt
$(AgNa_3(S_2O_3)_2)$ and suggested that these bacteria could be used for treating
photographic waste water [3,4]. Silver resistance of other microorganisms
in photographic industrial water was also reported [5]. A community of
bacteria, consisting of Pseudomonas maltophilia, Staphylococcus aureus and
a coryneform organism, was isolated from soil near a silver mine and was
found to be resistant to Ag concentration up to 100 mM; moreover, these Ag
adapted bacteria accumulated Ag [6]. Thiobacillus thiooxidans and
Thiobacillus ferrooxidans were also reported to accumulate small Ag
granules consisting of Ag_2S on the surface of cell membranes [7].

I now report that a photosynthetic sulfur bacterium, Chromatium
vinosum, cultured in the presence of silver thiosulfate complex salt accu-
mulates Ag^0 granules in the membrane. Conditions and some suggested
mechanisms for the accumulation will be discussed.

3. METHODS AND MATERIALS

3.1. Culture conditions. Chromatium vinosum was grown in a standard medium
containing per liter: NaCl, 3.0g; NH_4Cl, 1.0g; $CaCl_2$, 0.05g; KH_2PO_4, 0.5g;

$FeCl_3$ $6H_2O$, 0.05g; K_2HPO_4, 0.5g; and $NaHCO_3$, 2.0g. To this solution were added 2g of $Na_2S_2O_3 \cdot 5H_2O$ for control experiments or various concentrations of $AgNa_3(S_2O_3)_2$ prepared from photographic fixing solutions used for processing photographic film. The pH of the final solution was 5.8 to 6.0. Narrow-neck, 125 ml glass bottles were filled with the medium containing about 100mg dry cells per l and were sealed with wax. Oxygen in the solutions was consumed for oxidation of S^{2-} or $S_2O_3^{2-}$, and thus anaerobic conditions were maintained during the incubation. The bottles were kept at $30^\circ C$ and were illuminated with a set of incandescent lamps while stirring. The light intensity at the surface of the bottle was $2mW/cm^2$. The cells were collected on membrane filters, and dry cell weight was measured after drying the filters in an oven at $110^\circ C$ for 1 hr.

3.2. Quantitative analysis of Ag. The amount of Ag accumulated in the cells was determined by two different procedures: decrease of Ag in the culture solutions and direct measurement of Ag in the cells. Ag concentrations in culture solutions were measured with an atomic absorption spectrophotometer (Hitachi Z-7000) before inoculation and after removal of the cultured cells following various incubation times. The amount of Ag accumulated in the cells was determined directly by the following procedure. All cells in a bottle were collected by centrifugation and were redispersed in 5ml of 5% gelatin solution. The whole amount of the solution was then coated homogeneously on a $50cm^2$ glass plate and dried. The amount of Ag per unit area of the plate was measured with an X-ray fluorophotometer (Philips PW 1410). The results were corrected with a calibration curve prepared by using solutions containing predetermined concentrations of $AgNO_3$.

X-ray microanalysis of each cell was performed with a Shimadzu EPM 810 X-ray microanalyzer. One drop of the cell culture from a high concentration thiosulfate solution was spotted on a glass plate, dried and placed on the microanalyzer. An electron photomicrograph of one cell was taken, and then the X-ray diffraction pattern of a granule in the membrane was measured.

4. RESULTS AND DISCUSSION

Chromatium vinosum was found to be resistant to silver thiosulfate complex salt $(AgNa_3(S_2O_3)_2$. It grew continuously in a medium containing Ag at concentrations lower than 10mg/l. The culture reached a steady state at 350 to 600mg of dry cells per 125ml in 2 to 3 days. When the Ag concentration was higher, the color of the cells changed from red to dark red and finally black. The cell suspension was stable at the beginning of the incubation, but when the bacteria accumulated a substantial amount of Ag, they sedimented easily upon termination of stirring.

As Chromatium accumulated Ag from the medium, Ag concentrations in the medium were reduced. When the initial Ag concentraion was 4.6mg/l, almost all Ag was removed from the medium in 18 hr. When the initial Ag concentration was 76mg/l, 43% was removed in 18 hr and 98% in 62 hr.

The amount of Ag accumulated in the cells were determined directly by an X-ray fluorophotometer. It was found that the accumulation of Ag in the cells depended on the initial Ag concentrations in the culture medium. The results are shown in Table 1. When the initial Ag concentration was 45mg/l, the cells accumulated Ag up to 10% of the dry cell weight. When the initial concentration was 266mg/l, 29% of the dry cell weight was Ag.

In order to identify the site of Ag accumulation, a single cell was observed and analyzed with an X-ray microanalyzer. Figure 1 is an electron photomicrograph of Chromatium cultured in a medium with high Ag

Table 1. Amount of silver accumulated by <u>Chromatium vinosum</u> from solutions containing various concentrations of silver thiosulfate.

Initial Ag Conc. (mg/l)	Dry cell weight (mg/125ml)	Ag found in cells (mg/125ml)	Ag content (%)
0	342	0	0
3.0	541	3.0	0.6
45.2	430	43.3	10
226	561	135	29

concentration. It may be seen from the figure that the Ag granules (electron dense spots) are located mostly in the cell membrane. The number of the granules is not very large but they look to almost fill the membrane. The size of the granules varies but many of them are as large as 0.1 to 0.3 μm.

Figure 2 is a result of an X-ray diffraction pattern of one cell. A distinctively high peak assigned as Ag^o and small peaks assigned as Ag_2S and AgCl are observed. This investigation indicates that Ag is accumulated by <u>Chromatium</u> mainly in metallic form, rather than Ag_2S as reported by Pooley [7] with <u>Thiobacillus</u>.

Some experiments were carried out to understand more about the mechanisms of Ag accumulation by <u>Chromatium</u>. <u>Chromatium vinosum</u> is a facultative autotrophic bacterium. It utilizes reducing energy of S^{2-} or $S_2O_3^{2-}$ in strictly inorganic medium. But when it is cultured in the presence of organic proton acceptor such as malate, it utilizes the organic reductant preferably to the inorganic reductants. In order to confirm whether the Ag accumulation process is correlated to an energy uptake mechanism of the bacterium, the effect of malate on the Ag accumulation was studied. It was found that <u>Chromatium</u> grew more rapidly in the presence of malate but accumulation of Ag diminished markedly as compared with

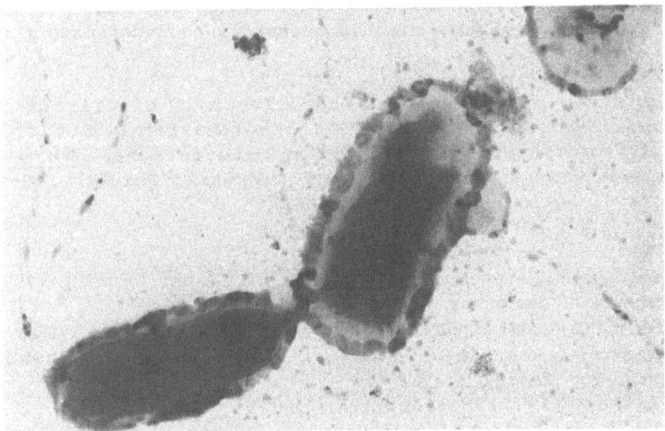

FIGURE 1. Electron photomicrograph of <u>Chromatium vinosum</u> containing silver granules in the membrane.

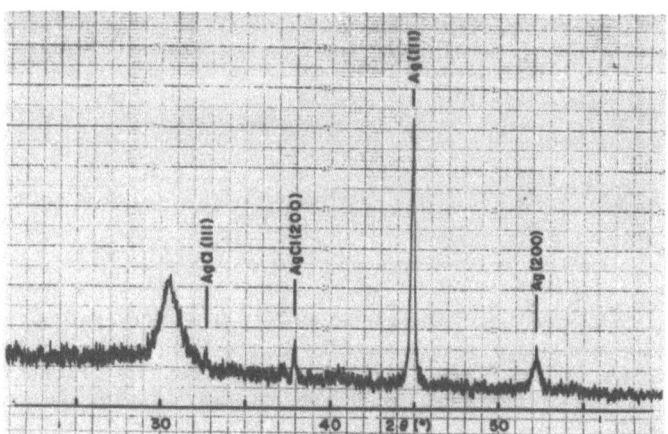

FIGURE 2. X-ray diffraction pattern of a silver granule in the cell membrane measured by an X-ray microanalyzer.

experiments done in the inorganic medium. Effect of illumination on the Ag accumulation was also investigated. It was found that the bacterium accumulated Ag under dark conditions, although it grew much slower than with illumination. The effect of sodium pentachlorophenolate, a strong uncoupler of phosphorylation [8], was also examined. The bacterium was found to accumulate Ag in the presence of the uncoupler. The amount of the accumulated Ag was about 2/3 of the control.

$AgNa_3(S_2O_3)_2$ is stable in solution and its solubility is quite high. Chromatium may not distinguish $Na_2S_2O_3$ and $AgNa_3(S_2O_3)_2$ at the membrane surface. It utilizes $S_2O_3^{2-}$ as energy source and may leave Ag in the intramembrane structure. As the intramembrane region is maintained in reductive atmosphere, Ag^+ is reduced and may deposit as Ag^o and form granules. Preliminary studies do not provide much detail about the accumulation mechanism, and more work is necessary to understand it fully.

ACKNOWLEDGEMENTS

This study was started at the University of California, San Diego under the auspices of the late Professor Warren L. Butler. The author wishes to thank him for his support and suggestions. Chromatium vinosum was obtained through the courtesy of Professor Shigehiro Morita at Tokyo University of Agriculture and Technology.

REFERENCES

1. Resenkranz HS, Coward JE, Wloodkowski TJ and Carr HS (1974) Antimicrob Antigents & Chemother 5: 199-201
2. Haefeli CC, Franklin C and Hady K (1984) J Bacteriol 158: 389-392
3. Kitajima M, Abe A and Tomotsu T (1976) In: Proc Ann Meeting Photographic Soc Japan, pp.107-109
4. Kitajima M (1977) U S Patent 4,155,810
5. Belly RT and Kydd GC (1982) Develop Industr Microbiol 23: 567-577
6. Charley RC and Bull AT (1979) Arch Microbiol 123: 239-244
7. Pooley FD (1982) Nature 296: 642-643
8. Krogmann DW, Jagendorf AT and Avron M (1959) Plant Physiol 34:272-275